DYNAMICS OF PLANETS AND SATELLITES
AND THEORIES OF THEIR MOTION

ASTROPHYSICS AND
SPACE SCIENCE LIBRARY

A SERIES OF BOOKS ON THE RECENT DEVELOPMENTS

OF SPACE SCIENCE AND OF GENERAL GEOPHYSICS AND ASTROPHYSICS

PUBLISHED IN CONNECTION WITH THE JOURNAL

SPACE SCIENCE REVIEWS

VOLUME 72
PROCEEDINGS

DYNAMICS OF PLANETS AND SATELLITES AND THEORIES OF THEIR MOTION

PROCEEDINGS OF THE 41ST COLLOQUIUM OF THE
INTERNATIONAL ASTRONOMICAL UNION HELD IN
CAMBRIDGE, ENGLAND, 17–19 AUGUST 1976

Edited by

VICTOR SZEBEHELY

The University of Texas at Austin

D. REIDEL PUBLISHING COMPANY

DORDRECHT : HOLLAND / BOSTON : U.S.A.

Library of Congress Cataloging in Publication Data
Main entry under title:

Dynamics of planets and satellites and theories of their motion

(Astrophysics and space science library; v. 72)
Bibliography: p.
Includes indexes.
1. Planets—Congresses. 2. Satellites—Congresses. 3.
Mechanics, Celestial—Congresses. I. Szebehely, Victor, G., 1921–
II. International Astronomical Union. III. Title: Interna-
tional Astronomical Union colloquium no. 41. IV, Series.
QB603.M6D95   523.4   77–28322
ISBN-13: 978-94-009-9811-7   e-ISBN-13: 978-94-009-9809-4
DOI: 10.1007/978-94-009-9809-4

Published by D. Reidel Publishing Company,
P.O. Box 17, Dordrecht, Holland

Sold and distributed in the U.S.A., Canada and Mexico
by D. Reidel Publishing Company, Inc.
Lincoln Building, 160 Old Derby Street, Hingham,
Mass. 02043, U.S.A.

TABLE OF CONTENTS

INTRODUCTION

P.J. MESSAGE
University of Liverpool

The papers which comprise this volume were presented at Colloquium
No. 41 of the International Astronimical Union, which was held in
Cambridge, England, from the 17th to the 19th of August, 1976, and
had as its subject 'Dynamics of Planets and Satellites and Theories
of their Motion'. The Colloquium was held just prior to the XVIth
General Assembly of the Union (which was held from 24th August to
2nd September, in Grenoble, France) to provide an opportunity for the
presentation of research papers on a number of active and lively
branches of Celestial Mechanics to a gathering of experts in the field,
and for the stimulus of discussion of research problems of interest to
participants. A number of papers testify to the progress being made
in General Planetary Theory, the theories of motion of the minor
planets, the Moon, and the satellites of Jupiter and Saturn, and to
significant advances in both the general and restricted gravitational
problems of three bodies.

The Organizing Committee of the Colloquium was comprised of J.
Chapront, R.L. Duncombe, J. Hadjidemetriou, Y. Kozai, B. Morando,
J. Schubart, V. Szebehely, and P.J. Message (Chairman). The local
Organizer was D.C. Heggie, to whose tireless efforts the success of
the arrangements is due.

LIST OF PARTICIPANTS

N. Abu-el-Ata, Bureau des Longitudes, 77 Avenue Denfert Rochereau,
    75014 Paris, France
K. Aksnes, Center for Astrophysics, 60 Garden Street, Cambridge,
    Massachusetts 02138, U.S.A.
G. Antonacopoulos, Department of Astronomy, Patras, Greece
M. Antonacopoulos, Department of Mathematics, Patras, Greece
J. Baumgarte, Lehrstuhl A für Mechanik, Technische Universität,
    Pockelsstrasse 4, D-3300 Braunschweig, West Germany
D. Benest, Observatoire de Nice, Le Mont Gros, F-06007 Nice
    Cedex, France
D.G. Bettis, Munich
K.B. Bhatnagar, IA/47-C Ashok Vihar, Delhi-110052, India
R. Bien, Astronomisches Recheninstitut, Munchhofstrasse 12-14,
    69 Heidelberg 1, West Germany
V.R. Bond, NASA Johnson Space Center, FM5, Houston, Texas 77058, U.S.A.
P. Bretagnon, (see Abu-el-Ata)
C. Brookes, Department of Mathematics, University of Aston in
    Birmingham, Gosta Green, Birmingham B4 7ET, U.K.
O. Calame, CERGA, Avenue Nicolai Copernic, 06130 Grasse, France
J. Chapront, (see Abu-el-Ata)
T. Christides, Department of Theoretical Mechanics, University of
    Thessaloniki, Thessaloniki, Greece
H. Claes, Facultés Universitaires de Namur, Rempart de la Vierge 8,
    B-5000 Namur, Belgium
A. Deprit, Department of Mathematics, University of Cincinnati,
    Cincinnati, Ohio 45221, U.S.A.
L. Duriez, Laboratoire d'Astronomie, 1 Impasse de l'Observatoire,
    59000 Lille, France
R. Dvorak, Universitätssternwarte Graz, Universitätsplatz 5, 8010
    Graz, Austria
B. Garfinkel, Department of Astronomy, Yale University, New Haven,
    Connecticut 06520, U.S.A.
J.D. Hadjidemetriou, (see Christides)
D.C. Heggie, Department of Mathematics, University of Edinburgh,
    King's Buildings, Mayfield Road, Edinburgh, EH9 3JZ, U.K.
J. Henrard, (see Claes)
P. Herget, Cincinnati Observatory, Observatory Place, Cincinnati,
    Ohio 45208, U.S.A.
H.G. Hertz, 2301 E St. NW, Apartment A608, Washington, D.C. 20037,
    U.S.A.
S. Hughes, University of Leicester, Department of Astronomy and
    History of Science, University Road, Leicester, LE1 7RH, U.K.
G. Janin, ESOC, Robert-Bosch-Strasse 5, D-6100, Darmstadt, West

Germany
A.H. Jupp, Department of Applied Mathematics, University of Liverpool,
     P.O. Box 147, Liverpool, L69 3BX, U.K.
D.G. King-Hele, Royal Aircraft Establishment, Farnborough, Hampshire,
     U.K.
H. Kinoshita, (see Aksnes)
Y. Kozai, Tokyo Astronomical Observatory, Tokyo, Japan 181
J.H. Lieske, Jet Propulsion Laboratory, 4800 Oak Grove Drive,
     Pasadena, California 91103, U.S.A.
V.V. Markellos, Department of Astronomy, The University, Glasgow,
     G12 8QQ, U.K.
J.J. Martinez Benjamin, University of Texas at Austin, Department
     of Aerospace Engineering and Engineering Mechanics, Austin,
     Texas 78712 U.S.A
P.J. Message, (see Jupp)
M. Michalodimitrakis, (see Christides)
M. Moons, (see Claes)
J.D. Mulholland, Department of Astronomy, 15.212 R.L. Moore Hall,
     College of Natural Sciences, University of Texas at Austin,
     Austin, Texas 78712, U.S.A.
C. Oesterwinter, NSWC/DL Code DK-10, Dahlgren, Virginia 22448, U.S.A.
G. Ratier, Observatoire du Pic du Midi, 65200 Bagneres de Bigorre,
     France
W.J. Robinson, Department of Mathematics, University of Bradford,
     BD7 1DP, U.K.
E.A. Roth, (see Janin)
J.L. Sagnier, (see Abu-el-Ata)
J. Schubart, (see Bien)
P.K. Seidelmann, U.S. Naval Observatory, Washington, D.C. 20390,
     U.S.A.
R. Sergysels, Ecole Polytechnique, Faculté des Sciences Appliquées,
     Université Libre de Bruxelles, Avenue F.D. Roosevelt 50, 1050
     Bruxelles, Belgium
A.T. Sinclair, Royal Greenwich Observatory, Herstmonceux Castle,
     Hailsham, Sussex, U.K.
N. Spyrou, Institute of Astronomy, Madingley Road, Cambridge,
     CB3 OHA, U.K.
D. Standaert, (see Claes)
V. Szebehely, (see Martinez Benjamin)
D.B. Taylor, (see Markellos)
J.P. Vinti, Massachusetts Institute of Technology, Building W91-202,
     Cambridge, Massachusetts 02139, U.S.A.
J. Waldvogel, Eidgenossische Technische Hochschule, Seminar für
     Angewandte Mathematik, Clausiusstrasse 55, 8006 Zurich,
     Switzerland
C.A. Williams, Department of Astronomy, University of South Florida,
     Tampa, Florida 33620, U.S.A.
C. Zagouras, Department of Mechanics, University of Patras, Patras,
     Greece
K. Zare, Farah Park, Afarin Street No. 19, Isfahan, Iran

PART I

PLANETARY THEORY AND ANALYTICAL METHODS

PLANETARY THEORIES AND OBSERVATIONAL DATA

R. L. Duncombe   and   P. K. Seidelmann
University of Texas   U. S. Naval Observatory

ABSTRACT

   A brief review is given of planetary theories from
Leverrier to Newcomb to the age of computers. The presently
used planetary theories are discussed and the process of
replacing these theories with new ones is described. Some
difficulties in preparing new planetary theories and the observational discrepancies which have been encountered previously are discussed.

HISTORY

   The first systematic application of dynamical principles
to the motions of planetary bodies was made by Leverrier for
the theories of the motions of the planets Mercury, Venus,
Earth, and Mars for the epoch 1850. (Leverrier 1858, 1859,
1861a, 1861b). These theories suffered from two defects:
(1) they were not based on a consistent set of planetary
masses, and (2) they did not satisfy the observed longitudes
of the planets. Leverrier used only Newtonian mechanics and
was unable to account for the secular motions of the perihelia
Therefore, he augmented his dynamical theories by nondynamical
terms in order to achieve agreement with the observations.
Leverrier's theories formed the basis for the tables of Mercury, Venus, Earth and Mars in the various national ephemerides until 1900 and in the Connaissance de Temps until 1960.

   In 1900, the planetary theories of Simon Newcomb were
generally adopted. Newcomb's theories of Mercury, Venus,
Earth and Mars (Newcomb 1895) were also based on Newtonian
dynamics, but they incorporated a consistent system of
planetary masses. Again the theories did not satisfy the
observational data with respect to the observed motions of
the perihelia, and so Newcomb was forced to assume that the

3

V. Szebehely (ed.), Dynamics of Planets and Satellites and Theories of Their Motion, 3-14.

gravitational forces toward the Sun did not vary exactly
according to the inverse square law of the distance. Thus,
in the expression for the gravitation, $f = mm'/r^n$, where m
and m' are the masses of the two bodies and r is the distance
between them, the exponent n of r was assumed to be of the
form $n = 2 + \Delta$, where Δ is a small quantity. In this case
the perihelion of each planet will have a direct motion
found by multiplying its mean motion by $\frac{1}{2}$ Δ. Newcomb used
the value of n = 2.0000001612 and augmented the secular
motion of the perihelia of the four planets by the corresponding amount. Shortly after the introduction of Newcomb's
tables of Mars into the national ephemerides, it was realized
that they failed to represent the observations of the planet
by amounts large enough to indicate errors in the tables.
F. E. Ross in 1912 concluded that the eccentricity of the
orbit adopted by Newcomb was in error. He compared the observations of Mars with Newcomb's tables and obtained new
values of the elements and calculated corrections to
Newcomb's tables, which have been incorporated in the ephemerides of the planet since (Ross 1917). In 1950 Eckert,
Brouwer, and Clemence (1951) produced a simultaneous numerical integration of the orbits of the five outer planets,
which replaced the ephemerides based on the tables of Hill
for Jupiter and Saturn and Newcomb for Uranus and Neptune.
By that time the observations of Saturn and Neptune had de-
parted significantly from the ephemerides. In 1960 these
special perturbation theories were introduced as the basis
for the tables of the outer planets. To this day the tables
of the inner planets printed in the national ephemerides are
based on the theories of Newcomb with Mars revised by Ross,
and the integration by Eckert, Brouwer and Clemence is the
basis for the tables of the outer planets.

PRESENT STATUS

    In the meantime Clemence (1949, 1961) has used Hansen's
method to calculate a new general theory for Mars and
R. E. Laubscher (1971) has derived definitive constants for
this theory based on the observations of Mars from 1750 to
1971. Improved elements for Mercury, Venus and Earth have
also been determined by comparison of Newcomb's theories
with observations (Clemence 1943, Duncombe 1958, Morgan 1933).
Sharaf (1955, 1964) has prepared a numerical general theory
of the motion of the planet Pluto, while Cohen, Hubbard and
Oesterwinter (1967) performed a numerical integration to
calculate an improved ephemeris of Pluto. Numerical integrations have been used by numerous individuals and groups to
calculate improved ephemerides, particularly for special
purposes or limited periods of time (e.g. Ash et al, 1971,
Oesterwinter and Cohen 1972, Standish et al, 1976).

It is recognized that before planetary theories can be significantly improved some of the underlying constants must be corrected. For this reason in 1970 Commission 4 of the IAU appointed Working Groups on units and time scales, precession, and planetary ephemerides. A Joint Report of these working groups will be submitted to the IAU for adoption in Grenoble this year. This report includes recommendations for a new set of planetary masses, a new value for the precessional constant, a definition of time scales for dynamical theories, and a consistent set of basic constants. Once these recommendations are adopted by the IAU it is anticipated that new planetary theories will be prepared at various institutions and by several different methods. These new theories will permit the preparation of ephemerides which can be fitted to the observations and which can be intercompared to ensure their accuracy. For this purpose at the U. S. Naval Observatory, in cooperation with other organizations, the preparation of new general theories, particularly for the inner planets, are being undertaken. The methods of Musen and Carpenter are being applied to determine numerical general theories (Musen and Carpenter 1963, Carpenter 1963, 1965, 1966, 1966). Numerical integrations of all the planets of the solar system are being planned and observations for all the planets are being collected and reduced to the FK4 system. The observations will be compared with the general theories and the numerical integrations in order to rectify the constants. It is hoped that these efforts will result in accurate ephemerides for all the planets that will satisfy the observations over an extended period of time.

However, before the impression is given that all the problems with respect to the formation of planetary theories are solved and that the process is perfectly straightforward, it is advisable to describe some of the difficulties which have been experienced with the theories and some of the observational problems that have been encountered.

PROBLEMS

In calculating a general theory for a planet there is an ever present problem of convergence. With classical methods of calculating general theories, as higher orders are calculated, the amplitude of the changes of the coefficients reduces, and it was assumed that convergence was being achieved. The practical considerations of hand computations limited the process to somewhere between second and third order and the numerical contributions at that order were small. While there was no method of testing for convergence, it was assumed that contributions from higher orders would be smaller and that they would not be cumulative.

With the availability of computers for calculating general
theories, either numerical or analytical, it has become feasible and advisable to depart from the concept of orders and
instead to use an iterative approach. With iterative methods
the expressions for the motion of the planet are substituted
into the fundamental equations of motion and a new theory
for the planetary motion is generated. This new theory can
then be substituted back into the equations of motion and
the process repeated. The assumption is made that eventually, when the theory is substituted into the equations of
motions, identical expressions are derived and thus the
theory has converged. Experience has shown that this process
apparently works well except when dealing with resonant terms.
When resonance is involved, several possibilities arise;
either the values of the coefficients may oscillate about
some value, or the coefficients change from iteration to
iteration, normally finding a minimum or maximum and then
diverging from those values with each successive iteration
in a parabolic manner, or the values of the coefficients in
each successive iteration assume an asymptotic approach to
some value. (Figure 1) Where there is a single predominant
resonant term, such as in the case of the Earth perturbed by
Venus, it appears possible to isolate the term and determine
its value independently. When this value of the resonant
term and the other coefficients are substituted into the
equations of motion, the results are equivalent to the input
values. Thus a form of convergence is achieved. In more
complicated cases such as the Jupiter-Saturn resonance, it
is not certain that convergence by such manipulations can be
achieved.

The second problem arising in the calculation of general
theories, and possibly related to the first problem, is the
question of accuracy. General planetary theories have in
most cases failed to deliver the positional accuracy to
which they were calculated. This could be due to a lack of
convergence, neglected terms, or other causes and indicates
that extreme care must be exercised to guard against these
sources of error. Therefore, careful investigation of the
accuracy of a general theory should be made, in addition to
comparing the theory to observations.

The basic purpose of a planetary theory is to represent
the motion of the planet, particularly as observed from the
Earth, so the third problem relates to the comparison of the
theories with the observational data. This is particularly
evident in regard to the latitude residuals of Uranus and
Neptune. Having fitted the numerical integrations to observations, in the case of Uranus to normal points and in the
case of Neptune to the individual observations, a systematic
effect remains in the residuals. (Figures 2 and 3) For Uranus

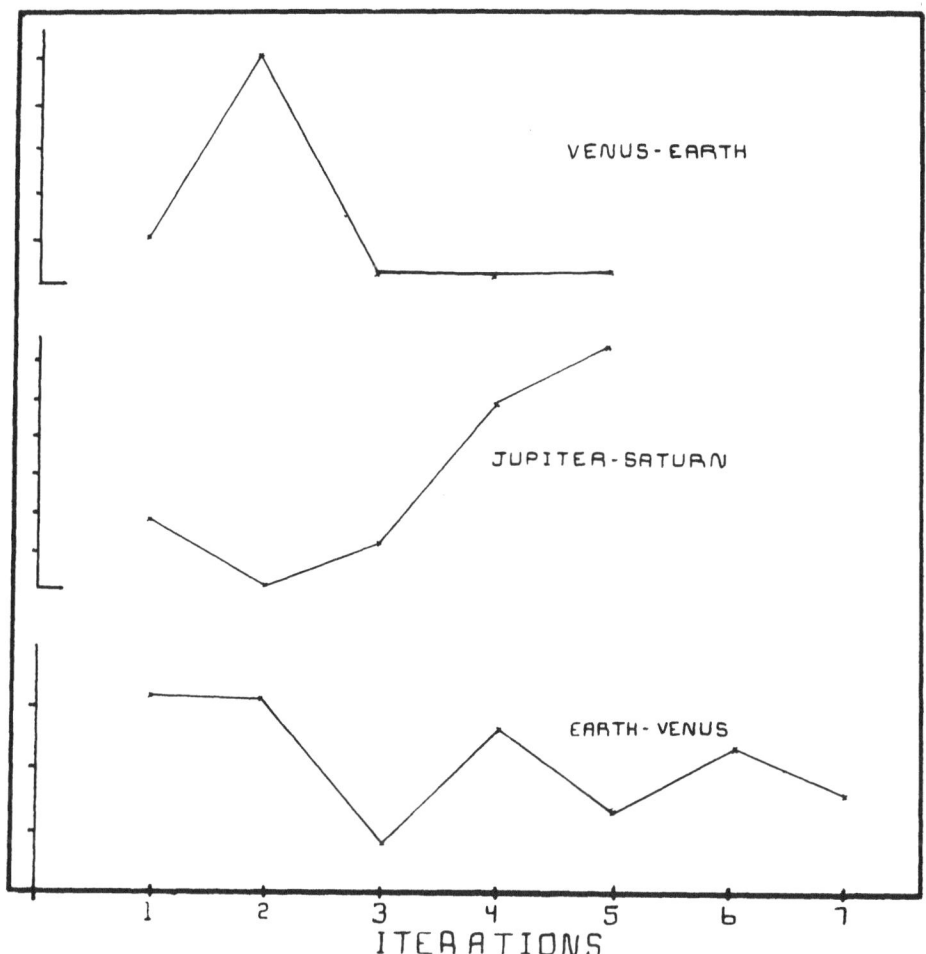

Figure 1.   Plot of convergence characteristics of resonance
            terms.

more than two orbital periods are covered by the observations,
and the residuals in latitude have the characteristic of a
secular change. The effect was observed by Newcomb (1898),
Wylie (1947) and Seidelmann, Duncombe and Klepczynski (1969).
The latitude residuals of Neptune show a periodic effect,
but it must be remembered that a complete period of Neptune's
orbit has not been observed (Seidelmann et al 1971). The
residuals, after a least squares fitting to data with a
secular effect but not covering the whole orbital period,
would have the periodic signature observed. Thus, it appears
that there is a secular effect in the observations of both
Uranus and Neptune, which is either due to an omitted effect
in the calculation of the theories, or due to a systematic
error present in the observations of these planets. Further,

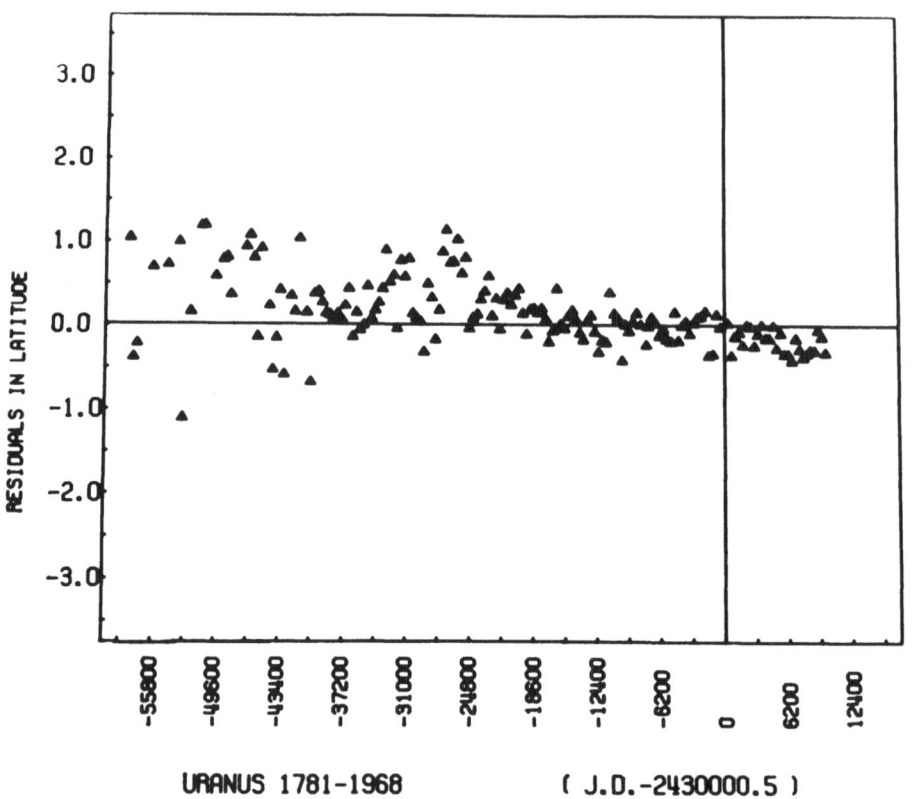

Figure 2.   Latitude residuals of Uranus.

when the best fit of the observations to the ephemerides of
the outer planets is performed and the residuals are analyzed,
significant periodic variations are found in the residuals.
Unfortunately no logical hypothesis appears to satisfy the
data.   There is also the disturbingly large difference between the mass determinations for Uranus.   The reciprocal
mass of Uranus, determined from the observations of Saturn,
is 22693 ± 33 (Klepczynski et al 1970), while the determination from the satellite observations is 22945 ± 15 (Dunham,
1971).

The fourth problem might be described as the opposition, or phase, effect.   When the optical observations of
the brighter planets are compared with ephemerides, a systematic effect is present which is a function of the time before and after opposition (Standish et al 1976). (Figures 4
and 5) This systematic effect in the residuals is found to
be present when observations are compared to several different theories.   Thus, this is most likely an observational
error, perhaps due to an incomplete phase correction to the

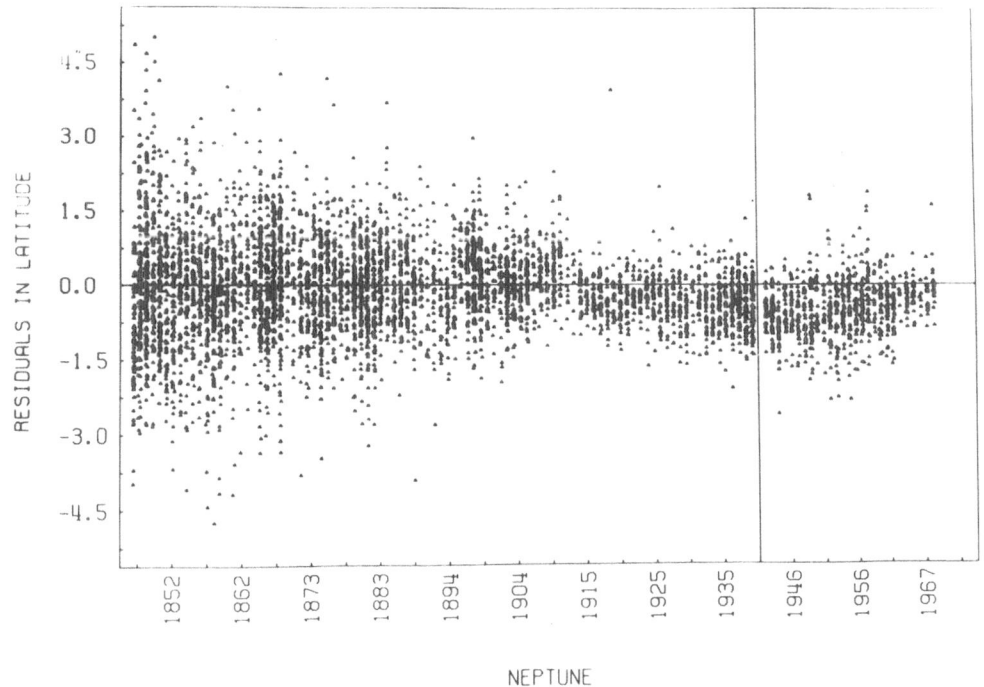

NEPTUNE

Figure 3.   Latitude residuals of Neptune.

observations, or the effect of irradiation in the observer's
personal equation, or some effect which is a function of the
brightness of the object.

     The fifth problem is the secular variation of the obli-
quity of the ecliptic as discussed by Duncombe and Van
Flandern (1976).   The planetary observational data has indicated a correction to Newcomb's secular variation of the
obliquity of (-0.277+.18) T where T is in centuries from
1900.0.   Further there seems to be a difference between the
observations made before 1900 and those made after 1900,
which may indicate how rigorously the observations have been
reduced to the FK3, FK4 system. (Figure 6) If this is truly
an observational problem it could be related to the observational effects observed in the residuals of Uranus and Neptune.   The evidence seems to indicate that the theoretical
value for the secular variation for the obliquity of the
ecliptic is correct, but that care must be exercised in the
use of pre-1900 observations, particularly of the inner
planets which are observed as day objects.

     The sixth problem may be described as the many aspects
of that peculiar planet Pluto.   The knowledge of Pluto is

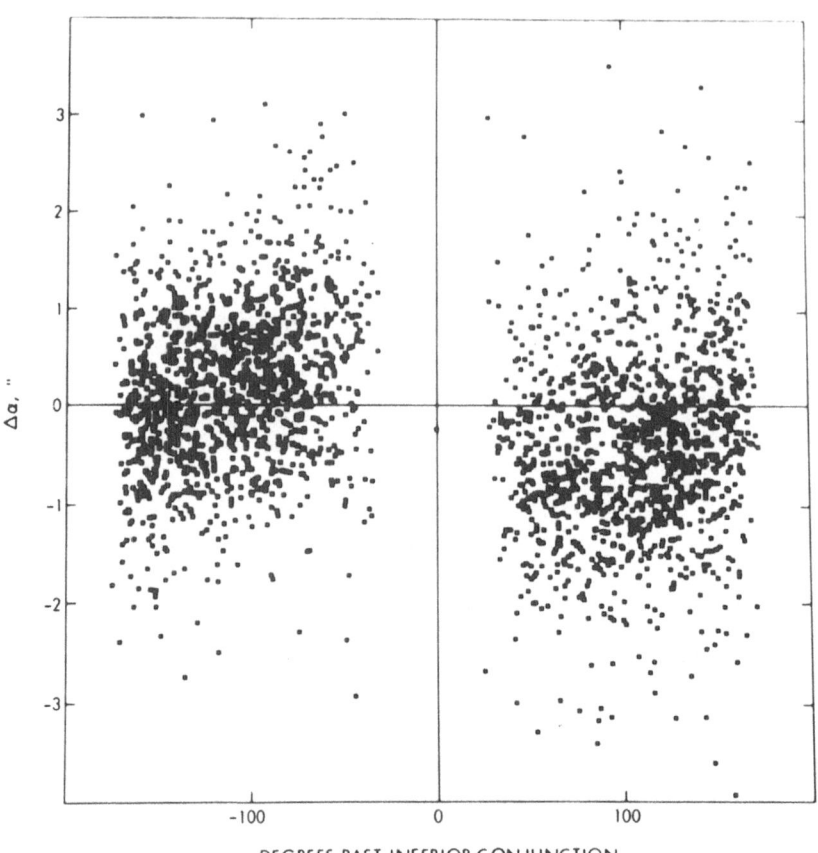

Figure 4.   Optical residuals illustrating "opposition effect"
             for Mercury. (Standish et al, 1976)

very limited in part due to the fact that observations for
only about one fourth of its orbital period are available.
There are estimates of its mass and radius, but they lead to
an uncomfortably large density. Even so the mass estimates
are so low that the prediction of the existence of Pluto
based solely on its effects on the orbit of Uranus is impossible. Yet Pluto was discovered very close to its predicted
position. Either its discovery was a case of serendipity or
perhaps only one of many objects in that part of the solar
system has been discovered. Assuming that the mean elements
of Pluto are sufficiently well known to merit the preparation
of a general theory, there remains the problem that the orbit of Pluto crosses the orbit of Neptune. This makes it
impossible to use harmonic analysis on the distances between
those two bodies and thus prohibits the calculation of a
general theory of Pluto, except by techniques such as that

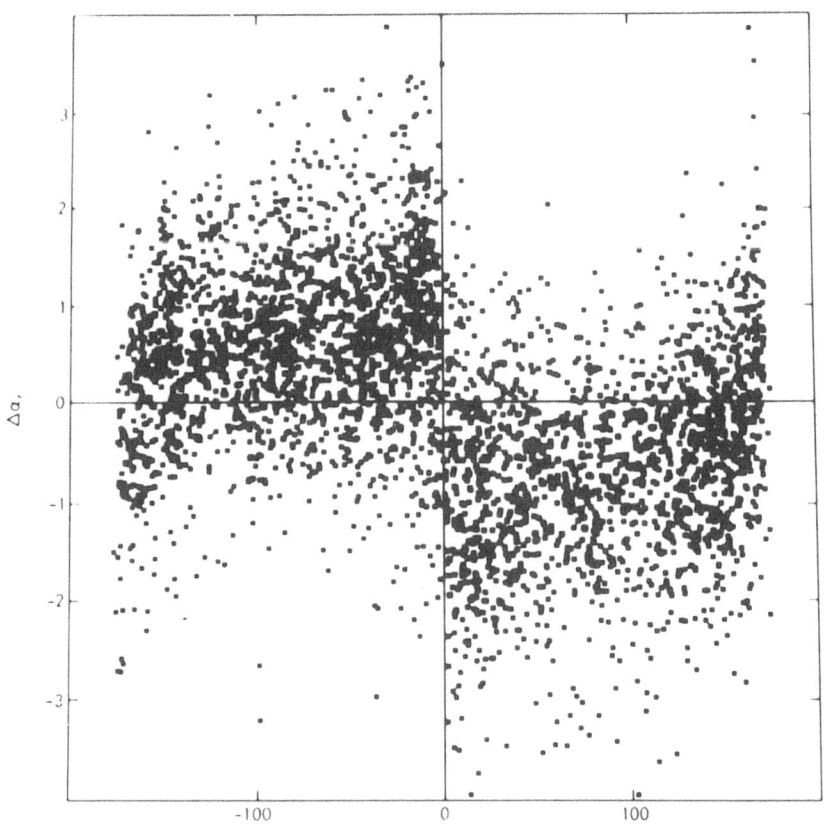

DEGREES PAST INFERIOR CONJUNCTION

Figure 5.   Optical residuals illustrating "opposition effect"
            for Venus.   (Standish et al, 1976)

of Goodrich and Carpenter (1966) where an exact resonance
is assumed, or that of Sharaf (1955, 1964) where numerical
integration is used for the Neptune-Pluto terms, or that of
Petrovskaya (1972) where special expansions are used.

    The seventh problem concerns the use of mixtures of the
various types of observational data; namely optical, radar
and spacecraft observations. Optical observations provide an
angular determination in two coordinates of the position of
the object as projected on the celestial sphere. With tran-
sit circle instruments one observation per day can be made
at a given observatory. Radar observations provide an
accurate observation of the distance from the Earth to the
object and back in a short span of time, and many observations can be made on a given day at a given observatory.
Spacecraft data also provide accurate range or range-rate

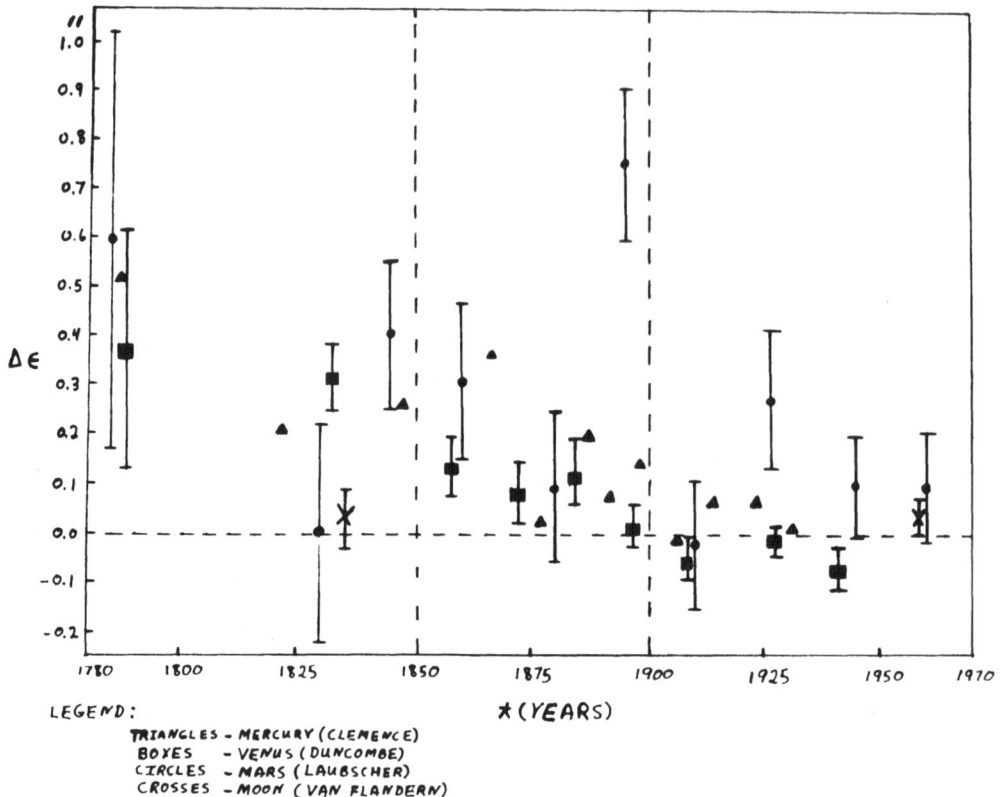

Figure 6.   Determinations of the obliquity of the ecliptic.

determinations and many observations are possible in a
short period of time from a given observatory. The optical
data covers a long period of time with some improvements in
the accuracy of the observations over that period, although
the possibility remains of the presence of systematic errors,
particularly in the older observations. The spacecraft and
radar data cover a shorter time period with significant
changes in the accuracy of the observations over that period.
The important question here concerns the manner in which
these observations are combined. What is the best way of
determining the relative weights to be applied to these
different types of observations in order that the best possible ephemeris for the object may be determined?

        The eighth problem concerns the implications of the
lunar observational data. Van Flandern (1975, 1976) has
hypothesized a secular change in the gravitational constant
based on occultation observations of the Moon. Whether the
hypothesis of such a change in the gravitational constant
is correct or not, there is some systematic effect present

in the lunar observations. The Moon, due to its short period, tends to magnify effects which may also be present in plane- tary data. Thus it can be anticipated that whatever is af- fecting the observations of the Moon will also affect plane- tary observations with a smaller amplitude, or over a longer period of time.

These problems, and perhaps others as yet undetected, face us as we commence the task of improving the theories of motion of the principal planets. It will require the cooperation and effort of many investigators in the fields of astrometry, dynamical astronomy and celestial mechanics to bring this project to a successful conclusion.

REFERENCES

Ash, M. E., Shapiro, I. I., and Smith, W. B., 1971, Science 174, 551.

Carpenter, L., 1963, NASA Tech. Note D-1898.

Carpenter, L., 1965, NASA Tech. Note D-2852.

Carpenter, L., 1966, NASA Tech. Note D-3078.

Carpenter, L., 1966, NASA Tech. Note D-3168.

Clemence, G. M., 1943, Astron. Pap. Amer. Ephemeris 11, part 1

Clemence, G. M., 1949, Astron. Pap. Amer. Ephemeris 11, part 2 1961, 16, part 2.

Cohen, C. J., Hubbard, E. C., and Oesterwinter, C., 1967, Astron. J., 72, 973.

Duncombe, R. L., 1958, Astron. Pap. Amer. Ephemeris 16, part 1

Duncombe, R. L., Seidelmann, P. K., and Janiczek, P. M., 1974, Highlights of Astronomy, 223.

Duncombe, R. L. and Van Flandern, T. C., 1976, Astron. J. 81, 281.

Dunham, D. W., 1971, Dissertation, Yale University.

Eckert, W. J., Brouwer, D., and Clemence, G. M., 1951, Astron. Pap. Amer. Ephemeris 12.

Goodrich, E. F. and Carpenter, L., 1966, Goddard Space Flight Center X-643-66-133.

Klepczynski, W. J., Seidelmann, P. K. and Duncombe, R. L., 1970, Astron. J. 75, 739.

Laubscher, R. E., 1971, Astron. & Astrophys. 13, 426.

Leverrier, U. J. J., 1858, Ann. de l'Obs. Paris 4, 1.

Leverrier, U. J. J., 1859, Ann. de l'Obs. Paris 5, 1.

Leverrier, U. J. J., 1861a, Ann. de l'Obs. Paris 6, 1.

Leverrier, U. J. J., 1861b, Ann. de l'Obs. Paris 6, 185.

Morgan, H. R., 1933, Astron. J. 42, 149.

Musen, P. and Carpenter, L., 1963, J. Geophys. Res. 68, 2727.

Newcomb, S., 1895, Astron. Pap. Amer. Ephemeris 6.

Newcomb, S., 1898, Astron. Pap. Amer. Ephemeris 7.

Oesterwinter, C. and Cohen, C. J., 1972, Celestial Mechanics 5, 317.

Petrovskaya, M. S., 1972, Celestial Mechanics 6, 328.

Ross, F. E., 1917, Astron. Pap. Amer. Ephemeris 9, 251.

Seidelmann, P. K., Duncombe, R. L., and Klepczynski, W. J.,
    1969, Astron. J. 74, 776.

Seidelmann, P. K., Klepczynski, W. J., Duncombe, R. L., and
    Jackson, E. S., 1971, Astron. J. 76, 488.

Sharaf, Sh. G., 1955, Trans. of Inst. of Theor. Astr. 4, USSR
    Academy of Sciences Press, Moscow-Leningrad (NASA Tech.
    Translation F-490).

Sharaf, Sh. G., 1964, Trans. of Inst. of Theor. Astr. 10,
    USSR Academy of Sciences Press, Moscow-Leningrad (NASA
    Tech. Translation F-491).

Standish, E. M., Keesey, M. S., and Newhall, XX, 1976, JPL
    Tech. Report 32-1603.

Van Flandern, T. C., 1975, Month. Not. R. Astr. Soc. 170,
    333.

Van Flandern, T. C., 1976, Scientific American 234, 44, Feb.

Wylie, L. R., 1947, Publ. U. S. Naval Obs. XV, 2nd Series III.

CORRESPONDANCES ENTRE UNE THEORIE GENERALE PLANETAIRE EN VARIABLES
ELLIPTIQUES ET LA THEORIE CLASSIQUE DE LE VERRIER

L. DURIEZ
Université des Sciences et Techniques de Lille (France)

ABSTRACT. In order to improve the determination of the mixed terms in
classical theories, we show how these terms may be derived from a
general theory developed with the same variables (of a keplerian nature).
We find that the general theory of the first order in the masses already
allows us to develop the mixed terms which appear at the second order in
the classical theory. We also show that a part of the constant perturbation of the semi-major axis introduced in the classical theory is
present in the general theory as very long-period terms; by developing
these terms in powers of time, they would be equivalent to the appearance
of very small secular terms (in t, t^2, ...) in the perturbation of the
semi-major axes from the second order in the masses. The short period
terms of the classical theory are found the same in the general theory,
but without the numerical substitution of the values of the variables.

INTRODUCTION

Les résultats récents obtenus par Bretagnon et Simon (1976) lors de la
comparaison de leur théorie planétaire de type classique (amélioration
de celle de Le Verrier), avec une intégration numérique sur 1000 ans
(type A.P.A.E.) montrent schématiquement que la précision atteinte sur
les perturbations à courtes périodes des éléments elliptiques est très
satisfaisante; par contre, il subsiste encore des écarts lentement
variables avec le temps et importants vis à vis de la précision atteinte
sur les termes à courtes périodes. Ces écarts peuvent être réduits
notablement en ajoutant à la théorie certains termes séculaires et mixtes
empiriques. D'ailleurs, les termes mixtes ainsi introduits ne sont relatifs qu'à certains termes à longues périodes (par exemple il s'agit du
terme $2\lambda_J-5\lambda_S$ dans le cas de Jupiter ou de Saturne).
    Dans le but d'améliorer la connaissance de ces quelques termes
mixtes, nous avons cherché à les mettre en évidence à partir d'une
théorie générale développée avec les mêmes variables que la théorie classique. En procédant ainsi, nous verrons que pour obtenir ces termes, la
théorie générale a besoin d'être poussée à un ordre des masses moins
élevé que la théorie classique. Nous avons également mis en évidence le

15

V. Szebehely (ed.), Dynamics of Planets and Satellites and Theories of Their Motion, 15-32.
All Rights Reserved. Copyright © 1978 by D. Reidel Publishing Company, Dordrecht, Holland.

fait qu'une partie de la perturbation constante du demi-grand axe (celle
qui dépend des excentricités et des inclinaisons) se retrouve dans la
théorie générale sous forme de termes à très longues périodes; nous verrons que le développement de ces termes au voisinage d'un instany initial
serait équivalent à l'apparition dans la théorie classique, d'un terme
séculaire en t à l'ordre deux des masses dans la perturbation du demi-
grand axe.

On utilise les notations suivantes:
k la constante de gravitation, $m_0 = 1$ la masse du Solei, m la masse
d'une planète, puis les éléments osculateurs héliocentrique de cette
planète à un instant τ, rapportés à un repère héliocentrique de direction fixe: a le demi-grand axe, e l'excentricité, i l'inclinaison, l la
longitude moyenne de l'époque, $\bar\omega$ la longitude du périhélie, Ω celle du
noeud.

On note encore:

$$n = \sqrt{k(1 + m)}\ a^{-3/2} \qquad \text{le moyen mouvement}$$

$$\lambda = n(t - \tau) + l \qquad \text{la longitude moyenne à l'instant t}$$

$$z = e \exp \sqrt{-1}\ \bar\omega \qquad \text{et } \bar z \text{ son conjugué}$$

$$\zeta = \sin \frac{i}{2} \exp \sqrt{-1}\ \Omega \qquad \text{et } \bar\zeta \text{ son conjugué}$$

La théorie générale et la théorie classique partent l'une et
l'autre des mêmes êquations (de Lagrange), écrites ainsi pour une planète d'indice i perturbée par une planète d'indice j:

$$\frac{1}{a_i} \frac{da_i}{dt} = \mu\ \mathcal{L}_{ij}^{(1)} \qquad\qquad \frac{dl_i}{dt} = \frac{d\lambda_i}{dt} - \eta_i = \mu\mathcal{L}_{ij}^{(2)}$$

$$\frac{dz_i}{dt} = \mu\ \mathcal{L}_{ij}^{(3)} \qquad\qquad \frac{d\bar z_i}{dt} = \mu\overline{\mathcal{L}_{ij}^{(3)}} \qquad\qquad (1)$$

avec: $\quad \dfrac{d\zeta_i}{dt} = \mu\ \mathcal{L}_{ij}^{(4)} \qquad\qquad \dfrac{d\bar\zeta_i}{dt} = \mu\ \bar{\mathcal{L}}_{ij}^{(4)}$

$$m_j = \mu\varepsilon_j$$

$$\mathcal{L}_{ij}^{(e)} = \sqrt{-1}\ \frac{n_i}{1+m_i}\ \varepsilon_j \sum_{k_i,k_j} L_{k_i,k_j}^{(e)} (a_i,a_j,z_i,z_j,\bar z_i,\bar z_j,$$

$$\zeta_i,\zeta_j,\bar\zeta_i,\bar\zeta_j) \exp \sqrt{-1}(k_i\lambda_i + k_j\lambda_j) \qquad\qquad (2)$$

où $L_{k_i,k_j}^{(e)}$ est une fonction de $\alpha = \dfrac{\min(a_i,a_j)}{\max(a_i,a_j)}$, développable en série
entière de z_i, z_j, $\bar z_i$, $\bar z_j$, ζ_i, ζ_j, $\bar\zeta_i$, $\bar\zeta_j$; k_i et k_j sont des entiers
realtifs. (cf. Duriez, 1976).
Chacune des théories représente une solution des êquations (1),
obtenue par approximations successives sous forme d'un développement
ordonné suivant les puissances croissantes des masses, c'est à dire

encore suivant les puissances de μ. Pour bien mettre en évidence les
différences et aussi les relations existant entre les deux théories, et
pour introduire les notations, rappelons d'abord brièvement les princi-
pales caractéristiques de la théorie classiques; nous verrons ensuite
comment cette théorie peut être "généralisée".

LA THEORIE CLASSIQUE

Elle consiste principalement à développer les seconds membres des équations (1) en séries de Taylor par rapport aux 6 éléments osculateurs de
chaque planète, au voisinage d'un instant initial fixé $\tau=\tau_0$. La première
approximation (ordre zéro en μ) est alors le mouvement képlérien, avec
les valeurs numériques des éléments osculateurs à cet instant: a_{i_0}, l_{i_0},
z_{i_0}, $\bar{z}_{i_0}$, ζ_{i_0}, $\bar{\zeta}_{i_0}$ et avec n_{i_0} calculé par la 3eme loi de Kepler:

$$n_{i_0}^2 \, a_{i_0}^3 = k \, (1 + m_i).$$

On recherche ensuite une solution de la forme:

$$a_i = a_{i_0} + \mu \Delta_1 a_i + \mu^2 \Delta_2 a_i + \ldots$$

$$l_i = l_{i_0} + \mu \Delta_1 l_i + \mu^2 \Delta_2 l_i + \ldots$$

$$z_i = z_{i_0} + \mu \Delta_1 z_i + \mu^2 \Delta_2 z_i + \ldots \qquad (3)$$

$$\zeta_i = \zeta_{i_0} + \mu \Delta_1 \zeta_i + \mu^2 \Delta_2 {}_i + \ldots$$

Pour cela, on identifie les termes de même ordre en μ dans la dérivée
des expressions (3), et dans le développement de Taylor de la relation
(2):

$$\mu \mathcal{L}_{ij}^{(e)} = \mu [\mathcal{L}_{ij}^{(e)}]_0 + \mu^2 \sum_{k=i \text{ et } j} \{ \Delta_1 a_k \, [\frac{\partial \mathcal{L}_{ij}^{(e)}}{\partial a_k}]_0 + \Delta_1 l_k \, [\frac{\partial \mathcal{L}_{ij}^{(e)}}{\partial l_k}]_0$$

$$+ \Delta_1 z_k \, [\frac{\partial \mathcal{L}_{ij}^{(e)}}{\partial z_k}]_0 + \Delta_1 \bar{z}_k [\frac{\partial \mathcal{L}_{ij}^{(e)}}{\partial \bar{z}_k}]_0 + \Delta_1 \zeta_k [\frac{\partial \mathcal{L}_{ij}^{(e)}}{\partial \zeta_k}]_0 + \Delta_1 \bar{\zeta}_k [\frac{\partial \mathcal{L}_{ij}^{(e)}}{\partial \bar{\zeta}_k}]_0 \} + \ldots$$

$$(4)$$

où la notation $[\ldots]_0$ signifie que $\mathcal{L}_{ij}^{(e)}$ et ses dérivées partielles sont
calculées numériquement pour la valeur des éléments osculateurs à
l'instant τ_0; ce sont ainsi des développements de la forme:

$$\mu \sum_{k_i, k_j} A_{k_i, k_j}^{(e)} \quad \exp \sqrt{-1} \, (k_i \eta_{i_0} + k_j \lambda_{j_0}) \qquad (5)$$

où $A_{k_i k_j}^{(e)}$ est une constante numérique, et où λ_{i_0} et λ_{j_0} sont de la forme d'une fonction connue du temps:

$$\lambda_{i_0} = n_{i_0} (t - \tau_0) + l_{i_0}. \tag{6}$$

On sait qu'à l'ordre 1 en μ, $\mu[\mathcal{L}_{ij}^{(e)}]_0$ contient un terme constant: $\mu A_{00}^{(e)}$, nul pour e=1, qui par intégration donne un terme séculaire linéaire en t (noté suivant le cas $\mu\Delta_1^{(s)} l_i$, $\mu\Delta_1^{(s)} z_i$, $\mu\Delta_1^{(s)} \zeta_i$), tandis que les autres termes donnent des termes périodiques qu'on note: $\mu\Delta_1^{(p)} a_i$, $\mu\Delta_1^{(p)} l_i$, $\mu\Delta_1^{(p)} z_i$ et $\mu\Delta_1^{(p)} \zeta_i$.

On remarque l'absence de terme séculaire par les a_i et les n_i. D'un point de vue analytique, $\Delta_1^{(s)} l_i$ admet un développement pair par rapport aux excentricités et aux inclinaisons commençant au degré zéro, tandis que $\Delta_1^{(s)} z_i$ et $\Delta_1^{(s)} \zeta_i$ admettent des développements impairs de ces variables commençant au degré 1.

A l'ordre 2, on doit effectuer dans le développement (4), le produit de la solution du 1er ordre avec les dérivées partielles de $\mathcal{L}_{ij}^{(e)}$: le terme constant de ces dérivées partielles se combine à l'éventuel terme séculaire du 1er ordre pour donner par intégration un terme en t^2, et aux inégalités périodiques du 1er ordre pour donner de nouvelles inégalités périodiques de même période. Ensuite, chaque terme périodique des dérivées partielles se combine d'une part au terme séculaire du 1er ordre pour donner après intégration un terme mixte en t exp $\sqrt{-1}(k_i\lambda_{i_0} + k_j\lambda_{j_0})$ et une inégalité périodique de même période; il se combine d'autre part avec les inégalités périodiques du 1er ordre pour donner des inégalités périodiques avec de nouvelles périodes, dont éventuellement un terme constant qui donne par intégration un terme séculaire en t d'ordre 2 des masses.

A cet ordre, les éléments a_i et n_i ne possèdent toujours pas de terme séculaire en t et t^2, mais comportent des termes mixtes (théorème de Poisson).

Aux ordres supérieurs en μ, on obtient d'autres inégalités périodiques, ainsi que d'autres termes séculaires et mixtes, mais pour les a_i et n_i, le problème de la présence de termes séculaires est encore ouvert.

Ainsi, la théorie classique en variables elliptiques est surtout caractérisée par une certaine "stablité" des seuls éléments a_i et n_i (mis à part les termes mixtes). On peut cependant obtenir le même type de stabilité (absence de terme séculaire) pour les éléments l_i en utilisant dès l'ordre 1 en μ l'expression suivante à la place de (6):

$$\lambda_{i_0} = (n_{i_0} + \Delta^{(s)} l_i) (t - \tau_0) + l_{i_0} \tag{7}$$

où $\Delta^{(s)} l_i t$ représente le terme séculaire de la longitude moyenne de l'époque τ_0, qui ne doit alors plus figurer dans l'expression (3) de l_i. Pour cela, le terme constant de l'équation relative à l_i sert à former une perturbation constante dans le demi-grand axe. Dans ces conditions, $\Delta_1 a_k$ et $\Delta_1 l_k$ étant dépourvus de termes séculaires, les termes mixtes ne peuvent provenir à l'ordre 2 en μ que des termes séculaires de $\Delta_1 z_k$, $\Delta_1 \bar{z}_k$, $\Delta_1 \zeta_k$ et $\Delta_1 \bar{\zeta}_k$. En particulier, les termes mixtes qui apparaissent dans l'expression des a_i et des l_i sont uniquement introduits par les termes séculaires qui existent pour les autres éléments. Le but principal de la théorie générale est justement d'introduire à la place de ces termes séculaires, des termes à très longues périodes, avec comme conséquence la suppression des termes mixtes.

Remarque: En utilisant l'expression (7) de λ_{i_0}, la solution n'est plus exactement ordonnée suivant les puissances de μ, mais on a l'avantage de bien connaitre dès le départ la valeur de la somme $n_{i_0} + \Delta^{(s)} l_i$, sensiblement égale à la valeur moyenne du moyen mouvement, fournie par les observations sur une durée suffisamment longue. On verra plus loin la signification de ce moyen mouvement à propos de la théorie générale.

Par ailleurs, à chaque ordre en μ, on fait généralement une comparaison de la théorie obtenue avec les observations; ceci conduit à modifier progressivement les constantes d'intégration initiales, pour passer finalement des éléments osculateurs à l'instant τ_0 aux éléments moyens pour cet instant. Ces éléments moyens sont ainsi relatifs à la théorie qui les a produits.

LA THEORIE GENERALE.

1. Préliminaires

Contrairement à la théorie classique, on s'efforce d'obtenir maintenant une solution bornée aux équations (1), sous forme de fonctions quasi-périodiques de t. L'existence d'une telle solution a été démonstrée par Jefferys et Moser (1966) et par Krasinsky (1969); Brumberg (1970), puis Brumberg et Chapront (1973) ont donné un algorithme pratique pour obtenir une telle solution au problème planétaire; Sagnier (1973) l'a aussi appliqué au problème des satellites galiléens de Jupiter.

Dans ce type de méthode, on s'efforce d'obtenir la solution analytiquement sous la forme la plus générale possible. Les théories déjà construites montrent que moyennant l'introduction pour chaque planète d'une quantité qui a la nature d'un moyen mouvement moyen, on peut exprimer la solution sous forme quasi-périodique de t; ceci permet de supposer qu'il existe pour chaque planète un certain moyen mouvement moyen N_i, constant, indépendant de tout instant initial, et tel qu'on puisse écrire:

$$\eta_i = N_i (1 + p_i) \tag{8}$$

$$\lambda_i = N_i (t - \tau_0) + L_{i_0} + q_i \tag{9}$$

où p_i et q_i sont des nouvelles variables (remplaçant n_i (ou a_i) et l_i) qu'on pourra exprimer sans termes séculaires ni mixtes, sous forme quasi-périodique de t. Dans (9), L_{i_0} est une constante dépendant de τ_0, qu'on peut laisser indéterminée avec τ_0 jusqu'à la comparaison de la théorie avec les observations.

En définissant pour chaque planète une autre quantité constante A_i par la relation:

$$N_i^2 A_i^3 = k(1 + m_i) = n_i^2 a_i^3 \tag{10}$$

on tire encore:

$$a_i = A_i(1 + p_i) - 2/3 \tag{11}$$

et on peut alors remplacer les équations (1) relatives à a_i et l_i par les suivantes:

$$\frac{dp_i}{dt} = -\frac{3}{2}(1 + p_i) \frac{1}{a_i} \frac{da_i}{dt} = \mu \mathcal{L}_{ij}^{*(1)} \tag{12}$$

$$\frac{dq_i}{dt} = N_i p_i + \frac{dl_i}{dt} = N_i p_i + \mu \mathcal{L}_{ij}^{(2)}. \tag{13}$$

Le fait d'avoir explicité la constante L_{i_0} dans l'expression (9) de λ_i permet de prendre nulle la constante d'intégration relative à la variable q_i.

Le fait de rechercher pour p_i et q_i des expressions quasi-périodiques de t, entraine que le terme constant de $\mu \mathcal{L}_{ij}^{(2)}$ dans (13) devra être éliminé par un choix convenable de la constante d'intégration introduite dans (12) pour la variable p_i.

Il n'y à donc plus d'arbitraire pour les variables p_i et q_i et on constate que dans ce cas, ces variables vont forcément contenir μ en factuer: ce sont donc des quantités d'ordre 1 au moins en μ. L'absence d'arbitraire des p_i et q_i revient encore à considérer les constantes N_i et L_{i_0} déjà introduites, comme constantes d'intégration pour ces variables. De ce fait, il suffira de rechercher pour les équations relatives à ces variables, une solution particulière sans constante d'intégration.

Inversement, si on permit l'introduction d'une constante d'intégration arbitraire pour p_i (même une constante de l'ordre de μ), on ne peut empêcher l'apparition d'un terme séculaire sur q_i et la solution n'est plus quasi-périodique de t. Il faut noter cependant qu'une telle constante d'intégration reviendrait à modifier N_i.

Toutefois, alors que les constantes L_{i_0} peuvent rester arbitraires tout au long de la théorie, il est indispensable de fixer numériquement dès le début les constantes N_i, ainsi que les A_i, par l'intermédiaire des constantes m_i, car les développements complétement analytiques par rapport aux a_i ou aux A_i seraient beaucoup trop volumineux; par ailleurs,

pour l'intégration, il est préférable d'avoir les valeurs numériques
des N_i à cause des particularités de la solution pouvant se produire
pour des combinaisons presque nulles des N_i, qui amènent des petits
diviseurs lors de l'intégration.

Dès lors se pose le problème du choix de la valeur de chaque constante N_i par rapport au moyen mouvement moyen tel qu'il est actuellement
connu pour chaque planète. Jusqu'à présent, c'est principalement par les
théories classiques de Le Verrier et de Newcomb qu'on connait des valeurs
du moyen mouvement moyen. Celles-ci résultent de la comparaison de leur
théorie avec les observations dont ils disposaient, et il est certain
que ces moyens mouvements moyens dépendent de l'instant initital τ_0
qu'ils ont choisi. En effet, ils sont de la forme $n_{i_0} + \Delta^{(s)} l_i$ déjà
mentionnée en (7), et dans cette expression, le terme $\Delta^{(s)} l_i$ dépend de
τ_0 car la partie constante de $\mu [\mathcal{L}_{ij}^{(2)}]_0$ dont il provient, est un développement en puissances de z et ζ de ζ dans lequel ces variables ont été
remplacées par leur valeur numérique à l'instant τ_0. De même n_{i_0}, dépend
de τ_0 car il est obtenu en retranchant du moyen mouvement moyen observé
la partie séculaire de l_i qu'on aura calculée au préalable avec ce moyen
mouvement moyen observé; par ailleurs la période pendant laquelle on
dispose d'observations étant limitée, et la théorie étant inévitablement
imparfaite surtout en ce qui concerne les termes de petit diviseur (qui
sont aussi à longues périodes), l'absence ou la mauvaise connaissance
de ces termes contribuent globalement à former des termes séculaires
empiriques lors de la comparaison de la théorie avec les observations.
Cette contribution empirique dépend encore de la date moyenne des observations, donc en quelque sorte d'un certain τ_0, et on ne sait pas la
séparer de la partie du moyen mouvement moyen qui est indépendante de τ_0.

On peut penser cependant que cette partie empirique est très faible,
eu égard à la qualité des travaux de Le Verrier et de Newcomb. En la
négligeant, et en supposant donc que la dépendance du moyen mouvement
moyen vis à vis de τ_0 découle principalement de celle de $\Delta^{(s)} l_i$, on voit
que le meilleur choix qu'on puisse faire pour N_i consiste à prendre le
moyen mouvement moyen obtenu par Le Verrier ou par Newcomb, auquel on
retranche la partie de $\Delta^{(s)} l_i$ qui dépend des variables z et ζ dans leur
théorie. Cela revient encore à ne conserver de $\Delta^{(s)} l_i$ que la partie
qui ne dépend que des demi-grands axes, eux-mêmes calculés avec ces N_i.

On peut noter cependant que cette correction des moyens mouvements
est au moins d'ordre 2 en excentricités et inclinaisons, et d'ordre 1
en masses; elle est donc assez minime (inférieure en pratique à $10^{-5} N_i$).

Avec ce choix des N_i, par analogie avec la théorie classique, on
peut encore prévoir que la perturbation constante du demi-grand axe
(liée à l'utilisation d'un moyen mouvement moyen au lieu d'un moyen
mouvement osculateur) sera elle aussi indépendante des excentricités et
des inclinaisons.

On voit ainsi la nécessité de posséder au départ une théorie classique la meilleure possible afin de déterminer au mieux les N_i qui sont
à la base de toute théorie générale. Cependant, si avec les N_i ainsi
choisis la comparaison de la théorie générale avec les observations ne
peut être satisfasante qu'en ajoutant un terme séculaire empirique, on
saura que c'est la valeur dont il faut modifier N_i pour le rendre indépendant de τ_0 et on pourra reprendre la théorie avec cette nouvelle

valeur. D'ailleurs il n'y a pas que les N_i qui puissent être sujets à
modification: les masses m_i interviennent aussi numériquement dans toute
la théorie et une modification de celles-ci entraine une modification
des N_i. Ceci justifie encore la nécessité de rendre la méthode de développement de la théorie générale aussi souple et adaptable que possible.

2. Méthode d'intégration

Nous envisageons de ne donner ici que les principes de la méthode,
en insistant surtout sur la façon dont s'opère la "généralisation" de
la théorie classique.

Nous partons donc toujours des équations (1), sauf en ce qui concerne les variables a_i et l_i, pour lesquelles on utilise les équations
(12) et (13). On a cependant, à la place du développement (2) un nouveau
développement:

$$\mathcal{L}_{ij}^{(e)} = \sqrt{-1} \, \frac{N_i}{1+m_i} \, \varepsilon_j \sum_{k_i, k_j} L_{k_i k_j}^{(e)} \, (A_i, A_j, p_i, p_j, q_i, q_j,$$

$$z_i, z_j, \bar{z}_i, \bar{z}_j, \zeta_i, \zeta_j, \bar{\zeta}_i, \bar{\zeta}_j) \, \exp \sqrt{-1} (k_i L_i + k_j L_j) \tag{14}$$

où $L_{k_i k_j}^{(e)}$ est une fonction de $\alpha_0 = \dfrac{\min(A_i, A_j)}{\max(A_i, A_j)}$ développable en série

entière des variables p_i, p_j, q_i, $\ldots$ $\bar{\zeta}_j$, et où L_i et L_j désignent des
expressions de la forme:

$$L_i = N_i (t - \tau_0) + L_{i_0}$$

(on a donc $\lambda_i = L_i + q_i$). On peut noter qu'en prenant à la place de z_i
et ζ_i les variables x_i et y_i ainsi définies:

$$x_i = \bar{z}_i \, \exp \sqrt{-1} \, L_i$$

$$y_i = \bar{\zeta}_i \, \exp \sqrt{-1} \, L_i \tag{15}$$

on obtient des équations formellement semblables à celles de Brumberg:

$$\frac{dx_i}{dt} = \sqrt{-1} \, N_i x_i + \mu \mathcal{E}_{ij}^{(3)}$$

$$\frac{dy_i}{dt} = \sqrt{-1} \, N_i y_i + \mu \mathcal{E}_{ij}^{(4)}$$

en précisant bien ici que, contrairement à celles de Brumberg, $\mu \varepsilon_{ij}^{(e)}$ est
au moins de l'ordre de μ. Avec les variables (15), la structure des
développements en z et ζ permet d'obtenir à la place de (14) un dévelop-

pement plus simple:

$$\mathcal{L}_{ij}^{(e)} = \sqrt{-1}\ \frac{N_i}{1+m_i}\ \varepsilon_j \sum_k \mathcal{L}_k^{(e)}\ (A_i,A_j,p_i,p_j,q_i,q_j,$$

$$x_i,x_j,\bar{x}_i,\bar{x}_j,y_i,y_j,\bar{y}_i,\bar{y}_j)\ \exp\ \sqrt{-1}\ k(L_i - L_j) \qquad (16)$$

et un développement du même type pour $\mathcal{E}_{ij}^{(e)}$.

Cependant pour mieux indiquer comment on passe de la théorie classique à la théorie générale, on va plutôt continuer à utiliser le
développement (14) en z et ζ, et les équations (1) pour ces variables.

Remarquons encore qu'à la place de (9), on pourrait utiliser
$\lambda_i = L_i - \sqrt{-1}q_i$ où q_i est alors imaginaire pure permettant un développement plus maniable des fonctions $\exp \sqrt{-1}\ k_i\lambda_i$.

Le développement (14) contient quatre variables supplémentaires par
rapport à (2): p_i, p_j, q_i et q_j. En fait, le développement de $L_{k_i k_j}^{(e)}$ par
rapport a ces variables représente un développement de Taylor de
$L_{k_i k_j}^{(e)}$ dans (2), au voisinage des valeurs A_i et A_j, et au voisinage des
fonctions L_i et L_j. D'ailleurs, comme les variables p_i, p_j, q_i et q_j
sont de l'ordre de μ au moins, la partie des développement de $\mu\mathcal{L}_{ij}^{(e)}$
qui les contient intéresse seulement la solution générale à partir de
l'ordre 2 des masses.

On peut distinguer dans le développement de $\mathcal{L}_{ij}^{(e)}$ deux sortes de
termes: d'une part ceux qui ne dépendent que des A_i, A_j et du temps, et
d'autre part les autres termes qui dépendent en plus des variables p_i,
p_j, q_i, q_j, z_i, z_j, $\bar{z}_i$, $\bar{z}_j$, ζ_i, ζ_j, $\bar{\zeta}_i$, $\bar{\zeta}_j$, qu'on regroupe dans un
vecteur noté V. On regroupe également ces deux sortes de termes dans
les notations: $\mathcal{L}_{ij}^{(e)}(t)$ et $\mathcal{L}_{ij}^{(e)}(V,t)$. Les équations ont alors la forme
suivante:

$$\frac{dp_i}{dt} - \mu\ \mathcal{L}_{ij}^{*(1)}(V,t) = \mu\ \mathcal{L}_{ij}^{*(1)}(t) \qquad (17)$$

$$\frac{dq_i}{dt} - N_i p_i - \mu\ \mathcal{L}_{ij}^{(2)}(V,t) = \mu\ \mathcal{L}_{ij}^{(2)}(t) \qquad (18)$$

$$\frac{dz_i}{dt} - \mu\ \mathcal{L}_{ij}^{(3)}(V,t) = \mu\ \mathcal{L}_{ij}^{(3)}(t) \qquad (19)$$

$$\frac{d\zeta_i}{dt} - \mu\ \mathcal{L}_{ij}^{(4)}(V,t) = 0 \qquad (20)$$

On rappelle cependant que $\mathcal{L}_{ij}^{(2)}(t)$ contient un terme indépendant de t
(constant), tandis que $\mathcal{L}_{ij}^{(1)}(t)$ et $\mathcal{L}_{ij}^{(3)}(t)$ en sont dépourvus.

Les seconds membres étant des fonctions connues du temps, on a ainsi

un système différentiel dit "avec second membre". On sait que la solution générale du s tel système est la somme de la solution générale du
système "sans second membre", et d'une solution particulière du système
complet (appelée orbite intermédiaire dans la m'thode de Brumberg). En
notant la solution particulière par le symbole δ, on aura ainsi:

$$p_i = p_i^{(0)} + \delta p_i$$

$$q_i = q_i^{(0)} + \delta q_i$$

$$z_i = z_i^{(0)} + \delta z_i$$ $\qquad (21)$

$$\zeta_i = \zeta_i^{(0)}$$

La solution particulière de ζ_i peut être choisie nulle car le développe-
ment de $\mathcal{L}_{ij}^{(4)}(V,t)$ est impair en ζ_i, ζ_j, $\bar{\zeta}_i$ et $\bar{\zeta}_j$.

On recherche ces solutions par approximations successives suivant
les puissances de μ, en faisant en sorte d'éviter l'introduction de
termes séculaires. A la première approximation on aura ainsi:

$$\frac{d\delta p_i}{dt} = \mu \, \mathcal{L}_{ij}^{*(1)}(t); \qquad \frac{d\delta q_i}{dt} = \mu \, \mathcal{L}_{ij}^{(2)}(t) + N_i \delta p_i$$

$$\frac{d\delta z_i}{dt} = \mu \, \mathcal{L}_{ij}^{(3)}(t) \qquad (22)$$

et:

$$\frac{dp_i^{(0)}}{dt} = \mu \, \mathcal{L}_{ij}^{*(1)}(V,t); \qquad \frac{dq_i^{(0)}}{dt} = \mu \, \mathcal{L}_{ij}^{(2)}(V,t) + N_i p_i^{(0)}$$

$$\frac{dz_i^{(0)}}{dt} = \mu \, \mathcal{L}_{ij}^{(3)}(V,t); \qquad \frac{d\zeta_i^{(0)}}{dt} = \mu \, \mathcal{L}_{ij}^{(4)}(V,t). \qquad (23)$$

L'absence de terme constant dans $\mathcal{L}_{ij}^{*(1)}(t)$ et $\mathcal{L}_{ij}^{(3)}(t)$ entraine l'absence
de terme séculaire dans δp_i et δz_i à l'ordre 1 en μ. La constante d'intégration de p_i est choisie pour annuler le terme constant de $\mathcal{L}_{ij}^{(2)}(t)$ ce
qui permet de trouver δq_i sans terme séculaire. Cette constante
d'intégration sur δp_i équivaut à la perturbation constante du demi-grand
axe du 1er ordre des masses. En annulant les constantes d'intégration de
δq_i et δz_i, la solution particulière obtenue ainsi sans terme séculaire
est d'ordre 1 en μ. Les approximations suivantes de la solution particulière peuvent alors s'obtenir en développant les équations (17) à (29)
au voisinage des $p_i^{(0)}$, $q_i^{(0)}$ et $z_i^{(0)}$; en développant δp_i en puissance de
μ: $\delta p_i = \mu \delta_1 p_i + \mu^2 \delta_2 p_i + \ldots$ on aura par exemple pour l'équation (17):

$$\frac{dp_i^{(0)}}{dt} + \mu \frac{d}{dt} \delta_1 p_i + \mu^2 \frac{d}{dt} \delta_2 p_i + \ldots = \mu \, \mathcal{L}_{ij}^{*(1)}(t) +$$

$$\mu(\mathcal{L}_{ij}^{*(1)}(V,t))^{(0)} + \mu^2\left(\frac{\partial\mathcal{L}_{ij}^{*(1)}(V,t)}{\partial v}\right)^{(0)}.(\delta_1 V + \mu\delta_2 V + \ldots)$$

(24)

où $_k V$ (k=1,2) désigne le vecteur composé des $\delta_k p_i$, $\delta_k p_j$, $\delta_k q_i$, $\delta_k q_j$, $\delta_k z_i$, $\delta_k z_j$, $\delta_k \bar{z}_i$, $\delta_k \bar{z}_j$ déterminés à l'ordre k en μ, et où $\left(\frac{\partial\mathcal{L}_{ij}(V,t)}{\partial v}\right)^{(0)}$ désigne le vecteur des dérivées partielles de $\mathcal{L}_{ij}$ par rapport aux variables qui composent le vecteur V, dans lequel V doit être remplacé par:

$$v^{(0)} = \{p_i^{(0)}, p_j^{(0)}, q_i^{(0)}, z_i^{(0)}, z_j^{(0)}, \bar{z}_i^{(0)}, \bar{z}_j^{(0)}, \zeta_i^{(0)}, \zeta_j^{(0)},$$

$$\bar{\zeta}_i^{(0)}, \bar{\zeta}_j^{(0)}\}.$$

On identifie $\mu^2\frac{d\delta_2 p_i}{dt}$ à la partie de $\mu^2(\partial\mathcal{L}_{ij}^{*(1)}(V,t))^{(0)}. \partial_1 V$ qui est fonction de t seulement (elle correspond à la partie de $\mathcal{L}_{ij}^{*(1)}(V,t)$ qui est de degré 1 par rapport à V). Les particularités qui avaient permis d'éviter les termes séculaires dans $\delta_1 V$, se retrouvent au deuxième ordre à condition d'avoir annulé les constantes d'intégration des $\mu\sigma_1 q_i$ et $\mu\sigma_1 z_i$, et d'avoir choisi celle de $\mu\sigma_1 p_i$ de façon à annuler le terme constante l'équation relative à $\mu\delta_1 q_i$. Aux ordres suivants, la possibilité d'obtenir une solution sans terme séculaire reste conditionnée à la non apparition de terme constant dans l'équation $\frac{dp}{dt}$. Il faut encore noter que des termes constants pourraient apparaitre dans des cas de résonnance stricte, lesquels sont ici exclus par hypothèse.

Il reste ensuite à identifier:

$$\frac{dp_i^{(0)}}{dt} = \mu (\mathcal{L}_{ij}^{*(1)}(V,t))^{(0)} + \mu^2(\frac{\partial\mathcal{L}_{ij}^{*(1)}(V,t)}{\partial v})^{(0)}.\delta_1 V$$

(25)

où le terme en μ^2 concerne la dérivée partielle des termes de degré au moins égal à 2 par rapport à V. Le second membre est ainsi déjà développé en puissances de μ; il est de degré 1 au moins par rapport aux variables qui constituent le vecteur V.

Avec des notations semblables, on obtient pour les autres variables:

$$\frac{dq_i^{(0)}}{dt} = N_i p_i^{(0)} + \mu(\mathcal{L}_{ij}^{(2)}(V,t))^{(0)} + \mu^2(\frac{\partial\mathcal{L}_{ij}^{(2)}(V,t)}{\partial V})^{(0)}.\partial_1 V+\ldots$$

(26)

$$\frac{dz_i^{(0)}}{dt} = \mu(\mathcal{L}_{ij}^{(3)}(V,t))^{(0)} + \mu^2(\frac{\partial\mathcal{L}_{ij}^{(3)}(V,t)}{\partial V})^{(0)}.\delta_1 V+\ldots$$

(27)

$$\frac{d\zeta_i^{(0)}}{dt} = \mu(\mathcal{L}_{ij}^{(4)}(V,t))^{(0)} + \mu^2(\frac{\partial\mathcal{L}_{ij}^{(4)}(V,t)}{\partial V})^{(0)}.\delta_1 V+\ldots$$

(28)

Finalement, la solution générale des équations sans second membre doit être tirée des équations (25) à (28). La méthode exposée par Brumberg (1970) dans sa théorie générale, revient à regrouper tout d'abord dans un système différentiel autonome une partie des équations, formée de termes ne dépendant pas explicitement de t; cette partie est exactement celle qui, dans la théorie classique, introduit les termes séculaires $\Delta^{(s)} z$ et $\Delta^{(s)}$, mais son intégration va maintenant être menée de façon à éviter les termes séculaires: la solution du système autonome sera une somme de termes à très longues périodes, et contiendra les constantes d'intégrations relatives aux variables z et ζ (on rappelle que les variables p et q doivent en être dépourvues).

En appelant V_s le vecteur solution du système autonome:

$$V_s = \{ p_{s\,i}^{(0)} \,,\, p_{s\,j}^{(0)} \,,\, q_{s\,i}^{(0)} \,,\, q_{s\,j}^{(0)} \,,\, z_{s\,i}^{(0)} \,,\, z_{s\,j}^{(0)} \,,\, \bar{z}_{s\,i}^{(0)} \,,\, \bar{z}_{s\,j}^{(0)} \,,$$

$$\zeta_{s\,i}^{(0)} \,,\, \zeta_{s\,j}^{(0)} \,,\, \bar{\zeta}_{s\,i}^{(0)} \,,\, \bar{\zeta}_{s\,j}^{(0)} \}$$

on recherchera ensuite une solution de la forme:

$$p_i^{(0)} = p_{s\,i}^{(0)} + \mu \mathscr{P}_i (V_s, t) \qquad (29)$$

$$q_i^{(0)} = q_{s\,i}^{(0)} = \mu \mathscr{Q}_i (V_s, t) \qquad (30)$$

$$z_i^{(0)} = z_{s\,i}^{(0)} + \mu \mathscr{X}_i (V_s, t) \qquad (31)$$

$$\zeta_i^{(0)} = \zeta_{s\,i}^{(0)} + \mu \mathscr{Y}_i (V_s, t) \qquad (32)$$

où les fonctions $\mathscr{P}_i$, $\mathscr{Q}_i$, $\mathscr{X}_i$ et $\mathscr{Y}_i$ seront des fonctions quasi-périodiques de t, à déterminer; la solution générale restera alors dans le voisinage de la solution du système autonome.

A l'ordre 1 en μ, il suffit de remplacer partout dans les équations (25) à (28) les éléments de $V^{(0)}$ par ceux de V_s; on isole le système autonome suivant:

$$\frac{d}{dt} p_{s\,i}^{(0)} = 0 \qquad (33)$$

$$\frac{d}{dt} q_{s\,i}^{(0)} = N_i p_{s\,i}^{(0)} + \mu [\mathcal{L}_{ij}^{(2)}] \,(p_i, p_j) \qquad (34)$$

$$\frac{d}{dt} z_{s\,i}^{(0)} = \mu [\mathcal{L}_{ij}^{(3)}] \qquad (35)$$

$$\frac{d\zeta_{i_s}^{(0)}}{dt} = \mu[\mathcal{L}(4)]_{ij}$$ (36)

où les notations signifient qu'on ne garde de $\mathcal{L}_{ij}^{(2)}$ que les termes explicitement indépendants de t et qui contiennent p_i ou p_j en facteur, tandis qu'on prend en plus dans $\mathcal{L}_{ij}^{(3)}$ et $\mathcal{L}_{ij}^{(4)}$ les termes qui ne dépendent que des variables z et ζ (les variables q ne peuvent pas apparaitre dans les termes explicitement indépendant de t).

De cette façon, les seconds membres de (33) et (34) s'annulaient avec les $p_{i_s}^{(0)}$, on a la solution particulière $p_{i_s}^{(0)} = 0$ et $q_{i_s}^{(0)} = 0$, qui convient parfaitement au fait que ces variables ne doivent pas contenir de constantes d'intégration, ni amener de terme séculaire dans les variables q.

Quant aux équations (35) à (36), avec la solution $p_{i_s}^{(0)} = 0$ il n'en reste qu'un système ne dépendant que des variables $z_{i_s}^{(0)}$ et $\zeta_{i_s}^{(0)}$ dont la partie linéaire par rapport à ces variables constitue exactement le système de Laplace-Lagrange. On sait que la solution générale de ce système est de la forme d'une some finie de termes à très longues périodes; on sait également que les termes de degré supérieur n'altèrent pas la forme de cette solution, qui reste à très longues périodes (méthode de Krilov-Bogolioubov, voir par exemple Bretagnon, 1173). Les constantes d'intégration sont à évaluer par comparaison avec les observations: les amplitudes de ces termes à très longues périodes sont de l'ordre des excentricités et des inclinaisons.

La détermination des fonctions $\mathscr{P}_i$, $\mathscr{Q}_i$, $\mathscr{X}_i$ et $\mathscr{Y}_i$ s'opère maintenent en dérivant les expressions (29) à (32) et en les identifiant aux équatinso (25)˜(28):

$$\frac{dp_{i_s}^{(0)}}{dt} + \mu\left(\frac{\partial\mathscr{P}_i(V_s,t)}{\partial V_s}\right)\cdot\frac{dV_s}{dt} + \mu\frac{\partial\mathscr{P}_i(V_s,t)}{\partial t} = \mu(\mathcal{L}^{*(1)}(V,t))_{ij}^{(0)}+\ldots$$ (37)

$$\frac{dq_{i_s}^{(0)}}{dt} + \mu\left(\frac{\partial\mathscr{Q}_i(V_s,t)}{\partial V_s}\right)\cdot\frac{dV_s}{dt} + \mu\frac{\partial Q_i(\mathscr{V}_s,t)}{\partial t} = N_i p_{i_s}^0 +$$

$$\mu N_i\mathscr{P}_i(V_s,t) + \mu(\mathcal{L}_{ij}^{(2)}(V,t))^{(0)} + \mu^2\ldots$$ (38)

$$\frac{dz_{i_s}^{(0)}}{dt} + \mu\left(\frac{\partial X_i(V_s,t)}{\partial V_s}\right)\cdot\frac{dV_s}{dt} + \mu\frac{\partial X_i(V_s,t)}{\partial t} = \mu(\mathcal{L}_{ij}^{(3)}(V,t))^{(0)} + \ldots$$ (39)

$$\frac{d\zeta_i^{(0)}}{dt} + \mu \left(\frac{\partial Y_i(V_s,t)}{\partial V_s}\right) \cdot \frac{dV_s}{dt} + \mu \frac{\partial Y_i(V_s,t)}{\partial t} = \mu (\mathcal{L}_{ij}^{(4)}(V,t))^{(0)} + \dots$$

(40)

Dans les seconds membres, il faut encore remplacer le vecteur V par le vecteur $V_s + \mu F(V_s,t)$, qui condense les notations (29) à (32). Comme $\dfrac{dV_s}{dt}$ est déjà d'ordre 1 en μ (éqs. (33) A 836)), il reste l'identification suivante, à l'ordre 1 en μ:

$$\mu \frac{\partial \mathcal{P}_i(V_s,t)}{\partial t} = \mu \, \tilde{\mathcal{L}}_{ij}^{(1)}(V_s,t)$$

(41)

$$\mu \frac{\partial \mathcal{Q}_i(V_s,t)}{\partial t} = \mu \, N_i \mathcal{P}_i(V_s,t) + \mu \tilde{\mathcal{L}}_{ij}^{(2)}(V_s,t)$$

(42)

$$\mu \frac{\partial X_i(V_s,t)}{\partial t} = \mu \tilde{\mathcal{L}}_{ij}^{(3)}(V_s,t)$$

(43)

$$\mu \frac{\partial y_i(V_s,t)}{\partial t} = \mu \tilde{\mathcal{L}}_{ij}^{(4)}(V_s,t)$$

(44)

où les expressions $\tilde{\mathcal{L}}_{ij}(V_s,t)$ représentant ce qui reste des seconds membres de (25) à (28), (avec V_s mis à la place de $V^{(0)}$) lorsqu'on leur enlève les seconds membres du système autonome (33) à (36); ce sont donc des expressions dépendant explicitement de t. Seul $\tilde{\mathcal{L}}_{ij}^{(2)}(V_s,t)$ contient encore des termes indépendants de t, ceux qui ne dépendent que des variables $z_i^{(0)}$ et $\zeta_i^{(0)}$. On peut les éliminer en les identifiant à des termes de la même forme dans l'expression de $\mathcal{P}_i(V_s,t)$; ainsi, ces termes qui contribuaient à la perturbation constante du demi-grand axe se retrouvent, dans la théorie générale, exprimés en fonction des variables V_s, c'est à dire qu'ils sont maintenant à très longues périodes.

A chacun des autres termes des $\tilde{\mathcal{L}}_{ij}(V_s,t)$ (qui sont quasi-périodiques de t), on associe un terme des fonctions $\mathcal{P}_i$, $\mathcal{Q}_i$, $\mathcal{X}_i$, ou $\mathcal{Y}_i$, qui a la même forme par rapport aux variables V_s et qui a la même période. L'intégration de ces termes par les équations (41) à (44) est l'équivalent de la détermination des inégalités à courtes périodes du 1er ordre en dans la théorie classique mise à part la substitution numérique de la valeur des variables z et ζ. Cependant, ces termes à courtes périodes sont ici en facteur d'éléments de V_s: ils doivent ainsi être composés à des termes à très longues périodes; si on développe ces derniers en séries entières de t au voisinage d'un instant initial τ_0, ces termes fourniront alors le terme à courte période de la théorie classique au 1er ordre en μ, puis les termes mixtes relatifs à cette même période.

Ainsi, les termes mixtes qui n'apparaissent qu'à l'ordre 2 en μ au moins dans la théorie classique, sont déjà présents dans la théorie générale à l'ordre 1 en μ. Ceci vient du fait que les basses fréquences de la

solution à très longues périodes sont de l'ordre de μ; le développement
en série entière de t de ces termes est donc aussi un développement en
puissances de μ, et on obtient par la théorie générale du 1er ordre les
termes mixtes en $\mu^2 t \exp \sqrt{-1}(k_i L_i + k_j L_j)$, puis en $\mu^3 t^3 \exp \sqrt{-1}(k_i L_i +$
$k_j L_j)$, ...; les termes mixtes en $\mu^3 t \exp \sqrt{-1}(k_i L_i + k_j L_j)$,
$\mu^4 t^2 \exp \sqrt{-1}(k_i L_i + k_j L_j)$ ... ne seraient fournis qu'à l'ordre 2 en
de la théorie générale.

La solution générale des équations (25) à(28), à l'ordre 2 et aux
ordres supérieurs en μ s'obtiendrait de façon analogue par les équations
(37) à(40), en y considérant que les $\mathscr{P}_i$, $\mathscr{Q}_i$, $\mathscr{X}_i$ et $\mathscr{Y}_i$ se développent en
puissances de μ (par exemple):

$$\mu \mathscr{P}_i = \mu \mathscr{P}_1 + \mu^2 \mathscr{P}_2 + \ldots, \quad \mu \mathscr{F} = \mu \mathscr{F}_1 + \mu^2 \mathscr{F}_2 + \ldots$$

on sépare alors à chaque ordre en μ, un système autonome du même ordre,
selon les mêmes principes qu'à l'ordre 1: termes indépendants de t pour
les équations (39) à (40) et seulement ceux qui s'annulent avec les
$p_s^{(0)}$ et $q_s^{(0)}$ pour les équations (37) et (38); la solution du système
autonome conserve alors la même forme qu'à l'ordre 1. Les termes rsstant
déterminent ensuite forcément, pour chaque ordre en μ, le vecteur F sous
forme quasi-périodique de t.

On a également à chaque ordre en μ un équivalent de la perturbation
constante du demi-grand axe qui se développe en termes à longues périodes par l'intermédiaire des variables.

Remarquons encore que l'intégration du système autonome peut se
faire indépendemment de la recherche de la solution quasi-périodique
(d'où son nom); la détermination des constantes d'intégration de la
solution générale peut ainsi intervenir seulement quand on estime avoir
déterminé avec suffisamment de précision la solution quasi-périodique.

APPLICATIONS ET CONCLUSION

D'après ce qui précède, on peut constater que le développement en μ
n'a pas la même signification dans la théorie générale et dans la théorie
classique, notamment en ce qui concerne les termes séculaires et mixtes:
ainsi, la théorie générale semble devoir être poussée à un ordre moins
élevé en μ que la théorie classique en ce qui concerne ces termes; c'est
un résultat intéressant car c'est presque uniquement pour ces termes
séculaires et mixtes que la théorie classique nécessite un développement
d'ordre élevé en μ, les termes mixtes étant alors toujours relatifs à
des longues périodes. Nous allons montrer plus précisement sur un exemple comment peut s'effectuer ce calcul des termes séculaires et mixtes
à partir de la théorie générale d'ordre 1 en μ

Ainsi par exemple, pour calculer les termes mixtes dans le demi-
grand axe de Saturne, relatifs à la grande inégalite, on devra d'abord
déterminer le développement analytique de cette inégalité dans $\mathscr{L}_{65}^{*(1)}$;
on trouve (cf. Duriez, 1976) que ce développement comporte en facteur de

$\exp \sqrt{-1}(2L_5 - 5L_6)$, 10 termes de degré 3:

$$- \frac{3}{2} N_6 (0,01208 \; z_6\zeta_6^2 - 0,02417 \; z_6\zeta_5\zeta_6 + 0,01208 \; z_6\zeta_5^2$$

$$- 0,00672 \; z_5\zeta_6^2 + 0.01255 \; z_5\zeta_5\zeta_6 - 0,00627 \; z_5\zeta_5^2 \qquad (45)$$

$$+ 0,02492 \; z_6^3 - 0,04560 \; z_5 z_6^2 + 0,02752 \; z_5^2 z_6 - 0,00550 \; z_5^3)$$

puis 70 termes de degré 5, dont la contribution globale dans une théorie classique est environ 100 fois moindre que celle du degré 3 (on compte encore 10+70 termes conjugués changés de signe des précédents, car $\sqrt{-1}\mathcal{L}_{ij}^{(1)}$ est imaginaire pur, en facteur de $\exp -\sqrt{-1}(2L_5 - 5L_6)$). La solution quasi-périodique $\mathscr{P}_6$ relative à cette inégalité s'obtient en intégrant l'équation (41): cela revient à diviser chaque terme de (45) par $\sqrt{-1}(2N_5 - 5N_6)$, et c'est donc exactement la même intégration que dans la théorie classique, dans la substitution numérique des z et ζ. C'est en fait la substitution de la solution du système autonome qui donne des termes mixtes cette dernière solution est de la forme, pour z ou pour ζ:

$$z = \sum_i A_i \exp \sqrt{-1}(c_i t + \phi_i)$$

où A_i, c_i et ϕ_i sont constants, c_i de l'ordre de μ, et A_i de l'ordre des excentricités (cf. Bretagnon, 1974).

Les termes mixtes s'obtiennent en substituant à z et ζ dans l'expression intégrée de (45), des expressions développées en puissances de t, de la forme:

$$z = \sum_i A_i \exp \sqrt{-1} \; \phi_i [1 + \sqrt{-1} \; c_i t - \frac{c_i^2}{2} t^2 + ...] \qquad (46)$$

On pourrait aussi utiliser une solution du système autonome exprimée sous forme polynomiale de t (Brumberg, 1975). Ce schéma de calcul est d'ailleurs valable pour toutes les inégalités périodiques.

La partie indépendante de t dans les expressions de la forme (46), substituée aux variables z et ζ dans la solution générale (45), donne la grande inégalité du demi-grand axe de Saturne tel qu'on l'aurait dans la théorie classique au 1er ordre en μ. Les termes suivants dans le développement (46) donneraient les termes mixtes en $t \exp \sqrt{-1}(2L_5 - 5L_6)$, $t^2 \exp \sqrt{-1}(2L_5 - 5L_6)$ d'ordre 2,3,... en μ.

Les termes mixtes associés à d'autres inégalités périodiques s'obtiendraient de la même façon, mais sont en général négligeables sauf dans le cas de petits diviseurs (comme la grande inégalité).

Une substitution de la forme (46) dans certains termes de la solution générale peut également conduire à des termes séculaires au lieu de termes mixtes, si les termes à très longues périodes qu'on développe

ainsi ne sont en facteur d'aucune fonction de t. C'est notamment le
cas des termes de $\mathcal{L}_{ij}^{(2)}$ introduits dans P_i pour éviter les termes séculaires dans (42), et qui correspondent à la partie de la perturbation
constante du demi-grand axe, qui dépend des excentricités et des inclinaisons. Il nous a paru intéressant de calculer le terme séculaire qui
serait ainsi introduit dans la perturbation du demi-grand axe de Saturne
perturbé par Jupiter: le développement de la partie de $\mu\mathcal{L}_{65}^{(2)}$ indépendante
de t, et de degré 2 par rapport aux z et ζ, comprend 8 termes:

$$N_6(-848\ \zeta_6\bar{\zeta}_6 - 765\ \zeta_5\bar{\zeta}_5 + 806(\zeta_5\bar{\zeta}_6 + \zeta_6\bar{\zeta}_5)$$

$$+ 212\ z_6\bar{z}_6 + 191\ z_5\bar{z}_5 - 156(z_5\bar{z}_6 + z_6\bar{z}_5)) \times 10^{-6}$$

En y substituant les développements de z_5, z_6, ζ_5 et ζ_6 de la forme (46),
calculés à apartir des plus gros termes de la solution de Lagrange donnée
par Bretagnon (1974), nous avons trouvé que le terme séculaire en t,
d'ordre 2 des masses, qui en résulte pour la variable p_6, vaut 7,66 10^{-6}
(11/an)t, ce qui donnerait sur le demi-grand axe de Saturne 2,37 10^{-10}
($^{u.a.}$/an)t . Ces valeurs sont du même ordre de grandeur que les très
petits termes séculaires qu'on trouve numériquement dans la theorie classique en variables elliptiques héliocentriques dans le demi-grand axe,
au deuxième ordre des masses.

   Finalement, on ne peut que constater l'intérêt de la théorie générale pour déterminer les termes séculaires et mixtes de la théorie
classique. Cependant, comme les résultats qu'on attend de la théorie
générale dépendent de la qualité de la théorie classique qui fournit la
valeur des N_i, on constate aussi la nécessité de développer en même
temps ces deux types de théories. Actuellement, la théorie générale est
acquise à l'ordre 1 en μ, pour tous les termes de degré inférieurs à 5
en z et ζ, ainsi que la solution intermédiaire à l'ordre 3 en μ. L'ordre
2 en μ de la solution générale est en cours d'élaboration.

REMERCIEMENTS

Je remercie chaleureusement Monsieur J. Chapront pour les nombreux conseils qu'il m'a donnés, et les discussions fructueuses qu'il m'a permis
d'avoir avec lui. Je remercie éaglement Messieurs Bretagnon et Simon
pour les résultats de leurs travaux qu'ils ont eu l'amabilité de me
communiquer.

BIBLIOGRAPHIE

Bretagnon, P.: 1974, *Astron. & astrophys.* **30**, 141
Bretagnon, P. 1976:, communication au colloque n°41 de l'UAI à Cambridge
Brumberg, V.A.: 1970, dans *Periodic Orbits, Stability and Resonances*,
     G.E.O. Giacaglia (ed.), Reidel, Dordrecht, p. 409
Brumberg, V.A. et Chapront, J.: 1973, *Celes. Mech.* **8**, 335

Brumberg, V.A.: 1975, *Celes. Mech.* 11, 131

Chapront, J., Bretagnon, P. et Mehl, M. 1975, *Astron. & Astrophys.* 38, 57

Duriez, L. 1977, *Astron. & Astrophys.* 54, 93

Jefferys, W.H., Moser, J.: 1966, *Astron. J.* 71, 568

Krasinsky, G.A. 1969, *Trans. Inst. Theor. Astron.* (Leningrad) 13, 105

Sagnier, J.L.: 1973, *Astron. & Astrophys.* 25, 113

Simon, J.L. et Bretagnon, P.: 1975, *Astron. & Astrophys.* 42, 259

Simon, J.L., et Bretagnon, P.: 1975, *Astron. & Astrophys. Suppl.* 22, 107

MATHEMATICAL RESULTS OF THE GENERAL PLANETARY THEORY IN RECTANGULAR COORDINATES

V.A. BRUMBERG, L.S. EVDOKIMOVA and V.I. SKRIPNICHENKO
Institute of Theoretical Astronomy, Leningrad

ABSTRACT. Mathematical construction of the general planetary theory has led to the series of two forms for the coordinates of eight major planets (excluding Pluto). The series of the first form are Poisson series where all orbital elements except the semi-major axes occur in literal shape. The series of the second form are polynomial-exponential series with respect to the time and serve to calculate the ephemerides. The arbitrary constants of the theory are related to the Keplerian elements. The terms of the zero and first degree in eccentricities and inclinations have been found in the second approximation with respect to the disturbing masses. Among those of particular interest are the resonant terms caused by the commensurabilities of the mean notations of triplets of planets.

1. INTRODUCTION

This paper summarizes the results of the general planetary theory for eight major planets based on the semi-analytical method exposed earlier (Brumberg and Chapront, 1973). The actual construction of the theory started at the Bureau des Longitudes (Paris) on a computer IBM-360-65 has been continued at the Institute of Theoretical astronomy (Leningrad) on a computer BESM-6. At present our main results are as follows:
1. literal series of the first approximation theory for all eight planets;
2. polynomial solution of the secular system for the slow variables;
3. numerical series for ephemeris calculation;
4. linear second approximation theory.
The exact meaning of these results will be explained below.
The ephemerides based on this theory are erroneous due to the neglect of higher order terms (truncation errors) and the tentative estimates of the constants of the theory (estimation errors). Therefore, for completing our theory both in mathematical and astronomical sense it is necessary to add to our series some corrections influenced by the higher orders terms and to improve the numerical values of the constants by means of comparison of our ephemerides with observations.

V. Szebehely (ed.), Dynamics of Planets and Satellites and Theories of Their Motion, 33-48.

However, the results already obtained seem to be of interest by themselves. They may be used for comparison with the results of different
planetary theories now in progress. Moreover, they enable one to
estimate a size of the resulting series representing the motion of the
major planets and to realize the amount of necessary calculations. That
is why the presentation of our results seems to be useful, even though
the final completion of our theory may be available in the future.

2. TWO FORMS OF SERIES OF THE GENERAL PLANETARY THEORY

The theory has been built in p_i, w_i-variables related to the
heliocentric rectangular coordinates x_i, y_i, z_i ($i = 1, \ldots, N; N = 8$)
as follows:

$$x_i + \sqrt{-1}\, y_i = A_i\,(1 - p_i)\, \exp \sqrt{-1}\, \lambda_i, \qquad z_i = A_i\, w_i, \qquad (1)$$

$$\lambda_i = n_i t + \varepsilon_i, \qquad n_i^2 A_i^3 = k^2\,(m_0 + m_i). \qquad (2)$$

Expressions for p_i, w_i may be given in two forms. The first form
corresponds to the Poisson series (4N, N-1) where 4N polynomial variables have an order of planetary eccentricities and inclinations and
N-1 angular arguments are differences of the mean longitudes. The
coefficients of these series are real numbers depending on the adopted
numerical values of mean notions n_i and masses m_i. Using numerical
values of these constants dictates the semi-analytical form of our
theory. Nevertheless, taking into account that all remaining parameters
of the theory enter in literal form these series will be referred to as
literal series of the general planetary theory.

With a large computer it would be suitable to construct these
series by iterations without using the developments in powers of the
disturbing masses (Brumberg, 1974). We have tested this way to recal-
culate the intermediate solution independent of the planetary eccentricities and inclinations (Brumberg et al., 1975a). But for the iterative solution on the whole the capcaities of BESM-6 are unsufficient
and we have found the Poisson series using the expansions in powers of
the disturbing masses. Explicitly these series have the form

$$p_i = p_{0i} + \mu p_{1i} + \mu^2 p_{2i} + \ldots, \qquad (3)$$

$$w_i = w_{0i} + \mu w_{1i} + \mu^2 w_{2i} + \ldots, \qquad (4)$$

μ (=0.001) being a basic small parameter. Here p_{0i} and w_{0i} represent the

Poisson series (4,0) for the undisturbed motion

$$p_{0i} = \sum_{m=1}^{\infty} \sum\sum P_{pqrs}\, a_i^p\, \bar{a}_i^{-q}\, b_i^r\, \bar{b}_i^{-s}, \qquad (5)$$

$$w_i \atop o = \sum_{m=1}^{\infty} \sum w_{pqrs} \; a_i^p \; \bar{a}_i^{-q} \; b_i^r \; \bar{b}_i^{-s}, \qquad (6)$$

where p_{pqrs}, w_{pqrs} are absolute numerical constants having the same values for all planets. Polynomial variables a_i and b_i are of order of eccentricity and inclination of planet i respectively. A bar denotes a complex conjugate quantity. Here and everywhere below the inner summation without explicit limits is performed over all non-negative values of power indices, the sum of which is equal to m. Evidently, the value of m indicates an analytical order of smallness of the corresponding term with respect to the planetary eccentricities and inclinations.

The next terms in (3), (4) yield the series of the first approximation theory

$$p_i \atop 1 = \sum_{j=1}^{N(i)} p_{ij}, \qquad w_i \atop 1 = \sum_{j=1}^{N(i)} w_{ij}, \qquad (7)$$

where perturbations $p_{ij} \atop 1$, $w_{ij} \atop 1$ due to a specific couple of planets i and j represent Poisson series (8,1):

$$p_{ij} \atop 1 = \sum_{m=0}^{\infty} \sum p_{pqrsp'q'r's'}^{(ij)} \; a_i^p \, \bar{a}_i^{-q} \, b_i^r \, \bar{b}_i^{-s} \, a_j^{p'} \, \bar{a}_j^{-q'} \, b_j^{r'} \, \bar{b}_j^{-s'}, \qquad (8)$$

$$w_{ij} \atop 1 = \sum_{m=1}^{\infty} \sum w_{pqrsp'q'r's'}^{(ij)} \; a_i^p \, \bar{a}_i^{-q} \, b_i^r \, \bar{b}_i^{-s} \, a_j^{p'} \, \bar{a}_j^{-q'} \, b_j^{r'} \, \bar{b}_j^{-s'}. \qquad (9)$$

The coefficients of these series are exponential series with respect to the argument $\sqrt{-1}$ $(\lambda_i - \lambda_j)$ with real coefficients. These coefficients contain a multiplier x_{ij} determined by the relation

$$\mu x \atop ij = m_j / (m_0 + m_i). \qquad (10)$$

Therefore, if one changes the adopted values of planetary masses it is sufficient to multiply the obtained coefficients by a related correction factor (within the first approximation theory one may neglect the variations of coefficients due to the changes of semimajor axes related with the planetary masses by the third Kepler law (2)).

Series (9) is self-conjugate and does not contain pure exponentail terms with m = 0 (such terms in (8) yield the quasi-periodic intermediary).

Our work results mainly in actual constructing (8) and (9) for all couples of the major planets.

Literal series (5), (6), (8), (9) are important for the investigation of the analytical and numerical structure of the planetary inequalities. For ephemeris computation the second form of the series of the general planetary theory is more suitable. By this form only the

mean longitudes of the planets retain their literal shape whereas all
other planetary elements have numerical values. These series will be
called numerical series of the general planetary theory. They are represented by the exponential series in multiples of $\sqrt{-1}\,\lambda_i$ ($i = 1, \ldots,$
N), the coefficients being polynomials in powers of time. These polynomials result from the slow secular variation of the planetary elements. Their coefficients are complex numbers. Construction of the
numerical series is achieved by substitution into literal series

$$a_i = \alpha_i \exp \sqrt{-1}\,\lambda_i, \qquad b_i = \beta_i \exp \sqrt{-1}\,\lambda_i \qquad (11)$$

α_i, β_i being Lagrange elements satisfying the autonomous secular system
of differential equations. The polynomial solution of this system is of
the form (Brumberg et al., 1975b):

$$\alpha_i = \sum_{k=0}^{\infty} \alpha_k^{(i)}\, t^k, \qquad \beta_i = \sum_{k=0}^{\infty} \beta_k^{(i)}\, t^k \qquad (12)$$

where $\alpha_k^{(i)}$, $\beta_k^{(i)}$ are numerical complex coefficients. $\alpha_o^{(i)}$ and $\beta_0^{(i)}$ may
be regarded as arbitrary constants determining all subsequent coefficients.
In fact, solution (12) represents an expansion in powers of the dimensionless quantity $\mu n t$, n being a characteristic mean planetary motion
and this solution may be valid for interval of several centuries. The
substitution of (11) and (12) into literal series gives numerical series
described above. Due to the unsufficient memory capacity of BESM-6 we
had to change here the order of summation and to deal with these series
as power series, the coefficients being exponential polynomials in multiples of the mean longitudes with complex coefficients. There results

$$p_{0_i} = \sum_{k=0}^{\infty} p_k^{(i)}\, t^k, \qquad w_{0_i} = \sum_{k=0}^{\infty} w_k^{(i)}\, t^k, \qquad (13)$$

$$p_{1_{ij}} = \sum_{k=0}^{\infty} p_k^{(ij)}\, t^k, \qquad w_{1_{ij}} = \sum_{k=0}^{\infty} w_k^{(ij)}\, t^k. \qquad (14)$$

$p_k^{(i)}$ and $w_k^{(i)}$ in (13) are one-argument exponential series with
respect to $\sqrt{-1}\,\lambda_i$ and the terms of order m in (5) and (6) lead in these
series to multiples $-m$, $-m+2$, $\ldots$, $m-2$, m. $p_k^{(ij)}$ and $w_k^{(ij)}$ in (14) are
two-argument exponential series with regard to $\sqrt{-1}\,\lambda_i$ and $\sqrt{-1}\,\lambda_j$. The
terms of the power order m and multiple $\pm\sigma$ with respect to $\sqrt{-1}\,(\lambda_i - \lambda_j)$
in (8) and (9) result in (14) to multiples $\sqrt{-1}\,(k\lambda_i + l\lambda_j)$ with k and l
changing in limits $\pm\sigma$ $\pm m$ provided that the sum k+l may take only the
values $-m$, $-m+2$, $\ldots$, $m-2$, m.
    Numerical series (13) and (14) may be used for the computation of
the planetary ephemerides.

3. ARBITRARY CONSTANTS OF THE THEORY

Arbitrary constants of our theory are represented by n_i, ε_i, $\alpha_o^{(i)}$, $\beta_o^{(i)}$ ($i = 1, \ldots, N$). Taken along with masses m_i these quantities should be determined by comparison of the computed coordinates with the results of observations. The mean motions n_i may be supposed to be known from observations sufficiently accurately and their numerical values are fixed in our theory once and for all. For initial evaluation of ε_i, $\alpha_o^{(i)}$, $\beta_o^{(i)}$ they should be related with the analogous quantities of the classicaal Keplerian expansions. This problem was treated earlier (Brumberg and Chapront, 1973) up to the terms of the seventh degree in eccentricities and inclinations by means of the straightforward comparison of (5), (6) with the classical expansions. We give below a general algorithm to obtain this relationship up to any degree of accuracy. The actual realization of this algorithm on BESM-6 has been performed with the aid of the Poisson series processor by Dasenbrock (Dasenbrock, 1973).

Omitting the subscript i rewrite (5), (6) in the form

$$p = -\frac{1}{2} a + \frac{3}{2} \bar{a} + \delta p(a, \bar{a}, b, \bar{b}),\tag{15}$$

$$w = b + \bar{b} + \delta w(a, \bar{a}, b, \bar{b}),\tag{16}$$

δp, δw denoting the series in powers of a, $\bar{a}$, b, $\bar{b}$ starting with the second degree terms. With the aid of the Keplerian processor (Brunberg and Isakovich, 1975) we find the Keplerian power expansions

$$(A \exp \sqrt{-1} \Lambda)^{-1} (x + \sqrt{-1}\, y) = 1 + \frac{1}{2} K - \frac{3}{2} \bar{K} + S(K, \bar{K}, L, \bar{L}),\tag{17}$$

$$A^{-1} z = L + \bar{L} + T(K, \bar{K}, L, \bar{L}),\tag{18}$$

where S and T are series in powers of K, $\bar{K}$, L, $\bar{L}$ staring with the second degree terms while these variables themselves are

$$K + e \exp \sqrt{-1}\,(\Lambda-\pi), \quad L + \frac{1}{2\sqrt{-1}} \sin I \exp \sqrt{-1}\,(\Lambda-\Omega),\tag{19}$$

$$\Lambda = nt + E.\tag{20}$$

e, I, π, Ω, E represent classical Keplerian elements, i.e. eccentricity, inclination, longitude of the perihelion, longitude of the ascending node and mean longitude at the epoch. In order to relate a, b, ε with K, L, E put

$$a = K + F(K, \bar{K}, L, \bar{L}),\tag{21}$$

$$b = L + G(K, \bar{K}, L, \bar{L}),\tag{22}$$

$$\exp \sqrt{-1} \ (\varepsilon - E) = 1 + H(K, \bar{K}, L, \bar{L}), \tag{23}$$

F, G, H being unknown power series with respect to the indicated variables. Substitution of (15)-(18), (21)-(23) into (1) results in relations

$$\frac{1}{2} F - \frac{3}{2} \bar{F} + H = \Phi, \tag{24}$$

$$G + \bar{G} = \Psi, \tag{25}$$

enabling one to determine F, G, H by iterations. Here we have

$$\Phi = S + (1 + H)\delta p + (-\frac{1}{2} K + \frac{3}{2} \bar{K})H + (-\frac{1}{2} F + \frac{3}{2} \bar{F})H, \tag{26}$$

$$\Psi = T - \delta w. \tag{27}$$

It is easy to show that F and G-series contain only the forms of odd degree in K, $\bar{K}$, L, $\bar{L}$ (starintg with the third degree terms) while H-series consists of the even degree forms only (starting with the fourth degree terms). Moreover, with p, q, r, s denoting the powers of K, $\bar{K}$, L, $\bar{L}$ the combination p-q+r-s is always unity for every term of F and G and zero for every term of H. Taking all this into account we can easily separate the variables in Equation (24). The most difficult operation of this algorithm is to substitute (21), (22) into series for δp, δw. To do this the Dasenbrock's system was supplemented with a special subroutine NTAYLR permitting to expand a function of several variables, represented by a Poisson series, in powers of variations of these variables, which are represented by Poisson series too. In the result one obtains the Poisson series expressed in new variables.

Thus having found F, G, H up to the terms of some degree with respect of K, $\bar{K}$, L, $\bar{L}$ we substitute (21), (22) into (26), (27) and obtain the expressions for Φ, Ψ accurate to one degree more. This leads to the more accurate expressions for F, G, H. Let us note that the structure of (26), (72) shows a way of modification of the algorithm for finding the series (5), (6) (Brumberg and Chapront, 1973) so as to have identically F = G = H = 0.

We have obtained the series (5), (6) and the developments for F, G, H accurate to m = 11 inclusively. From this we deduce immediately the expansions of α-k, β-l in powers of k, $\bar{k}$, l, $\bar{l}$ where α, β are our arbitrary constants α_0, β_0 and k, l represent the Lagrange elements

$$k = e \exp(-\sqrt{-1} \ \pi), \ l = \frac{1}{2\sqrt{-1}} \sin I \exp (-\sqrt{-1} \ \Omega). \tag{28}$$

In virtue of the structure of H the expansion (23) has the same form expressed in K, $\bar{K}$, L, $\bar{L}$ as well as in k, $\bar{k}$, l, $\bar{l}$. Inverting these developments we find expressions of k-α, l-β in powers of α, $\bar{\alpha}$, β, $\bar{\beta}$. This is achieved by means of a special subroutine INVERS yielding the power series inversion

$$X = Y + P(Y) \tag{29}$$

in form of

$$Y = X + Q(X). \tag{30}$$

The algorithm of inversion is based on a iterative relation

$$Q(X) = - P(X + Q) \tag{31}$$

which again calls for the use of the subroutine NTAYLR. All functions occurring in (29)-(31) are vectors of arbitrary dimension.

In this manner all these expansions give a complete solution for the problem of relationship between our constants and classical Keplerian elements. This opens a way for improvement of our constants with the aid of usual methods based on Keplerian elements, i.e. to improve at first these elements and then to return to our constants.

4. CONSTRUCTION OF THE LITERAL SERIES

To begin with, we describe some technical characteristics of the computer system employed by us.

Calculations have been performed on BESM-6 with a software provided by monitoring system DUBNA. The unit of the memory capacity on BESM-6 is a page, consisting of 1024 machine words with 48 bits in each. In employing this computer system a program together with the induced system subroutines and tables of data may occupy no more than 32 pages of the operating storage. Therefore to store intermediate results we had to use magnetic drums with the mean access time of about 0.02 sec. The final results have been written on magnetic tapes, each tape containing no more than 512 pages (zones). 40 bits of the machine word are designed for a mantissa of a number and thus the single precision is adequate for our calculations. Our programs were written in FORTRAN and only several subroutines for the magnetic storage exchange were formulated in an assembler.

Computation of the series (5), (6) common to all planets is a subject of a separate program. Along with this the program gives some auxiliary developments common to all planets as well. All these expansions have been obtained with the aid of a set of subroutines for manipulation with four-argument power series. The series were calculated up to the eighth degree terms inclusively. The coefficients of these series ordered in a tabular manner were stored on a magnetic tape. The series (5) is the longest one and up to the terms of the eight degree inclusively it contains 254 terms. All these series occupy 30 zones on a tape. Limiting up to the seventh degree terms this information demands only 15 zones.

The second program is designed to compute the series (8), (9) representing the disturbing action of one planet on the other. All 56 possible combinations of couples of planets (excluding Pluto) have been considered. The accuracy of calculations is given first of all by

the maximal degree of terms to be retained in power expansions.
Depending on the magnitude of the mutual perturbations this maximal
degree specific for each couple of planets has been chosen from 5 to 7.
Only for the Jupiter-Saturn case we have computed some terms of the
eighth degree. The maximal degree of the retained terms for each couple
of planets is presented in Table I where i, j are referred to disturbed
and disturbing planets respectively and the tabular value for i = j
corresponds to the undisturbed motion.

Table I
Maximal degree of terms in Series (8), (9)

i \ j	1	2	3	4	5	6	7	8
1	8	7	7	5	7	5	5	5
2	6	8	6	5	6	5	5	5
3	5	6	8	6	6	5	5	5
4	5	7	7	8	7	6	5	5
5	5	5	5	5	8	8	7	6
6	5	5	5	5	7	8	7	6
7	5	5	5	5	7	7	7	7
8	5	5	5	5	6	6	7	8

As mentioned above, the coefficients of (8), (9) are exponential
series in multiples of the difference between mean longitudes of
disturbed and disturbing planets:

$$
{}_1^{(ij)}p_{pqrsp'q'r's'} = \sum_{k=-\infty}^{\infty} P_k \exp \sqrt{-1}\, k(\lambda_i - \lambda_j),
$$

$$
{}_1^{(ji)}w_{pqrsp'q'r's'} = \sum_{k=-\infty}^{\infty} W_k \exp \sqrt{-1}\, k(\lambda_i - \lambda_j).
$$

The number of terms retained in these series is the second accuracy characteristics of the presentation of perturbations in our theory.
For all couples of planets in calculating the perturbations of zero,
first and second degrees with respect to the polynomial variables we
have taken into account the exponential terms with index k varying from
-23 to +23. For degrees between 3 and 7 the range of k was from -11 to
+11. For the eighth degree only the terms with k ranging from -5 to +5
were retained. Such a fixed stepping changing of the exponential terms
number was chosen instead of a more logical smooth ranging due to some
technical peculiarities of our tape storage system.

        As a function of the degree m we give in Table II a number NP of
m-degree power terms in the series for p-coordinate and a number NT
of numerical coefficients P_k in the same series. The corresponding
numbers for w-coordinate are equal or less than those for p-series.

        In the described program we have dealt with a set of subroutines to
manipulate with Poisson series having eight polynomial variables and one

exponential argument.

Table II
Number of terms in Series (8), (9) and amount of storage
pages necessary for computing the perturbations of a given
degree

m	1	2	3	4	5	6	7	8
NP	4	20	60	170	396	868	1716	3235
NT	188	940	2820	7990	18612	40796	80652	152045
M	17	19	23	32	58	107	128	143

The terms of series (8), (9) have been calculated subsequently in
increased order with respect to the power variables. In the same order
they were transposed on the magnetic storage. In calculating the m-degree
terms one needs to use just as all terms of lower degrees (from zero to
m-1) so also the four-argument power expansions related to the undisturbed motion. Besides this, it is necessary to provide storage for the
series of right-hand members and for the series of the m-degree perturbations themselves. All this information for each couple of planets
been handled at the given moment is stored on magnetic drums in form of
the tables of coefficients. In the operative storage we have only the
series being in immediate processing at the given moment. In Table II
we indicate a number M of the storage pages on magnetic drums necessary
for computation of perturbations of different degrees m. A relatively
slight increase of the necessary storage when passing from m=6 to m=7
is explained by the fact that the seventh degree terms are transposed
only on tape without using drums since usually they are not employed in
the subsequent calculations. In the case of the eight degree terms calculations the coefficients of the necessary series (from m=0 to m=7)
are transposed from tape on drum but only for the exponential index
ranging from -5 to +5.

We give below separately the amounts of processor and commercial
time needed for the calculation of m-degree perturbations (for one
couple of disturbed and disturbing planets):

degree	1	2	3	4	5	6	7	8
processor time	-	-	-	-	-	-	2h	4h 40m
commercial time	12s	26s	1m	5m	25m	1h35m	6h	14h

Significant distinction between the values of the second and third
lines for m=7 and m=8 is due to the extensive information exchange with
the magnetic drum. But this did not adversely affect the computer charge
in virtue of the multiprogram regime of our computer system.

The actual calculations for all planets have been performed by
subsequent steps with provision for starting solution with the results
of the uncompleted previous step. Along with the coefficients P_k, W_k of
the resulting series we have stored on tape the coefficients of the

right-hand members of the secular system (11). Depending on the maximal degree of power terms (five or seven) the number of these coefficients for each couple of planets is 240 or 940 respectively. These
coefficients were used further in a separate program to compute secular
perturbations in form of (12).

5. CONSTRUCTION OF THE NUMERICAL SERIES

The obtained series (5), (6), (8), (9) have been used in the third
program. This program may carry out the following operations:
1. For any given couple of planets to read the coefficients of the
series stored on tape.
2. Instead of power variables a_i, b_i to substitute into the literal
series the expressions (11), (12) resulted from the solution of the
secular system. By this substitution the series (5), (6), (8), (9)
transform to (13) or (14) relatively.
3. To convert the polynomial-exponential series (13), (14) to the
polynomial-trigonometric series for the rectangular coordinates of the
planets. The series of such form are well suited for comparison with
the results of other theories presented in classical shape with trigonometric and power terms.
4. Based on (13), (14) to calculate for given moments of time the
tabular values of p, w and of the rectangular coordinates of any planet
in the arbitrary fixed heliocentric coordinate system.
Depending on given initial values this program permits to take into
account the total perturbations of the given planet from all others or
alternatively the perturbations from each planet separately. In addition
one may change the maximal power degree of the terms retained in (13)
and (14). It is possible to estimate a contribution due to the terms of
any fixed degree.
We have controlled our results just as in performing the calculations themselves so also by comparison of our final series for rectangular coordinates with those obtained by G.A. Krasinsky in elaborating
a general planetary theory using the method of von Zeipel. (Krasinsky,
1973). In performing our calculations an ideal control is the absence
in the series (5), (6), (8), (9) of critical terms with associated zero
divisors. This type of control was of much use particularly in testing
our programs.
As to comparison of our series with the results of G.Å. Krasinsky
we have stated their coincidence under the same accuracy limitations.
In the cases when our accuracy is lower than that of G.A. Krasinsky's
theory we have observed discrepancies mainly in the eighth and ninth
decimals. Only in the Mercury case where our restriction by the terms
of the seventh and eight degrees is clearly unadequate the discrepancy
has attained the magnitude of the sixth decimal.
To appreciate a size of the resulting numerical series of the type
(13), (14) we give in Table III the numbers of terms in the series for
p-coordinate with coefficients to less than 10^{-8} radians by absolute
value. Here i and j correspond to the disturbed and disturbing planets
respectively and k denotes the degree of t in (13), (14). Just as in

Table III
Number of terms in Series (13), (14) for p-coordinate with $\varepsilon = 10^{-8}$ radians

i	0	0	0	1	2	2	2	3	3	3	3	4	4	4	4	5	5	5	6	6	6	6	6	7	7	7	8	8	8
j \ k	0	1	2	3	0	1	2	0	1	2	3	0	1	2	3	0	1	2	0	1	2	3	4	0	1	2	0	1	2
1	17	15	12	6	7	2	5	31	6	5	3	19	7	2	6	9	8	32	92	59	34	19	3	24	3	7	1	2	4
2	60	12	1		7	6		7	6			50	26	3		109	76		10	10	8	6		48	22	5	2	3	2
3	22	2			43	8		36	8			11	11	10		16			48	22	7			10	8	5	1	14	
4					9	3		25	6			50	19	2		5			17	2				94	28		7	5	
5	26	4			16	3		7	2			25	6														13		
6	8				6	2						5															59		
7												1															7		
8																													

Table I the value i=j is associated to the undisturbed motion. The
indicated numbers are related to calculations with the maximal degree
defined by Table I. The time t in (13), (14) is reckoned in millenia.

6. THE LINEAR THEORY OF THE SECOND ORDER

So far we have considered only the first two terms in series (3)
and (4). However the main parts of p_i and w_i have been found too. These
terms may be expressed as follows: 2^i 2^i

$$p_i = \sum_{j=1}^{N(i)} p_{ij} + \frac{1}{2} \sum_{j=1}^{N(i)} \sum_{k=1}^{N(i,j)} p_{ijk} \qquad (32)$$

$$w_i = \sum_{j=1}^{N(i)} w_{ij} + \frac{1}{2} \sum_{j=1}^{N(i)} \sum_{k=1}^{N(i,j)} w_{ijk}. \qquad (33)$$

p_{ij} and w_{ij} are represented by the Poisson series of the type
(8,1) in the same manner as (8), (9). Being symmetric in j and k p_{ijk}
and w_{ijk} are the Poisson articles (12, 2) where the variables a, $\frac{2}{a}$, b,
$\bar{b}$ for planets i, j, k enter as polynomial variables and the differences
$\lambda_i - \lambda_j$ and $\lambda_i - \lambda_k$ are the trigonometric arguments. We have obtained
the initial terms of the Poisson series (32), (33) with value m=0 (the
quasi-periodic intermediate solution) and m=1 (the inequalities of the
first degree in eccentricities and inclinations). The terms of p_{ij} and
w_{ij} are not of much interest since they are analogous to the terms of
p_{ij} and w_{ij} but make a significantly lesser contribution to the general
solution (3), (4). Contributions of p_{ijk} and w_{ijk} are more essential due
to the appearance of the resonance terms caused by the close commensurabilities between mean motions of the triplets of planets. These terms
are analytically of the second order with respect to the planetary masses
but numerically they are comparable with the first order terms. In the
paper (Brumberg et al., 1975a) a detailed analysis of these terms in
the quasi-periodic intermediate solution (m=0) is presented and it is
established that many of such terms are omitted in the classical planetary theories*. Similar analysis may be carried out for the terms of
the first degree in eccentricities and inclinations (m=1). It is convenient to write these terms in the following manner:

$$p_{ijk} = c(i,j,k,0)a_i + d(i,j,k,0)\bar{a}_i + c(i,j,k,1)a_j +$$

---

*In the paper cited there are some discrepancies with our basic results
obtained by the series expansion method. The reason lies in the unsufficient accurate computation of the right-hand members by the iteration
method. We are indebted to M. Luc Duriez for this remark.

$$d(i,j,k,1)\bar{a}_j + c(i,k,j,1)a_k + d(i,k,j,1)\bar{a}_k, \qquad (34)$$

$$w_{\frac{1}{2}ijk} = f(i,j,k,0)b_i + \bar{f}(i,j,k,0)\bar{b}_i + f(i,j,k,1)b_j +$$

$$\bar{f}(i,j,k,1)\bar{b}_j + f(i,k,j,1)b_k + \bar{f}(i,k,j,1)\bar{b}_k. \qquad (35)$$

Here $c(i,j,k,\Delta)$, $d(i,j,k,\Delta)$, $f(i,j,k,\Delta)$ ($\Delta=0$, $\Delta=1$) are exponential series in two arguments: $\sqrt{-1}\ (\lambda_i - \lambda_j)$ and $\sqrt{-1}\ (\lambda_i - \lambda_k)$. These series have been constructed in the range from -7 to $+7$ in each argument. This construction has revealed many resonance terms caused by the close commensurabilities between mean motions of some triplets of planets. The terms are regarded to be resonant if the ratio $(sn)/n_i$ approaches to zero or ±1 (s is the N-vector of indices of the mean longitudes λ_1, ..., λ_n, n is the N-vector of planetary mean motions).

It is of interest to compare our results with the classical theories of Newcomb and Hill. The paper (Newcomb, 1895) is devoted to the investigation of the long period inequalities of the second order in the mean longitudes of the four inner planets. Among them we are interested in the terms with the arguments

$$s\lambda = s_1\lambda_1 + \ldots + s_N\lambda_N$$

for which

$$s_1 + \ldots + s_N = 0, \pm1, \pm2.$$

There are six such arguments:

$$\lambda_1 - 5\lambda_2 + 4\lambda_3,\ 3\lambda_2 - 7\lambda_3 + 4\lambda_4,\ \lambda_3 - 2\lambda_4 + \lambda_5,$$

$$4\lambda_3 - 8\lambda_4 + 3\lambda_5,\ 3\lambda_3 - 6\lambda_4 + 2\lambda_5,\ 5\lambda_3 - 10\lambda_4 + 4\lambda_5.$$

According to Newcomb the perturbations in the motion of Mercury, Venus and the Earth associated with the first argument are negligible small while those due to the terms with the third argument are perceptible only in the orbit of Mars. Perturbations caused by the terms with the fourth argument are taken into account in the orbits of the Earth and Mars. The inequality of Le Verrier related to the second argument is present in the orbits of Venus, the Earth and Mars. The perturbations themselves are determined by a complicated artificial method. As possible arguments Newcomb indicates the fifth and sixth arguments but he does not examine the related perturbations. In the theory of the motion of Mars Newcomb (Newcomb, 1898a) takes into account the fifth argument and the argument $\lambda_2 - 2\lambda_4 + 2\lambda_6$.

The tables of perturbations of Jupiter elaborated by Hill (Hill, 1898a) present only one argument with three planetary mean motions; $3\lambda_5 - 6\lambda_6 + 3\lambda_7$. For the perturbations of Saturn Hill (Hill, 1898b) gives the terms related to four arguments of the type considered:

$-2\lambda_5 + 5\lambda_6 - 3\lambda_7$, $-\lambda_5 + 2\lambda_6 - \lambda_7$, $\lambda_5 - 3\lambda_6 + 2\lambda_7$, $\lambda_5 - 5\lambda_6 + 4\lambda_7$. In the Newcomb's theory for the motion of Uranus (Newcomb, 1898b) the following arguments occur: $-2\lambda_5 + 6\lambda_6 - 4\lambda_7$, $\lambda_5 - 4\lambda_6 + 3\lambda_7$, $2\lambda_5 - 7\lambda_6 + 5\lambda_7$, $\lambda_5 - 3\lambda_6 + 2\lambda_7$, $-\lambda_5 + 2\lambda_6 - \lambda_7$, $-\lambda_5 - \lambda_6 + 2\lambda_7$. At last, there are no such terms in the Newcomb's theory for the motion of Neptune (Newcomb, 1898c).

In our theory all terms entering into the classical theories are presented but along with them we have revealed a set of terms having numerically the same or even greater magnitude. In Table IV we give for illustration a list of arguments of the series for $\mu^2 c(i,j,k,\Delta)$ with a restricting condition that the absolute magnitudes of the corresponding coefficients are no less than 10^{-7} radians. An asterisk indicates the presence of the related term in the corresponding classical theory of Newcomb or Hill. It is to be noted that there is a full agreement in arguments for the motion of Mars between our results and the theory of Clemence (Clemence, 1961).

Table IV
Main triple arguments of the linear theory

i j k	s_i	s_j	s_k	sn/n_i		i j k	s_i	s_j	s_k	sn/n_i
2 3 4	3	-7	4	0.002 *		4 5 6	-3	4	-1	-2.430
	-4	6	-2	-0.963		4 5 7	-2	7	-5	-1.002
2 3 5	-4	5	-1	-0.976		5 6 7	-3	6	-3	-1.007 *
	3	-5	2	0.028			-2	3	-1	-0.933
2 5 6	-2	4	-2	-1.834			-3	5	-2	-1.269
3 2 4	5	-3	-2	-0.940			-4	6	-2	-1.866
	-7	3	4	0.003 *			-1	3	-2	-0.074
3 2 5	9	-6	-3	-1.006			-4	5	-1	-2.128
	4	-3	-1	-0.961			-3	4	-1	-1.530
3 4 5	-4	6	-2	-0.978 *			-3	2	1	-2.053
3 5 6	-2	4	-2	-1.731		5 6 8	-3	5	-2	-1.130
	-2	3	-1	-1.781			-4	5	-1	-2.058
	-1	-1	2	-1.016		5 7 8	-1	-1	2	-0.997
	-2	-1	3	-1.984			-2	4	-2	-1.579
4 2 3	4	3	-7	0.006 *		6 5 7	5	-2	-3	-1.018 *
4 3 5	7	-4	-3	-0.999			4	-2	-2	-1.668
	5	-3	-2	-0.959 *			-4	1	3	-0.465
	3	-2	-1	-0.920			-5	1	4	-1.114 *
	9	-5	-4	-1.038			-8	2	6	-0.930
4 3 6	-3	1	2	-0.991			2	-1	-1	-0.834 *
4 5 6	-2	4	-2	-1.493			3	-2	-1	-2.317
	-1	-1	2	-1.031			-3	1	2	0.184 *
	-2	3	-1	-1.588			5	-3	-2	-3.151
	-3	5	-2	-2.335			8	-3	-5	-1.203
	-2	-1	3	-1.967			7	-3	-4	-1.852
	-2	5	-3	-1.399			6	-2	-4	-0.369
	-1	3	-2	-0.652			-6	1	5	-1.763
	-2	1	1	-1.778			6	-3	-3	-2.502

Table IV (Continued)

planets indices			planets indices		
i j k	s_i s_j s_k	sn/s_i	i j k	s_i s_j s_k	sn/n_i
6 5 7	10 -4 -6	-2.037	7 6 8	3 -1 -2	-0.872
6 5 8	-4 1 3	-0.980		4 -2 -2	-2.724
	10 -4 -6	-1.006		-1 -1 2	-2.832
	4 -2 -2	-1.324		-5 1 4	-0.109
	6 -3 -3	-1.986		-7 1 6	-1.089
	3 -2 -1	-2.145		1 -3 2	-6.536
	-3 1 2	-0.159		5 -3 -2	-4.576
	2 -1 -1	-0.662		5 -2 -3	-2.234
	7 -3 -4	-1.165		2 -4 2	-8.388
6 7 8	-2 4 -2	-0.955		7 -2 -5	-1.253
	-1 -1 2	-0.993		4 -1 -3	-0.382
	-3 5 -2	-1.604		6 -2 -4	-1.743
	-4 6 -2	-2.254		2 -1 -1	-1.362
	-2 3 -1	-1.127		8 -2 -6	-0.763
	-1 3 -2	-0.306		6 -4 -2	-6.428
	-3 1 2	-2.292		3 -5 2	-10.241
	-5 7 -2	-2.903		-6 1 5	-0.599
	-2 1 1	-1.471		3 -2 -1	-3.214
	-3 4 -1	-1.776		7 -5 -2	-8.280
	-4 7 -3	-2.082		4 -6 2	-12.093
7 5 6	3 1 -4	-1.326 *		-8 1 7	-1.579
	-4 -2 6	-1.053 *	8 5 6	2 1 -3	-0.890
	5 2 -7	-0.799 *		4 2 -6	-1.781
	4 1 -5	-3.178 *		-1 -1 3	0.890
	2 1 -3	0.526 *	8 7 6	5 1 -6	-1.174
7 5 8	-1 -1 2	-7.063		6 1 -7	-2.136
	4 -2 -2	-11.185		4 1 -5	-0.213
	3 -1 -2	-5.102		3 -2 -1	-10.150

7. CONCLUSION

The results exposed here summarize our work in mathematical constructing the general planetary theory.

Taking into account a rather small operative storage capacity of BESM-6 computer accessible for us it is not suitable to compute by the same method inequalities of higher degree in eccentricities and inclinations and of higher order with respect to the planetary masses. But independent of an employed computer it is more advantageous to improve a semi-analytical theory by numerical iterative methods. Our work will proceed on this line.

For investigation of the long-term evolution of planetary motions the polynomial solution of the secular system used here is invalid and should be replaced by a pure trigonometric solution. This solution will be obtained in the nearest future.

As far as comparison with observations and determination of constants are concerned, the following solution seems to be reasonable. Considering the coincidence of our results with those of Krasinsky's theory it is of no use to determine from observations the constants for both theories separately. It is more simple to do this for one theory and then to use the algorithm of the relationship between constants exposed in Section 3.

Finally, we should like to express our sincere gratitude to Dr J. Chapront and his colleagues at the Bureau des Longitudes who collaborated with us and did so much in the initial stage of this work.

REFERENCES

Brumberg, V.A.: 1974, in Y. Kozai (ed.), The Stability of the Solar
      System and of Small Stellar Systems, p. 139. Reidel, Dordrecht.
Brumberg, V.A., Evdokimova, L.S., and Skripnichenko, V.I.: 1975a,
      Astron. J. (USSR) 52, 420. (in Russian)
Brumberg, V.A., Evdokimova, L.S., and Skripnichenko, V.I.: 1975b,
      Celes. Mech. 11, 131.
Brumberg, V.A. and Chapront, J.: 1973, Celes. Mech. 8, 335.
Brumberg, V.A. and Isakovich, L.A.: 1975, Algorithms of Celestial Mecha-
      nics No. 4. Inst. Theoret. Astron. (Leningrad). (in Russian).
Clemence, G.M.: 1961, Astron. Papers 16, pt. 2, 261.
Dasenbrock, R.R.: 1973, Naval Research Lab. Rep. No. 7564.
Hill, G.W.: 1898a, Astron. Papers 7, pt. 3, 287.
Hill, G.W.: 1989b, Astron. Papers 7, pt. 4, 417.
Krasinsky, G.A.: 1973, in N.S. Samoylova-Yakhontova (ed.), Minor
      Planets, Ch. 5, 6. Nauka, Moscow. (in Russian)
Newcomb, S.: 1895, Astron. Papers 5, pt. 2, 49.
Newcomb, S.: 1898a, Astron. Papers 6, pt. 4, 383.
Newcomb, S.: 1898b, Astron. Papers 7, pt. 1, 1.
Newcomb, S.: 1898c, Astron. Papers 7, pt. 2, 145.

CONSTRUCTION OF PLANETARY THEORY BY ITERATIVE PROCEDURE

T.V. Ivanova
Institute for Theoretical Astronomy, Leningrad,   U.S.S.R.

In this paper the method of determination of the planetary per-turbations is proposed which is a modification of Dziobek-Brouwer's method [1,2]. For the simplicity the case of two mutually disturbing planets is considered. In the original version of the method the perturbations of rectangular planetary coordinates are presented by means of the formal integrals

$$\delta X_{ik} = \int (\sum_{j=1}^{3} a_{ikj} G_{ij})dt + c_{ij} \iint (\sum_{j=1}^{2} b_{ij} G_{ij})dtdt, \tag{1}$$

$$i = 1,2; \quad k = 1,2,3$$

where index i corresponds to the number of the planet; δX_{ik} are per-bations of X_{ik} coordinates; G_{ij} – components of the perturbating accelerations. The coefficients of a_{ikj}, c_{ik}, b_{ij} are the well-known functions of the coordinates of the elliptic motion which can be developed as double Fourier series in mean longitudes. The denomi-nators in Davis' formulas [3] for these coefficients contain the eccentricities. For this reason Musen [4] expressed an opinion that Brouwer's method would lose its effectiveness when small eccentric-ities are involved. These fictitious peculiarities are eliminated in the present paper by means of trivial transformations and the expres-sions for the coefficients are given in a simple symmetric form.

According to Brouwer's method for the determination of the mutual perturbations the equations of the motion for any planet are written down in a special coordinate system. Therefore the combined integra-tion of the whole set of equations becomes difficult. The unified frame of reference connected with the ecliptic is taken now for all the planets under consideration and the formal integrals (1) are pre-sented in the following vectorial form

$$\delta \bar{r}_i = \int [A_i + A_i'(\tilde{t}-t)] \bar{G}_i dt, \quad i = 1,2 . \tag{2}$$

49

V. Szebehely (ed.), Dynamics of Planets and Satellites and Theories of Their Motion, 49-52.

Here index i still corresponds to the number of the disturbed planet; $\delta \tilde{t}_i$ is the perturbation of the radius-vector of the i-th planet; $\tilde{t}=t$ but it is considered as a constant while integrating; A_i and A_i' are linear operations in three-dimensional vector space to which the square third-order matrixes composed of perturbating function coefficients in Brouwer's equations (1) are related.

Brouwer expands the components of the vectors $\bar{G}_i$ in six-argument Taylor series with respect to the perturbations of the coordinates of both planets which is usually made while integrating the differential equations of the motion. The corresponding form of the integrands in (1) is very complicated, new terms being not described by the uniform formulas and added at each next step of the approximation. The most principal distinction of the proposed modification consists in new expansions of the functions $\bar{G}_i$. The expressions for $\bar{G}_i$ are independent on the order of the perturbations and well suited for the computer calculations. These are presented as functions symmetric with respect to the undisturbed radii-vectors and their perturbations

$$\bar{G}_i = \sum_{\ell=i}^{3} \kappa_{i\ell} \bar{\sigma}_{i\ell} - \frac{3}{2} \kappa_{ii} r_i^{-5} \delta \bar{r}_i [2\bar{r}_i \delta \bar{r}_i + (\delta \bar{r}_i)^2]$$

$$- \frac{3}{2} \kappa_{ii} r_i^{-5} r_i (\delta \bar{r}_i)^2 , \quad i = 1,2 . \tag{3}$$

Here

$$\bar{\sigma}_{i\ell} = (\bar{r}_\ell + \delta \bar{r}_\ell) \sum_{j=s_{i\ell}}^{\infty} d_j r_\ell^{-2j-3} [2\bar{r}_\ell \delta \bar{r}_\ell + (\delta \bar{r}_\ell)^2]^j ,$$

$$d_j = (-1)^j \frac{(\frac{3}{2})_j}{(1)_j} ; \quad s_{ii} = 2, \quad s_{ip} = 0 \quad \text{when } p = 1, 2, 3, \, p \neq i;$$

$$\tag{4}$$

$$\kappa_{ii} = -k^2(1 + m_i), \quad \kappa_{ip} = -k^2 m_p \quad \text{when } p = 1,2, \, p \neq i;$$

$$\kappa_{13} = k^2 m_2, \quad \kappa_{23} = k^2 m_1 ,$$

where k is the Gaussian constant, m_i means the mass of the i-th planet r_3 is the distance between the planets.

The integration in (2) is fulfilled by iterations

$$\delta \bar{r}_i^{(n+1)} = \int [A_i + A_i'(\tilde{t} - t)] \bar{G}_i^{(n)} dt \tag{5}$$

which allows more complete automatisation of the calculations, some terms of order higher than n with respect to the disturbed mass being taken

into account at every n-th approximation. As the first approximation
the keplerian ellipse or the results of any earlier elaborated theory
may be used.

The main difficulty in the present method as well as in any perturbation theory is the expansion of the reciprocal mutual planetary
distance r_3^{-1} involved in the expansions of the functions $\bar{G}_i$. Lately
Newton's iteration algorithm is widely used in developing r_3^{-1}

$$\left(\frac{a}{r_3}\right)_{n+1} = \frac{1}{2}\left(\frac{a}{r_3}\right)_n\left[3 - \left(\frac{r_3}{a}\right)^2\left(\frac{a}{r_3}\right)^2_n\right] \tag{6}$$

where n is the iteration number, a is a major semi-axis of the external planetary orbit.

Newton's method could not be used as yet to solve the problem of
Neptune-Pluto type because the known ways of choosing the first
approximation failed to guarantee the convergence of the process of
iterations.

In this paper Newton's algorithm is also used. But as the first

approximation for $\left(\frac{a}{r_3}\right)_0$ the quantity $\frac{a}{M}$ is accepted where M is the

maximum distance between the orbits of bodies considered. As was
theoretically shown by Dr. M.S. Petrovskaya (Celestial Mechanics, in
print) such a choice of the initial approximation can ensure the
iteration convergence for any form and configuration of planetary
orbits. We followed the above method in expanding the value
$\frac{a}{r_3}$ into double Fourier series in mean longitudes for the problems of

Jupiter-Saturn and Neptune-Pluto types. The calculations were made
by BESM6 computer. In the Jupiter-Saturn case 8 iterations were
necessary to get the precision of 10^{-8} and 9 iterations to provide
the accuracy as high as 10^{-9}. The number of terms in series was 916
and 940 respectively. As follows from the calculations the iterational
procedure for the Neptune-Pluto problem is also convergent, though
the convergence being rather slow. Thus, at an intermediate stage of
calculations 14 iterations proved to be necessary to expand $\frac{a}{r_3}$ with
the precision of 10^{-4}.

The proposed modification of Brouwer's method may be applied to
solve the classical problems of celestial mechanics in the cases of
small eccentricities and inclinations of orbits. It may be used as
well for the study of planetary and cometary motions when the orbits
of the bodies intersect in projection, in particular when the mutual
perturbations in the Neptune-Pluto system are determined. The method

is suitable for the calculations of high-order planetary perturbations
since any order perturbations may be represented by the same algorithm.

REFERENCES

1. D. Brouwer, Astron. J., 51, 37, 1944

2. O. Dziobek, Mathematical theories of planetary motions, New
    York, 1962.

3. M. Davis, Astron. J., 56, 188, 1952.

4. P. Musen, Geophys. Res., 71, 5997, 1966.

5. T.V. Ivanova, Astronomichesky Journal, 52, 839, 1975 (in Russian).

QUALITATIVE DYNAMICS OF THE SUN-JUPITER-SATURN SYSTEM

V. Szcbchcly
University of Texas, Austin, Texas

ABSTRACT

The stability of the three-body problem formed by the Sun, Jupiter
and Saturn is investigated using surfaces of zero velocity. The results obtained with the models of the restricted and general problems
of three bodies are compared with numerical integration. The system
is found to be stable in the sense that Saturn will neither interrupt
the (perturbed) binary orbit of Jupiter around the Sun, nor will it
escape from the system. It is shown that the known classical triple
stellar systems are "more stable" than the solar system, which in turn
is "more stable" than the Earth-Moon system.

INTRODUCTION

A general trend for instability of three-body systems containing
masses of the same orders of magnitude was demonstrated ten years ago
by Agekian (1967) and Szebehely (1967-a). These results suggested to
Kuiper (1973) that the onset of instability of planetary systems may
be enhanced by increasing the masses of the participating planets. In
this way the method known as the K-N-S theory (Kuiper-Nacozy-Szebehely)
was born and announced by Nacozy (1976). Numerical integrations using
increased planetary masses were performed by Nacozy (1976), and others.
Nacozy could not detect secular terms in the orbital elements of Saturn
unless the masses of Jupiter and Saturn were increased by a factor
(γ) of 29. When this factor was below 29 the system was stable (as
found by numerical integration and as defined by the absence of secular
terms). When γ was larger than 29 instability set in very soon after
the beginning of the motion, as displayed by the appearance of secular
terms.

This paper uses analytic qualitative methods as opposed to numerical integration to find the value of the above mentioned factor γ
at which instability sets in. The raison d'être for such study is that
the establishment of stability or the detection of long-period secular

V. Szebehely (ed.), Dynamics of Planets and Satellites and Theories of Their Motion, 53-55.
All Rights Reserved. Copyright © 1978 by D. Reidel Publishing Company, Dordrecht, Holland.

terms by numerical means is always open to questions.

ANALYSIS

     First the admittedly weak model of the restricted problem is used
to evaluate the effect of γ as described for instance by Szebehely
(1967-b). The Jacobian constant for the orbit of Saturn is computed
using the Sun and Jupiter as the primaries. In this way the mass-parameter of the restricted problem becomes $\mu = m_J/(m_\Theta + m_J) = 9.539 \times 10^{-4}$,
where m_J and m_Θ are the masses of Jupiter and of the Sun. When the
masses of Jupiter and Saturn are increased by a factor of $\dot\gamma$ we have
for the new mass-parameter $\mu' = \gamma m_J/(m_\Theta + \gamma m_J) \cong \gamma\mu$. The assumption
is made at this point that Saturn's orbit is fixed while μ changes.
In this way the Jacobian constant (C) for Saturn's orbit may be computed as a function of the mass-parameter μ' or as a function of γ. In
fact, we have

$$C = 3.2516 - 0.00128\gamma .$$

     On the other hand, the topology of the permissible regions of
motion is controlled by the critical value of the Jacobian constant,
C_{cr}, corresponding to the above introduced value of μ'. The functional
relation $C_{cr} = C_{cr}(\gamma)$ is complicated but it may be approximated by

$$C_{cr} = 3.0831 + 0.00774\gamma$$

in the range of interest. The intersection of the above two straight
lines corresponds to $C_{cr} = C$ and $\gamma = 18.7$. Therefore, the system is
unstable (in the sense that Saturn may penetrate the Sun-Jupiter region)
if $\gamma > 18.7$. On the other hand, if $\gamma < 18.7$ Saturn cannot enter the
region occupied by the Sun and Jupiter.

     The second model is the general problem of three bodies when orbi-
tal eccentricies as well as Saturn's effect on Jupiter and on the Sun
are included in the analysis. The role of the Jacobian constant is
now played by the dimensionless stability parameter $s = -c^2 H/(G^2 \bar{m}^5)$
which controls the topology of the zero-velocity surfaces. Here c is
the angular momentum, H is the total energy, $\bar{m}$ is the average mass and
G the gravitational constant. Once again, we compute the actual and
the critical values of s for various values of γ and the intersection
of the curves $s = s(\gamma)$ and $s_{cr} = s_{cr}(\gamma)$ will furnish the separation
between stability and instability. In the range of interest we have

$$s - s_{cr} = 10^{-6}(14.83 - 1.09\gamma),$$

giving $\gamma = 13.6$ for the intersection. Therefore, instability sets in
sooner (at a lower value of γ) when the (more realistic) model of the
general problem of three bodies is used, while the model of the

restricted problem gives more tolerance.

The computation of the stability parameter is described by Szebehely and McKenzie (1977) with additional details.

Similar analysis may be used to study the stability of classical triple stellar systems. For observed systems the measure of stability is given by $S = (s - s_{cr})/s_{cr}$ and it is found to be of the order of one. The same measure of stability for the above described model of the solar system ($\gamma = 1$) is $S = 3.6 \times 10^{-2}$, consequently, the known triple stellar systems are "more stable" than the solar system.

The Sun-Earth-Moon system is stable according to Hill's (1878) computation as well as according to the restricted problem (Szebehely, 1967-b). The measure of stability found by using the model of the restricted problem for the moon's orbit is $S = (C - C_{cr})/C_{cr} \cong 10^{-4}$, consequently, the moon's orbit is much "less stable" than the solar system. Corresponding values for the moon's stability, using the general problem are not available yet, but the stability is expected to be reduced because of the eccentricities of the orbits and because of the presence of general three-body effects. The hierarchy of stability according to these results is: triple stellar systems, planetary systems and satellites, in order of decreasing stability.

ACKNOWLEDGEMENT

A research grant from the Scientific Affairs Division of NATO in partial support of this work is gratefully acknowledged.

REFERENCES

Agekian, T.A., and Anosova, Z.P., (1967), Astronomical Zhurnal, 44, 1261.

Hill, G., (1878), Am. J. Math., 1, 5, 129, 245.

Nacozy, P., (1976), "On the Stability of the Solar System", Astron. J., 81, 787.

Szebehely, V., (1967-a), Proc. U.S. Nat. Acad. Sc., 58, 60.

Szebehely, V., (1967-b), Theory of Orbits, Academic Press, New York, NY.

Szebehely, V., and McKenzie, R., (1977), "Stability of Planetary Systems with Bifurcation Theory", Astron. J. 82, 79.

A NEW APPROACH FOR THE CONSTRUCTION OF LONG-PERIODIC PERTURBATIONS

Rudolf Dvorak
Astronomisches Institut, Universität Graz
Universitätsplatz 5, A-8010 Graz, Austria

ABSTRACT. The aim of this work is to study perturbations of planets
of a period of some thousands of years. The use of an analytical
method allows us to separate all different influences, e.g. near
resonances and is combined with the very precise method of the
numerical integration. The truncation to low orders can be avoided
which is made by analytical methods in using developments with
respect to the small parameters inclinations and eccentricities.
For this purpose a special form of the Lagrange Equations is used
where the terms containing the inverse distancefrom the planet to
the perturbing one are separated as it is the most difficult to
compute. To develop this a specific formulation has been found
where the short periodic terms can precisely be determined. Although
the development seems to be of a certain complexity the small
numbers of quantities used can be tabulated once and for all in a
specific problem. It should be possible to integrate the new form
of the Lagrange Equations within a reasonable computer-time to
determine the long periodic perturbations.

1. INTRODUCTION

To formulate the problem of constructing the long periodic
perturbations of the planets with the aid of analytical methods
is a priori restricted to a low order of the inclinations i and
the eccentricities e. On the other hand the determination of the
motion of the nodes and the perihelia is also dependant on the
choice of numbers of the near resonances fixed in advance, which
contributes to the solution of the problem.

The first aim of this paper is to extend the realm of validity
of the series expansions with respect to the small parameters e
and i to higher orders as it has to be done in methods based on
trigonometric series. Another truncation is always made in looking
only for solutions which are of low order in the masses. Therefore
our knowledge of the long periodic terms in the motion of the
planets is very uncertain. Although work has been done to understand

57

V. Szebehely (ed.), Dynamics of Planets and Satellites and Theories of Their Motion, 57-64.

the longer periods of some thousands of years by Hill (1889),
Brouwer and van Woerkom (1950) and Anolik et al.(1969) the truncation
mentioned above limits our understanding. Recently P. Bretagnon
(1974) studied the long periodic variations of the planetary system
in introducing long periodic terms of fourth order in eccentricities
and inclinations. He was then taking into account the short periodic
terms in providing the perturbations of the first order with regard
to the masses. Another approach is made by Cohen et al.(1973) by
a numerical integration in using rectangular coordinates over the
period of one million years. This method mixes up all the different
influences on the periodic perturbations, e.g. the role of the
near resonances cannot precisely be determined. In the diagrams
represented by these authors one can very well separate some main
perturbations (e.g. the multiples of the great inequality Jupiter-
Saturn) but no analysis of the periods mentioned above is made.

Therefore we try to combine an analytical method - which
allows us to introduce any desired near resonance term - with a
numerical integration by using the special advantage of the high
degree of accuracy of the last mentioned. The advantage of our
method will be a triple one:

1st we avoid the truncation due to the low powers in the
inclinations and eccentricities
2nd we are not restricted to a low order of masses
3rd the single influence of a group of near resonances can
be estimated carefully.

For this purpose we have chosen the system of the variables
of Lagrange. In a first part the Lagrange Equations are formed
in a special closed manner separating the terms dependant on the
inverse distance $1/\Delta$ planet - perturbing planet from the other one,
because the most complicated development arises from evaluating
the quantity Δ^{-s}.

δ being the vector of the osculating elements the equations
can be written as follows

$$\frac{d\delta}{dt} = \frac{F(\delta,\delta')}{\Delta^3} + \frac{G(\delta,\delta')}{r'^3} \tag{1}$$

As we can see later the short periodic terms appear through the
quantities of the form $(\frac{r}{a})^n \tau^m$, r being the radius vector, a the
semimajor axes, $\tau = \exp \sqrt{-1}\ (v+\tilde{\omega})$, $v+\tilde{\omega}$ being the true longitude.
The most heavy work is done on finding a development of the inverse
distance to be able to evaluate the expressions mentioned above.
Starting from the development

$$F(\frac{r}{a},\frac{a'}{r'},\tau,\tau',h,u,\ (\alpha^2)) \tag{2}$$

with $\alpha=a'/a$, $h=h(e,e',r/a,a'/r')$ and $u=u(i,i',\bar{c},\bar{c}')$ we used the convenient solution of Abu el Ata and Chapront (1974) where the $\Psi(\alpha^2)$ have been calculated very precisely even for larger α. One could use the method of the series expansion of Δ^{-s} in Legendre polynomials having the disadvantage for the planetary problem to be not suitable for large values of α .

Instead of the quantities h and u which are of the order of e and i^2 we transformed the development F into the new one G. The establishement of the form

$$G(\Psi(\alpha^2),\Gamma_1(i,i'),\Gamma_2(i,i'),(r/a)^n\bar{c}^m,(a'/r')^{n'}\bar{c}'^{m'}) \tag{3}$$

put in evidence the short periodic terms through the quantities $(a/r)\bar{c}$. These short periodic terms can be computed with the aid of the Hansen coefficients with a great accuracy without truncation in the form introduced by Brumberg (1973). Note that the Γ_1 and Γ_2 are of the order of thesquare of the inclinations. The calculated expression for the inverse distance is now combined with the equation (1) and enables to compute the long periods by a numerical integration of the system of the Lagrange Equations for the osculating elements used in taking the mean values of equation (1).

2. THE LAGRANGE EQUATIONS

We used the variables of Lagrange for the perturbing function in the following form

a the semimajor axes
λ the mean longitude

$$\begin{aligned} z &= e \exp \sqrt{-1}\ \tilde{\omega} \\ \xi &= \sin(\tfrac{i}{2})\exp\sqrt{-1}\Omega \end{aligned} \tag{4}$$

introduced by Chapront et al. (1975). We were able to find an expression where the inverse distance is isolated by starting from the equations givenin the paper mentioned above

$$\frac{da}{dt} = \frac{2a^2}{\varphi}\ \mathrm{Im}\bar{z}(\wp R_1+\pi R_2)+\mathrm{Im}(\bar{c}\pi)R_2$$

$$\frac{d\lambda}{dt} = n-\mathrm{Re}\left[(2\bar{\jmath}+\alpha\varphi\Psi\bar{z})(\wp R_1+\pi R_2)\right]+\left[\tfrac{\Psi}{\varphi}\mathrm{Im}(\bar{z}\wp)\mathrm{Im}(\bar{c}\pi)+\tfrac{1}{\varphi}\,\sigma_\lambda\right]R_2 \tag{5}$$

$$\frac{dz}{dt} = -ja\varphi(\wp R_1+\pi R_2)+\left[\tfrac{1}{\varphi}(zr+\wp)\mathrm{Im}(\bar{c}\pi)+jz\tfrac{1}{\varphi}\,\sigma_\lambda\right]R_2$$

$$\frac{d\xi}{dt} = \left[\wp-\xi\mathrm{Re}(\bar{\xi}\wp)\right]\tfrac{1}{\varphi}\,\sigma_\xi R_2$$

All the quantities like ϑ, π, φ, ψ, σ_λ and σ_ξ are functions of the Lagrange variables used; j signifies $\sqrt{-1}$. The R_1 and R_2 are the terms of the disturbing function where one has to distinguish on one hand for an inner planet perturbed by an outer one

$$R_1 = -\frac{\mu}{\Delta^3} \quad ; \quad R_2 = \mu(\frac{1}{\Delta^3} - \frac{1}{r'^3}) \quad ; \quad \mu = \frac{nam'}{1+m} \tag{6}$$

and on the other hand for an outer planet perturbed from an inner one (see the paper by Chapront et al. (1975)).

$$R_2 = \mu'(\frac{1}{\Delta^3} - \frac{1}{r^3}) \quad ; \quad \mu' = \frac{n'a'm}{1+m'} \quad ; \quad R_1 = -\frac{u'}{\Delta^3} \tag{7}$$

The separation of the terms dependant on the inverse distance leads to

$$\frac{da}{dt} = \mu(\frac{A}{\Delta^3} - \frac{A'}{r'^3})$$

$$\frac{d\lambda}{dt} = n-\mu(\frac{\Lambda}{\Delta^3} - \frac{\Lambda'}{r'^3})$$

$$\frac{dz}{dt} = \mu(\frac{Z}{\Delta^3} - \frac{Z'}{r'^3}) \tag{8}$$

$$\frac{d\xi}{dt} = \mu Z''(\frac{1}{\Delta^3} - \frac{1}{r'^3})$$

The according functions A, A', Λ, Λ', Z, Z' and Z'' are

$$A = \frac{2a^2}{\varphi} \operatorname{Im}(\bar{c}\pi) - \operatorname{Im} \bar{z}l$$

$$A' = \frac{2a^2}{\varphi} \operatorname{Im}(\bar{z}\pi) - \operatorname{Im}(\bar{c}\pi)$$

$$\Lambda = \operatorname{Re}(ml) + p$$

$$\Lambda' = \operatorname{Re}(m\pi) - p \tag{9}$$

$$Z = ja\varphi l + q$$

$$Z' = ja\varphi\pi - q$$

$$Z'' = [\vartheta - \xi \operatorname{Re}(\bar{\xi}\vartheta)] \frac{1}{\varphi} \sigma_\xi$$

with $l = \pi - \vartheta$; $m = 2\bar{\vartheta} + a\varphi\psi\bar{z}$; $p = \frac{\psi}{\varphi}\operatorname{Im}(\bar{z}\vartheta)\operatorname{Im}(\bar{c}\pi) + \frac{1}{\varphi}\sigma_\lambda$;

$q = \frac{1}{\varphi}(zr + \vartheta)\operatorname{Im}(\bar{c}\pi) + jz\frac{1}{\varphi}\sigma_\lambda$.

3. THE INVERSE DISTANCE

For the analytical development of the inverse distance we took the original form of Abu el Ata and Chapront (1975)

$$\Delta^{-6} = r'^{-s}(2-\delta_j^0)\,\text{Re}\sum_{j=0}^{j_m}\sum_{n=0}^{E(\omega/2)}\sum_{m=0}^{\omega-2n}\frac{(6/2)_n}{(1)_n}\alpha^n\varphi_{n+\frac{6}{2},m}^{(j)}(\alpha^2)\,*$$

$$* \;(\frac{r}{a}\frac{a'}{r'}\,\bar{c}')^{j}J_h^m(\frac{ra'}{ar'}\,u)^n \tag{10}$$

The development above should be transformed to a new expression which separates the $(r/a)^n c^m$. It could be possible to use the method of development of Δ^{-1} in Legendre polynomials of the form

$$\Delta^{-1} = \frac{a}{a'}\sum\frac{r^k}{r'^{k+1}}\,P_k(\cos H) \tag{11}$$

where r and r' are the radius vectors (prime signifies the perturbing planet) and H is the angular distance between the two planets. This is very convenient for the satellite problem where α is rather small.

Our purpose was to have the advantage of the Laplace development in which the functions $\varphi(\alpha^2)$ can be computed with a high degree of accuracy. Therefore we took directly

$$\Delta^{-6}=r'^{-6}(2-\delta_j^0)\,\text{Re}\sum_{j=0}^{j_M}\sum_{n=0}^{E(\omega/2)}\sum_{m=0}^{\omega-2n}\psi_{n,m}^{(j)}(\alpha^2)P^{j+n}c^{j}c'^{j}J_h^m u^n \tag{12}$$

with $P=\frac{ra'}{ar'}$ and $\psi_{n,m}^{(j)} = \frac{(6/2)_n}{(1)_n}\alpha^n\varphi_{n-\frac{6}{2},m}^{(j)}$

Taking the binomial development for $h=P^2-1$

$$h^n = \sum_{v=0}^{n}(-1)^v\binom{n}{v}\,P^{2(n-v)}$$

leads to the form

$$\Delta^{-6} = r'^{-6}(2-\delta_j^0)\,\text{Re}\sum_{j=0}^{j_M}\sum_{n=0}^{E(\omega/2)}F_n^{(j)}\,u^n \tag{13}$$

with $F_n^{(j)} = \sum_{w=0}^{\omega-2n}\sum_{v=w}^{\omega-2n}(-1)^{v-w}\binom{v}{v-w}\psi_{n,v}^{(j)}\,P^{2w+n}$

As an illustration one can realize for the seventh order of the
eccentricities ($\omega=7$) for the coefficients of u^2 (fourth order
of i or i')

$$F_2^{(j)} = P^2(\Psi_{2,0} - \Psi_{2,1} + \Psi_{2,2} - \Psi_{2,3}) + P^4(\Psi_{2,1} - 2\Psi_{2,2} + 3\Psi_{2,3}) +$$
$$+ P^6(\Psi_{2,2} - 3\Psi_{2,3}) + P^8\Psi_{2,3}$$

This leads to the following development of the inverse radius
vector which seems to be convenient for the plane case ($i=o, i'=o$,
$u=o, n=o$):

$$\Delta^{-6} = r'^{-6}(2-\delta_j^o)Re \sum_{j=o}^{j_M} \sum_{w=o}^{\omega-2n} \sum_{v=w}^{\omega-2n} (-1)^{v-w} \binom{v}{v-w} \psi_{n,v}^{(j)} P^{2w+n+j} c^j \bar{c}'^{,j} \qquad (14)$$

For the development in inclinations one has to evaluate

$$u=2Re(\bar{c}(\Gamma_1 c' + \Gamma_2 \bar{c}')) \text{ with } \Gamma_1 = \Gamma_1(i^2, i'^2); \Gamma_2 = \Gamma_2(i^2, i'^2)$$

Putting $\beta = \Gamma_1 c' + \Gamma_2 \bar{c}'$ we can express u by $u=\bar{c}\beta + c\bar{\beta}$ where $\bar{\beta}$ signifies
the conjugate complex of β. We find for β^n

$$\beta^n = \sum_{p=o}^{n} \binom{n}{p} \Gamma_1^{n-p} c'^{n-p} \Gamma_2^p \bar{c}'^p \qquad (15)$$

The power of u can now be developed by

$$u^n = \sum_{r=o}^{n} \sum_{p=o}^{n-r} \sum_{q=o}^{r} V_{r,p,q} c^{2r-n} c'^{n+2(q-r-p)} \qquad (16)$$

with $V_{r,p,q} = \dfrac{n!}{(n-r-p)!p!q!(r-q)!} \Gamma_1^{n-r-p} \Gamma_2^p \bar{\Gamma}_1^{r-q} \bar{\Gamma}_2^q$

giving the final expression for Δ^{-6}

$$\Delta^{-6} = (\frac{a'}{r})^6 (2-\delta_j^o)Re \sum_{j=o}^{j_M} \sum_{n=o}^{E(\omega/2)} \sum_{w=o}^{\omega-2n} \sum_{v=w}^{\omega-2n} \sum_{r=o}^{n} \sum_{p=o}^{n-r} \sum_{q=o}^{r} \qquad (17)$$
$$(-1)^{v-w} \binom{v}{v-w} V_{r,p,q} \psi_{n,v}^{(j)} (\frac{ra'}{ar'})^{2w+n+j} c^{j+2r-n} c'^{-j+n+2(q-r-p)}$$

In those expressions the $(r/a)^n c^m$ are developped with Hansen
coefficients as follows

$$(\frac{r}{a})^n c^m = \sum_{q=-oo}^{+oo} X_q^{n,m} (e)\exp \sqrt{-1} \, qM \qquad (18)$$

This elegant method for computing them has been elaborated by
Brumberg (1967). Although it seems to be a very cumbersome
calculation because of the great number of series many coefficients
disappear in the practical computation due to specific properties

of the Hansen coefficients which are the symmetry

$$X_o^{n,m} = X_o^{n,-m}$$

and the disappearance of terms when n e $[-2,-|m|-1]$. Note that in the equation (17) it remains the coefficients with index O when we take the mean values that is to say

$$(\frac{r}{a})^n_c m = X_o^{n,m}$$

The next work to be done is the numerical integration of the system of the Lagrange Equations to find the long periods in the variables used. This can be done by integrating only the system of long periodic terms

$$\frac{d\sigma^{(n)}}{dt} = \sum_{k \neq n} G_k(a^{(n)}, a^{(k)}, z^{(n)}, z^{(k)}, \varsigma^{(n)}, \varsigma^{(k)}) \tag{19}$$

for k perturbing planets; the index (n) signifies the regarded planet. The short periodic terms in λ are absent in taking the mean values of the equation (17) as it is explained above.

4. CONCLUSIONS

The new approach in this paper for constructing the long periodic terms is done in combining the classical method of analytical development with the very useful instrument of a numerical integration. As mentioned above in the classical works one does not know the truncation errors made in restrictions

1) of developments with respect to the small parameters e and i
2) in taking into account only low order of masses.

It should be mentioned that to analyze the effects order by order with respect to the perturbing masses it is possible to integrate numerically the system in the same way as it is done by the analytical methods in an iterative process. The method of using only a numerical integration is mixing up all different influences and does not allow to draw conclusions on the specific effects which play an important role in regard to the long periodic perturbations. We established a strict form of the Lagrange Equations and have the whole advantage of the powerful instrument of variation of constants. If we select some specific arguments to add to the system (8), e.g. the great inequality Jupiter-Saturn and integrate it together we should be able to evaluate rigorously the role of the near resonances which is always a fundamental one. On the other hand the numerical integration permits to integrate the system with the desired precision, withous taking into account truncation problems which are neglectable in the periods regarded, when we can take a suitable step length.

The special form of the equations (17) on regarding the small parameters e and i has allowed to work only on certain problems, but with slight changes it should be possible to establish also formulas for larger eccentricities and inclinations, as it has been done by Kozai (1962). Although the complexity of the Lagrange Equations established in this paper seems to lead to a cumbersome work for integrating the system numerically there is only a small number of quantities like $\Psi(\alpha^2)$, $\Gamma(i,i')$ and $(a/r)^n c^m$. These specific quantities can be tabulated in a certain regarded problem once and for all and at least should give a very precise determination of the periodic perturbations of some thousands of years in our planetary system .

ACKNOWLEDGEMENTS

First of all I have to thank Dr.J.Chapront from the Bureau des Longitudes in Paris for many fruitful discussions and his most valuable advice. I should also express my thanks to the"Goerres Gesellschaft zur Pflege der Wissenschaft" for the grant, which made this work possible.

REFERENCES

Abu-El-Ata, N., Chapront, J.: 1974, Astron. Astrophys. 38, 57.
Anolik, M.V., Krassinsky, B.A., Pius, L.J.: 1969, Trudy Inst. Theor. Astron. Leningrad 14, 3-14.
Brouwer, D., Van Woerkom, A.J.J.: 1950, Astron. Pap. 13, Part 2.
Bretagnon, P.: 1974, Astron. Astrophys. 30, 141.
Brumberg, V.: 1967, Bull. Inst. Theor. Astron. T 11, 125.
Chapront, J., Bretagnon, P., Mehl, M.: 1975, Celes. Mech. 11, 379.
Chapront, J., Dvorak, R.: 1977, Mitt. Astron. Ges. 40, 214.
Cohen, C.J., Hubbard, E.C., Oesterwinter, C.: 1973, Astron. Pap. 22, Part 1.
Hill, G.W.: 1889, Astron. J. 9, 89-91.
Kozai, Y.: 1962, Astron. J. 67, 591.

CONSTRUCTION D'UNE THEORIE PLANETAIRE AU TROISIEME ORDRE DES MASSES

J. L. Simon
Bureau des Longitudes, Paris, France

1. INTRODUCTION

Les théories planétaires avec termes séculaires ne sont valables,
par suite de la présence de termes augmentant avec le temps dans les
éléments métriques, que dans un intervalle de temps limité (de l'ordre
de mille ans) mais on peut les construire avec une bonne précision sur
cet intervalle. Les recherches récentes sur la construction de formulai-
res pour le développement de la fonction perturbatrice bien adaptés au
calcul sur ordinateur (Iszak 1964, Brumberg 1967, Chapront 1970) et les
progrès réalisés dans les méthodes de calcul utilisées (Chapront et al.
1974) nous permettent d'espérer une amélioration sensible des théories
existantes et des éphémérides qui en résultent. Les théories avec termes
séculaires actuellement en construction au Bureau des Longitudes sont
semi-numériques. Les éléments métriques et angulaires s'expriment sous
la forme :

$$x = x_0 + t(a_0 + a_1 t + a_2 t^2 + \ldots) + \left(\sum_{i_1 i_2} A_{i_1 i_2} \cos(i_1 \lambda_1 + i_2 \lambda_2 + \ldots) \right) \left(b_0 + b_1 t + \ldots \right) \tag{1}$$

$$+ \left(\sum B_{i_1 i_2} \sin(i_1 \lambda_1 + i_2 \lambda_2 + \ldots) \right) \left(b_0' + b_1' t + \ldots \right)$$

Les coefficients a_i, b_i, A_{ij}, B_{ij} etc... sont des coefficients numériques; les développements sont analytiques par rapport aux longitudes
moyennes des planètes ($\lambda_i = \lambda_{i0} + n_i t$), nous appellerons "arguments à

longue période" ceux qui font apparaître de petits diviseurs dans l'intégration, c'est à dire ceux pour lesquels on a $i_1 n_1 + i_2 n_2 + \ldots \ll n_i$ (ex-

emple : la grande inégalité $2\lambda_j - 5\lambda_s$ de période 1000 ans). Les théories

sont entreprises avec des valeurs modernes des masses et des constantes
d'intégration (qu'on améliorera par la suite éventuellement) ce qui
donne déja, indépendemment des progrès en précision que l'on souhaite
apporter aux théories anciennes, une actualisation de ces théories. La

V. Szebehely (ed.), Dynamics of Planets and Satellites and Theories of Their Motion, 65-75.
All Rights Reserved. Copyright © 1978 by D. Reidel Publishing Company, Dordrecht, Holland.

manière dont nous avons choisi nos constantes d'intégration est décrite
par Simon et Bretagnon (1975).

La construction de théories avec termes séculaires est actuellement
entreprise, au Bureau des Longitudes, par deux méthodes différentes.
L'une procède ordre par ordre par rapport aux masses et l'autre par approximation successive. Nous allons donner brièvement le principe de ces
deux méthodes avant de décrire plus en détail la construction d'une théorie planétaire à l'ordre trois des masses et de discuter les premiers
résultats obtenus.

2. MÉTHODES DE CONSTRUCTION D'UNE THÉORIE PLANÉTAIRE A TERMES SÉCULAIRES

2. 1. Théorie ordre par ordre par rapport aux masses (type Le Verrier)

Le système à intégrer est celui des équations de Lagrange, il a la
forme:

$$\frac{dx}{dt} = \mu F(x_i) \tag{2}$$

où μ est un paramètre de l'ordre des masses.
$F(x_i)$ se calcule à partir de la fonction perturbatrice et de ses dérivées
par rapport aux éléments. On cherche à exprimer la solution sous la
forme:

$$x = x_0 + \mu \Delta^1 x + \mu^2 \Delta^2 x + \mu^3 \Delta^3 x + \ldots \tag{3}$$

En substituant (3) dans (2), en développant suivant la formule de Taylor
et en identifiant les deux membres en μ, on obtient:

$$\frac{d\Delta^1 x}{dt} = F(x_i^0) \tag{4}$$

d'où, par intégration, les perturbations du premier ordre $\Delta^1 x$,

$$\frac{d\Delta^2 x}{dt} = \frac{\partial F}{\partial x_i} \Delta^1 x_i \tag{5}$$

d'où le deuxième ordre $\Delta^2 x$,

$$\frac{d\Delta^3 x}{dt} = \frac{\partial F}{\partial x_i} \Delta^2 x_i + \frac{1}{2} \frac{\partial^2 F}{\partial x_i \partial x_j} \Delta^1 x_i . \Delta^1 x_j \tag{6}$$

d'où le troisième ordre $\Delta^3 x$.
$F(x_i^0)$, $\dfrac{\partial F}{\partial x_i}$, $\dfrac{\partial^2 F}{\partial x_i \partial x_j}$ sont calculés analytiquement pour chaque argument, à
partir des développements analytiques de R et de ses dérivées; on substi-
tue ensuite les valeurs numériques des paramètres. On obtient donc dans
cette méthode:

 - la solution, semi-numérique, sous la forme (3)

 - les expressions des séries $\dfrac{\partial F}{\partial x_i}$, $\dfrac{\partial^2 F}{\partial x_i \partial x_i}$ qui permettront de recal-
culer facilement les perturbations pour une petite modification des
constantes. Notons, en particulier, que l'obtention de $\Delta^1 x$ et des $\dfrac{\partial F}{\partial x_i}$

est équivalente à un premier ordre analytique.

2.2 Théorie par approximation successive

Cette méthode est mise au point par P. Bretagnon. Le système à intégrer a la forme (2) mais on utilise un formulaire qui permet d'intégrer les équations de Lagrange sous une forme fermée par rapport aux variables sans qu'il soit besoin de développer la fonction perturbatrice (Chapront et al., 1974). Il faut résoudre l'équation de Képler et les formules habituelles du mouvement képlérien pour revenir de l'anomalie vraie v à l'anomalie moyenne l. L'intégration se fait par itération, notons x^n la solution de l'itération n, nous aurons, pour la première itération:

$$\frac{dx^1}{dt} = \mu F(x_i^0) \tag{7}$$

d'où $x^1 = x^0 + \mu\Delta^1 x$ avec: $\mu\Delta^1 x = \mu\int F(x_i^0)dt$.

Le résultat de la première itération est rigoureusement identique au premier ordre de la méthode précédente. Pour la deuxième itération nous aurons:

$$\frac{dx^2}{dt} = \mu F(x^1) = \mu F(x_i^0) + \mu^2 \frac{\partial F}{\partial x_i} \Delta^1 x_i + \frac{\mu^3}{2} \frac{\partial^2 F}{\partial x_i \partial x_j} \Delta^1 x_i \Delta^1 x_j + \ldots \tag{8}$$

On voit que le résultat de la deuxième itération contient des termes d'ordre trois et plus par rapport aux masses et est donc différent du deuxième ordre de la méthode exposée en 2.1. A l'itération n nous aurons:

$$\frac{dx^n}{dt} = \mu F(x^{n-1}) \tag{9}$$

Nous discuterons plus loin les avantages et inconvénients respectifs de ces deux méthodes dont l'application simultanée permet, soulignons le, de précieuses comparaisons et vérifications.

3. CONSTRUCTION D'UNE THEORIE A L'ORDRE TROIS DES MASSES

Nous appliquons la méthode décrite en 2.1. La manière dont nous avons développé la fonction perturbatrice et les dérivées premières des équations de Lagrange ainsi que la façon dont a été constitué la liste d'arguments ont été décrites en détail par Simon et Chapront (1974). Le calcul des dérivées secondes des équations de Lagrange qui sert à la construction du troisième ordre sera exposé dans un article ultérieur. Nous allons nous contenter, ici, de préciser les notations et donner les caractéristiques essentiels de nos développements avant de décrire, ordre par ordre, la forme de la solution et de discuter la précision des calculs

3.1. Notations. Caractéristiques des développements

Les variables métriques sont le demi-grand axe a, l'excentricité e et le sinus de la demi-inclinaison γ; les variables angulaires, la longitude moyenne λ, la longitude du périhélie $\bar{\omega}$ et la longitude du noeud h.

On utilisera aussi l'anomalie moyenne l et l'argument du périhélie g
(λ = l + g + h; ῶ = g + h). Considérons deux planètes P de masse m (planète intérieure) et P' de masse m' (planète extérieure) et un argument
donné:

$$\Phi = ql + q'l' + sg + s'g' + j(h-h')\qquad(10)$$

q, q', s, s', j sont des entiers et l', g', h' se rapportent à la planète
extérieure. On développe la fonction perturbatrice R à partir du formulaire de Brumberg (1967), suivant les méthodes de Chapront (1970) sous
la forme:

$$R = \mu\sum_{qq'ss'j}A_{qq'ss'j}\cos\Phi\qquad(11)$$

$A_{qq'ss'j}$ est une fonction analytique du rapport des demi-grands axes α et

de e, e', γ, γ'; μ est un coefficient numérique égal à (nαm'/l+m) dans
le cas d'une planète intérieure perturbée par une planète extérieure et
à (n'm/l+m') dans le cas contraire.

Nous avons constitué une liste d'environ 8000 arguments Φ de la
forme (10) répartis en 250 "blocs (q,q')" d'arguments correspondant à
une combinaison q,q' donnée des anomalies moyennes, de façon à obtenir
toutes les inégalités du premier ordre supérieures à 0",001. Pour chaque argument nous calculons les développements analytiques de $A_{qq'ss'j}$
et de ses dérivées premières, secondes et troisièmes. Les
développements sont effectués à l'ordre 9 par rapport à e, e', γ, γ'
pour un certain nombre d'arguments considérés comme petits diviseurs,
à l'ordre 7 pour les autres. On substitue ensuite numériquement, dans
les développements, les paramètres pour toutes les combinaisons de planètes (m,m') correspondant aux quatre grosses planètes et on rassemble
les résultats pour chaque bloc (q,q'). L'ensemble du calcul analytique
du premier ordre, des dérivées premières et secondes des équations de
Lagrange, pour les quatre grosses planètes, a demandé quatre heures de
calcul (dont la moitié pour les dérivées secondes) sur l'IBM 360-65 de
l'INAG.

3.2. Forme de la solution

Nous allons décrire la solution en nous bornant, pour simplifier
l'écriture, au cas d'une planète P de masse (m) perturbée par une planète P' de masse (m'). Dans le problème réel de l'une des quatre
grosses planètes perturbée par les trois autres, la forme de la solution
est identique, les séries étant à quatre arguments.

a. Le premier ordre. On intègre le système (4). F se présente sous
la forme d'une série trigonométrique à deux arguments λ et λ' qui contient, sauf pour l'équation en $\Delta^1 a$, un terme constant. Après intégration
le premier ordre $\Delta^1 x$ se présente donc sous la forme:

$$\Delta^1 x = x_1^1 t + X^1(\lambda,\lambda')\qquad(12)$$

t représente le temps, X^1 est une série trigonométrique à deux arguments

x_1^1, qui est le terme constant de F, est la variation séculaire du premier ordre de x. Il est nul pour le demi-grand axe a.

b. Le deuxième ordre. On intègre le système (5). $\Delta^1 x_i$ a la forme
(12) et $\frac{\partial F}{\partial x_i}$ est une série trigonométrique à deux arguments qui contient,
éventuellement, un terme constant. L'expression à intégrer se présente
sous la forme:

$$\frac{d\Delta^2 x}{dt} = b + ct + S(\lambda,\lambda') + tS'(\lambda,\lambda') \tag{13}$$

t représente le temps, b et c sont des coefficients numériques et S et
S', des séries de Fourier. L'intégration d'expressions du type
$\int tS'(\lambda,\lambda')dt$ donne une nouvelle série trigonométrique et des termes
mixtes de la forme $tT(\lambda,\lambda')$ où T est une série de Fourier. Ces termes
seront appelés "termes en tsint". Après intégration, le deuxième ordre
s'écrit:

$$\Delta^2 x = x_1^2 t + x_2^2 t^2 + X^2(\lambda,\lambda') + tY^2(\lambda,\lambda') \tag{14}$$

x_1^2 est la variation séculaire du deuxième ordre de l'élément x, x_2^2 le
terme en t^2, X^2 représente les perturbations périodiques du deuxième
ordre, Y^2 les perturbations en tsint.

c. Le troisième ordre. On intègre (6). $\Delta^1 x_i$ et $\Delta^1 x_j$ ont la forme
(12), $\Delta^2 x_i$, la forme (14); $\frac{\partial F}{\partial x_i}$ et $\frac{\partial^2 F}{\partial x_i \partial x_j}$ sont des séries de Fourier
avec, éventuellement, un terme constant. L'expression à intégrer est:

$$\frac{d\Delta^3 x}{dt} = b + ct + dt^2 + S(\lambda,\lambda') + tS'(\lambda,\lambda') + t^2 S''(\lambda,\lambda') \tag{15}$$

L'intégration d'expressions du type $\int t^2 S''(\lambda,\lambda')dt$ donne des séries de
Fourier, des séries en tsint et des séries du type $t^2 T(\lambda,\lambda')$ où T est
une série de Fourier. Ces derniers termes seront appelés "termes en
t^2sint". Le troisième ordre a donc la forme:

$$\Delta^3 x = x_1^3 t + x_2^3 t^2 + x_3^3 t^3 + X^3(\lambda,\lambda') + tY^3(\lambda,\lambda') \\ + t^2 Z^3(\lambda,\lambda') \tag{16}$$

3.3. Précision des calculs

Le calcul du deuxième et du troisième ordre s'opère donc en multi-
pliant entre elles des séries de Fourier. Les méthodes de programmation
utilisées ont été décrites par Chapront et al. (1974). Les calculs s'ef-
fectuent avec une précision variable suivant les arguments, de la manière suivante:

a. au deuxième ordre. Le calcul du deuxième membre de l'équation
(5) revient à effectuer une somme de produits de deux séries de Fourier
$\sum A_i B_i$. Un terme a_i de la série A_i est égal à:

$$a_i = \alpha_i \cos\phi_i + \alpha_i' \sin\phi_i \tag{17}$$

α_i et α_i' sont des coefficients numériques et l'argument ϕ_i est égal à:
$\phi_i = \sum_{j=1,4} p_j \lambda_j$ où les p_j sont des entiers. Un élément b_i de B_i s'écrit:

$$b_i = \beta_i \cos\psi_i + \beta'_i \sin\psi_i \qquad (18)$$

avec $\psi_i = \sum_{j=1,4} q_j \lambda_j$. Nous noterons:

$$|a_i| = |\alpha_i| + |\alpha'_i|, \quad |b_i| = |\beta_i| + |\beta'_i| \qquad (19)$$

Dans la série résultat un argument est de la forme $\sum_{j=1,4} r_j \lambda_j$ les r_j
étant des entiers. Si le diviseur qu'il donne dans
l'intégration ($\sum_i r_i n_i$ où les n_i sont les moyens mouvements) est inférieur
à une certaine limite arbitraire (fixée à 6000"/an) il est considéré
comme "petit diviseur". Les séries A_i et B_i sont, avant produit, rangés
par valeurs décroissantes de $|a_i|$ et $|b_i|$. Le produit s'effectue en uti-
lisant une précision "standard" ε_0 et une précision ε_1, plus petite,
pour les diviseurs. Soit a_h le h^e élément de la série A_i ordonnée sui-
vant $|a_i|$, on effectuera tous les produits $a_h b_j$ pour lesquels on a:
$|a_h||b_j| \geq \varepsilon_0$. Soit b_k le premier élément de B_i pour lequel on a:
$|a_h||b_k| < \varepsilon_0$, on regardera ensuite si les arguments résultants corres-
pondant aux produits $a_h b_l$ ($l \geq k$) et pour lesquels $|a_h||b_l| \geq \varepsilon_1$ sont des
petits diviseurs. C'est seulement dans ce cas qu'on effectuera les pro-
duits. L'exploration est terminée quand on rencontre un élément b_m tel
que $|a_h||b_m| < \varepsilon_1$. Cette méthode permet de calculer les petits diviseurs
avec une très bonne précision, sans conserver, avant intégration, un
nombre de termes trop important. Nous avons construit notre deuxième ordre avec une précision, avant intégration, de 10^{-7}" pour les petits
diviseurs, 10^{-5}" pour les autres.

b. <u>au troisième ordre</u>. Le calcul des $\dfrac{\partial^2 F}{\partial x_i \partial x_j} \Delta^1 x_i . \Delta^1 x_j$ revient à

faire une somme de produits de trois séries. Une fois les $\Delta^1 x_i . \Delta^1 x_j$ cal-
culés on pourra appliquer la méthode précédente aux produits $\left(\dfrac{\partial^2 F}{\partial x_i \partial x_j}\right)$
$(\Delta^1 x_i . \Delta^1 x_j)$ mais il faut calculer les $\Delta^1 x_i . \Delta^1 x_j$ avec toute la précision
nécessaire. Toutefois on pourra limiter le nombre de termes à conserver
dans les $\Delta^1 x_i . \Delta^1 x_j$ en utilisant une intéressante propriété des séries
$\dfrac{\partial^2 F}{\partial x_i \partial x_j}$ dans le cas "planète extérieure perturbée par planète intérieure" (ex. Uranus-Jupiter où Uranus-Saturne). Dans ce cas en effet ces
séries ont généralement deux où trois termes "extraordinaires" beaucoup
plus importants, numériquement, que les autres (de dix à cent fois).
Ceci est une conséquence de la présence de termes en $(1/\alpha^2)$ dans la
fonction perturbatrice pour ce type de couples. On calculera donc, dans
ce cas, les $\Delta^1 x_i . \Delta^1 x_j$ avec deux précisions, l'une ε'_0 liée aux termes
"ordinaires" de la série correspondante $\dfrac{\partial^2 F}{\partial x_i \partial x_j}$ et l'autre ε'_1 de dix à
cent fois plus petite. On calculera tous les termes du produit supérieurs

à ε_0' et on ne calculera les termes compris entre ε_0' et ε_1' qu'après avoir vérifié qu'ils donneront, par combinaison avec les termes "extraordinaires" de la série $\dfrac{\partial^2 F}{\partial x_i . \partial x_j}$ correspondante, des petits diviseurs dans le produit final. Dans le cas "planète intérieure perturbée par planète extérieure" (Uranus-Neptune) les séries $\partial^2 F/\partial x_i \partial x_j$ n'ont pas de

coefficients numériques trop gros et le calcul des $\Delta^1 x_i . \Delta^1 x_j$ s'effectue sans problème avec une seule précision. Cette méthode permet de limiter considérablement le nombre de termes à conserver dans les produits $\Delta^1 x_i . \Delta^1 x_j$ d'où un gain important en temps de calcul et en place pour stocker les séries.

Avec ces méthodes les produits sont effectués à une précision suffisante pour que les pertes de précision dans nos calculs soient dues uniquement aux limitations en ordre de nos développements et aux limites en nombre de notre liste d'arguments de départ.

4. RESULTATS

Dans ce paragraphe les indices 1,2,3,4 se rapportent respectivement aux planètes Jupiter, Saturne, Uranus, Neptune.

A l'heure actuelle le calcul des perturbations du premier et du deuxième ordre des quatre grosses planètes est terminé. Le premier ordre a été comparé avec les résultats obtenus par P. Bretagnon lors de sa première itération. L'accord est total à la précision de 0",001 (Simon et Bretagnon, 1975). Le deuxième ordre, lui, n'est pas tout à fait identique à la deuxième itération de P. Bretagnon, néanmoins les écarts, tant sur les termes périodiques que sur les termes en tsint restent faibles (quelques dixièmes de ") et de l'ordre des perturbations d'ordre supérieur sauf pour quelques arguments à longue période (en particulier $\lambda_3 - 2\lambda_4$) pour lesquels se posent, pour le moment, des problèmes de convergence dans la méthode itérative (Bretagnon, 1977). Ces résultats seront prochainement intégralement publiés. Le tableau 1 donne, à titre d'illustration, pour les variables λ et e, les deux plus gros termes (plus, eventuellement, le plus important terme à courte période) des perturbations du premier ordre $\Delta^1 x$, du deuxième ordre périodique $\Delta^2 x_p$, du deuxième ordre en tsint $\Delta^2 x_T$; (q_1, q_2, q_3, q_4) désigne l'argument $q_1 \lambda_1 + q_2 \lambda_2 + q_3 \lambda_3 + q_4 \lambda_4$; c_1 et s_1 sont les coefficients du cosinus et du sinus exprimés en ". Pour les termes en tsint c_1 et s_1 représentent la contribution au bout de 1000 ans. Le premier ordre $\Delta^1 \lambda_p$ comprend environ 400 termes supérieurs à 0",001 dont une centaine supérieurs à 0",1; $\Delta^2 \lambda_p$ comprend 650 termes supérieurs à 0",001 dont 80 supérieurs à 0",1; $\Delta^2 \lambda_T$ comprend 150 termes donnant, au bout de mille ans, une contribution supérieure à 0",001 et 20 donnant une contribution supérieure à 0",1. (La planète que l'on considére ici est la planète Uranus.)

Nous construisons actuellement les perturbations du troisième ordre de la planète Uranus. Les théories existantes de cette planète (Gaillot, 1910) sont, en effet, moins précises que celles de Jupiter et Saturne

et on peut espérer qu'une théorie au troisième ordre apportera une amélioration importante. Le calcul des perturbations périodiques n'a été,
pour le moment, conduit qu'à une précision limitée de 0'',05. Nous avons
comparé la théorie au deuxième ordre des quatre grosses planètes puis
cette théorie corrigée du troisième ordre provisoire d'Uranus avec une
intégration numérique des quatre grosses planètes faite à partir de valeurs initiales données par la théorie,à l'aide du programme de Schubart
et Stumpff (1966). Sur un intervalle de temps d'environ 200 ans la contribution des termes du troisième ordre en tsint et t^2sint, négligés
pour le moment n'est pas trop importante et on a constaté, en ce qui
concerne les éléments d'Uranus, que les écarts entre la théorie et l'intégration numérique étaient divisés par un facteur cinq environ lorsqu'
on tenait compte du troisième ordre provisoire. Ce résultat est satis-
faisant et permet de penser, en particulier, qu'aucune erreur n'a été
commise dans le calcul des dérivées des équations de Lagrange.

Depuis cette comparaison nous avons calculé les termes en tsint et
t^2sint du troisième ordre. Il reste à calculer la partie périodique à
la précision finale et à effectuer une nouvelle comparaison à l'intégration numérique sur un intervalle de temps plus large (1000 ans). Le tableau 2 donne, pour la planète Uranus avec les mêmes conventions que le
tableau 1, les valeurs des deux plus importants termes (plus, éventuellement, le plus grand terme à courte période) des séries du troisième
ordre périodique, en tsint et en t^2sint pour les variables λ et e. Pour
les séries en t^2sint, $\Delta^3\lambda_{T2}$ et $\Delta^3 e_{T2}$, nous donnons la valeur de la contribution de ces termes au bout de mille ans. Les vérifications n'étant
pas terminées, ces résultats ne sont pas définitifs mais donnent néanmoins une bonne idée de l'importance des perturbations du troisième ordre par rapport aux ordres précédents. On notera, en particulier, l'importance des termes en tsint d'ordre trois (la série $\Delta^3\lambda_T$ contient 25
termes donnant une contribution supérieure à 0'',1 au bout de mille ans)
et le fait que les plus gros termes périodiques, à courte période, d'ordre trois sont de l'ordre de 1''.

Nous terminerons par la remarque suivante: pour certains arguments
à longue période, les contributions au troisième ordre venant de
$\frac{\partial F}{\partial x_i} \Delta^2 x_i$ d'une part et de $\frac{\partial^2 F}{\partial x_i \partial x_j}$ d'autre part sont importantes mais
voisines et de signe contraire. Ainsi pour le terme en $\lambda_3 - 2\lambda_4$ de la par-
tie périodique du troisième ordre du demi-grand axe, ces deux contribu-
tions sont respectivement 15'',8 $\cos(\lambda_3 - 2\lambda_4)$ + 0'',6 $\sin(\lambda_3 - 2\lambda_4)$ et
-15'',7 $\cos(\lambda_3 - 2\lambda_4)$ - 0'',7 $\sin(\lambda_3 - 2\lambda_4)$ soit un total de 0'',14 $\cos(\lambda_3 - 2\lambda_4)$
- 0'',08 $\sin(\lambda_3 - 2\lambda_4)$. On peut expliquer cette propriété. Notons $S(q_1, q_2, q_3, q_4)$ le terme d'argument $q_1\lambda_1 + q_2\lambda_2 + q_3\lambda_3 + q_4\lambda_4$ d'une série S. L'équation
de Lagrange en a est du type:

$$\frac{d}{dt}\left(\frac{a}{a_0}\right) = \mu F \tag{20}$$

où μ est un paramètre de l'ordre des masses. Nous poserons: $\Delta^p a = \frac{1}{a_0}\delta^p a$
où $\delta^p a$ désigne les perturbations d'ordre p de a et nous noterons

TABLEAU 1. Résultats du premier et du deuxième ordre d'Uranus

	(q_1, q_2, q_3, q_4)				c_1	s_1
$\Delta^1 \lambda_P$	0	0	1	-2	233,48	-2954,99
	1	0	-1	0	-0,06	704,72
$\Delta^2 \lambda_P$	0	0	1	-2	-19,12	-133,00
	2	-6	+3	0	7,31	31,80
	1	0	-2	+2	-0,40	-5,10
$\Delta^2 \lambda_T$	0	0	1	-2	-43,53	10,35
	0	1	-3	0	-11,78	-8,87
	0	1	-2	0	1,60	2,78
$\Delta^1 e_P$	1	0	0	0	-561,20	76,02
	0	0	1	-2	417,98	56,15
$\Delta^2 e_P$	1	0	-1	+2	12,13	-3,32
	1	0	+1	-2	-12,46	0,02
$\Delta^2 e_T$	1	0	0	0	-1,05	-7,99
	0	0	1	-2	0,81	-5,96

TABLEAU 2. Résultats du troisième ordre d'Uranus

	(q_1, q_2, q_3, q_4)				c_1	s_1
$\Delta^3 \lambda_P$	0	0	1	-2	4,98	6,99
	0	0	2	-4	2,49	3,33
	1	0	-2	+2	0,03	-0,25
$\Delta^3 \lambda_T$	2	-6	+3	0	5,57	-4,17
	0	0	1	-2	-5,14	2,27
	0	1	-2	0	0,44	0,57
$\Delta^3 \lambda_{T2}$	0	1	-3	0	-0,06	1,01
	0	0	1	-2	0,59	0,03
	0	1	-2	0	0,14	-0,11
$\Delta^3 e_P$	0	0	2	-4	-0,91	-0,16
	1	0	0	0	0,80	-0,11
$\Delta^3 e_T$	0	1	-3	0	-1,09	0,56
	2	-6	+3	0	0,68	0,35
	1	0	0	0	-0,06	-0,68
$\Delta^3 e_{T2}$	0	1	-3	0	0,17	0,31
	0	0	1	-2	-0,05	0,05
	1	0	0	0	0,06	0,04

{1} la contribution provenant de $\partial F/\partial x_i . \Delta^2 x_i$, {2} la contribution venant

de $\partial^2 F/\partial x_i \partial x_j . \Delta^1 x_i . \Delta^1 x_j$. On constate, numériquement, que {1} provient

essentiellement du produit: $\{1\} \simeq \dfrac{\partial F^1}{\partial a_3} (1,0,-1,0) \big(\Delta^2 a_3(1,0,0,-2) + \Delta^2 a_3$

$(1,0,-2,2) \big)$, la quantité entre crochets provenant, elle même, essentiellement de l'intégration:

$$\int \frac{\partial F^1}{\partial \lambda_3} (1,0,-1,0) . \Delta^1 \lambda_3(0,0,1,-2) dt$$

{2} provient principalement du produit $\{2\} \simeq \dfrac{\partial^2 F^1}{\partial a_3 \partial \lambda_3} (1,0,-1,0) . \Delta^1 a_3(1,$

$0,-1,0) \Delta^1 \lambda_3(0,0,1,-2)$. L'argument $\lambda_3 - 2\lambda_4$ variant lentement avec le temps

on peut admettre que: $\{1\} \simeq \dfrac{\partial F^1}{\partial a_3} (1,0,-1,0) . \Delta^1 \lambda_3(0,0,1,-2) . \int \dfrac{\partial F^1}{\partial \lambda_3}(1,0,-1,0) dt$

De (20) on déduit: $\int \dfrac{\partial F^1}{\partial \lambda_3} (1,0,-1,0) = \dfrac{\partial}{\partial \lambda_3} \Delta^1 a_3(1,0,-1,0)$. Posons:

$A = \dfrac{\partial F^1}{\partial a_3} (1,0,-1,0)$, $B = \Delta^1 a_3(1,0,-1,0)$, nous pouvons écrire:

$\{1\} \simeq \Delta^1 \lambda_3(0,0,1,-2) . A \dfrac{\partial B}{\partial \lambda_3}$, $\{2\} \simeq \Delta^1 \lambda_3(0,0,1,-2) . B \dfrac{\partial A}{\partial \lambda_3}$

On vérifie aisément que $A \dfrac{\partial B}{\partial \lambda_3}$ et $B \dfrac{\partial A}{\partial \lambda_3}$ ont des termes constants opposés

et, par suite, $\{1\} \simeq -\{2\}$.

Pour la variable λ, à cause de la double intégration, ce phénomène
donne des résultats remarquables, le terme d'ordre trois en $(\lambda_3 - 2\lambda_4)$
qui est de l'ordre d'une dizaine de " s'obtenant par différence de deux
termes de l'ordre de 1200". Ce phénomène explique les difficultés de
convergence pour l'argument $\lambda_3 - 2\lambda_4$ rencontrées dans la méthode itérati-
ve; en effet on voit d'après (8) que les termes en $\partial^2 F/\partial x_i \partial x_j$ $\Delta^1 x_i . \Delta^1 x_j$
apparaissent dès la deuxième itération, sans évidemment, les termes
en $\partial F/\partial x_i$ $\Delta^2 x_i$. Cette propriété se reproduit pour d'autres arguments à
longue période et pour toutes les variables en utilisant, éventuellement,
d'autres dérivées (en particulier les dérivées de F^1 par rapport à e et
$\bar{\omega}$). Notons que la vérification numérique de ce phénomène donne une bonne
présomption d'exactitude du calcul de nos dérivées.

5. CONCLUSION

Les résultats obtenus nous permettent d'espérer obtenir avec une
bonne précision, tant sur les arguments à courte période que sur les
petits diviseurs, les perturbations du troisième ordre d'Uranus. Les dérivées secondes des équations de Lagrange ont été calculées pour tous
les couples correspondant aux quatre grosses planètes et il sera facile
de calculer ensuite les troisièmes ordres des autres grosses planètes.
A ce stade les perturbations à courte période seront connues avec une
précision de quelques centièmes de ". En revanche, sur les termes à
longue période, il manquera des termes d'ordre quatre qui peuvent attein-

dre quelques " sur la variable λ. Un quatrième ordre, par cette méthode, est tout à fait envisageable mais risque d'être assez lourd. On peut penser que la méthode itérative, dont la mise en oeuvre n'est pas plus compliquée d'itération en itération, donnera plus rapidement ces termes à longue période avec la même précision que les autres.

REMERCIEMENTS

Je remercie vivement Monsieur J. Chapront qui m'a prodigué conseils et encouragements tout au long de ce travail et Monsieur P. Bretagnon dont les travaux m'ont permis de précieuses vérifications et de fructueuses discussions. Je remercie également Monsieur J. Schubart qui nous a donné son programme d'intégration numérique et Monsieur R. Dvorak qui nous a aidé à utiliser ce programme.

'Construction of a Planetary Theory to the Third Order of the Disturbing Masses' by J.L. Simon

ABSTRACT. We recall briefly the sense which is given to a planetary theory of the Le Verrier type (i.e., with secular variations in the metrical elements. The solutions are developed with elliptical coordinates. We discuss the results obtained, order by order till the third order of the masses.

REFERENCES

Bretagnon, P. 1977, same issue
Brumberg, V. 1967, Bull. Inst. Theor. Astron. T11, 125
Chapront, J. 1970, Astronom. & Astrophys. 7,175
Chapront, J., Chapront, M., Simon, J.L. 1974, Astron. & Astrophys. 31,151
Chapront, J., Bretagnon, P., Mehl, M. 1975, Celes. Mech. Vol 11, 379
Gaillot, A. 1910, Ann. Obs. Paris, Vol 28
Iszak, I.G. 1964, Smithsonian Astrophys. Observ. Special report 140
Le Verrier, U.J.J. 1855, Ann. Obs. Paris, Vol 1
Schubart, J., Stumpff, P. 1966, Veröffentlichungen des Astronomischen
                            Rechen-instituts. Heidelberg. Nr 18
Simon, J.L., Bretagnon, P. 1975, Astron. & Astrophys. 42,259
Simon, J.L., Bretagnon, P. 1975, Astron. & Astrophys. Suppl. 22,107
Simon, J.L., Chapront, J. 1974, Astron. & Astrophys. 32,51

DISCUSSION SUR LES RESULTATS DE THEORIES PLANETAIRES

P. Bretagnon
Bureau des Longitudes, Paris, France

La construction de théories planétaires précises est un travail long et, pour ce qui nous concerne, n'est pas encore achevée. Aussi, les résultats dont nous parlerons ici ne sont que des résultats provisoires. Nous développerons quelques aspects des difficultés que nous avons rencontrées et les conséquences de ces difficultés sur l'orientation des méthodes utilisées pour résoudre les problèmes suivant qu'il s'agit des planètes intérieures ou des grosses planètes.

Dans la construction de théories planétaires que nous avons entreprise, l'objectif est d'atteindre une grande précision sur un intervalle de 1000 ans afin d'améliorer les éphémérides. Nous entendons par là conserver, sur cet intervalle, une précision de $0\overset{"}{.}001$ pour les planètes Mercure, Vénus, la Terre et Mars et de $0\overset{"}{.}01$ pour Jupiter, Saturne, Uranus et Neptune.

Les termes à longues périodes (périodes des périhélies et des nœuds qui sont comprises entre 50 000 ans et 2 000 000 d'années) sont exprimés sous la forme de développements par rapport au temps. Nos solutions ont donc la forme des solutions de Le Verrier (1855), c'est-à-dire qu'elles contiennent des termes de Poisson. Un élément quelconque s'écrit sous la forme :

$$x = x_0 + x_1 t + x_2 t^2 + x_3 t^3 + \ldots$$
$$+ \sum_i (s_0^i + s_1^i t + s_2^i t^2 + \ldots) \sin \Phi_i \qquad (1)$$
$$+ \sum_i (c_0^i + c_1^i t + c_2^i t^2 + \ldots) \cos \Phi_i$$

où t est le temps et les arguments Φ_i des combinaisons linéaires des 8 longitudes moyennes. Les coefficients x_0, x_1, x_2, x_3, ..., s_0^i, s_1^i, s_2^i, ..., c_0^i, c_1^i, c_2^i, ... sont des fonctions des éléments métriques et angulaires autres que la longitude . Ils sont conservés sous forme numérique

V. Szebehely (ed.), Dynamics of Planets and Satellites and Theories of Their Motion, 77-85.
All Rights Reserved. Copyright © 1978 by D. Reidel Publishing Company, Dordrecht, Holland.

et sont calculés à partir de valeurs moyennes prises pour 1950.0
(Simon J.L., Bretagnon P. 1975).

Deux systèmes de variables sont utilisés : le premier se compose
du demi-grand axe a, de la longitude λ, de l'excentricité de l'orbite e,
de la longitude du périhélie ῶ, de γ = sin i/2 où i est l'inclinaison
de l'orbite, de la longitude du noeud Ω ; le second se compose de a, λ,
h = e sin ῶ, k = e cos ῶ, p = γ sin Ω, q = γ cos Ω. Le premier système faci-
lite les comparaisons directes semi-analytiques avec la théorie de Le
Verrier, le second convient mieux dans le cas d'excentricités ou d'inclinaisons faibles.

Pour résoudre le système des équations de Lagrange pour l'ensemble
des huit planètes nous avons envisagé deux méthodes : l'une itérative
à travers un formulaire fermé dans les variables osculatrices, l'autre
opérant ordre par ordre par rapport aux masses perturbatrices à partir
d'un développement de Taylor des seconds membres des équations. L'une
et l'autre méthodes conduisant à des calculs volumineux, il est nécessaire, à chaque étape, de vérifier les résultats et de juger de la précision atteinte. Deux moyens efficaces sont la comparaison à d'autres
théories et la comparaison à l'intégration numérique. Une analyse terme
à terme avec les résultats de Le Verrier et les résultats actuels de
J.L. Simon met en évidence les écarts sur les coefficients des inégalités périodiques ou séculaires, sur les courtes ou les moyennes périodes.
L'intégration numérique permet une comparaison globale de la théorie
pour un temps fixé. Nous utilisons pour cela le programme d'intégration
numérique de J. Schubart et P. Stumpff (1966). La comparaison à l'intégration numérique se fait en substituant un temps fixé dans la théorie.
Ceci définit un point de départ pour l'intégration numérique, mais ce
point de départ est entaché de l'imprécision de la théorie utilisée. La
trajectoire engendrée par l'intégration numérique est donc légèrement
différente de la solution du problème que l'on cherche à résoudre. Toutefois, dans la comparaison entre la théorie et cette intégration numérique, la plus grande partie des écarts est due aux insuffisances de la
théorie et cette comparaison met bien en évidence la précision atteinte.
On peut également comparer la théorie à un modèle indépendant dont les
conditions initiales ne sont pas liées à notre théorie : l'intégration
numérique des APAE (1950), par exemple, ou plus précisément un prolongement sur 1000 ans de cette intégration.

Nous allons maintenant distinguer le cas des planètes intérieures :
Mercure, Vénus, la Terre et Mars et celui des grosses planètes : Jupiter, Saturne, Uranus et Neptune.

LES PLANETES INTERIEURES

Nous avons tout d'abord appliqué la méthode itérative (Chapront et
al., 1975) à l'ensemble des huit planètes. Les équations de Lagrange
s'écrivent :

$$\frac{dX_{n+1}}{dt} = F(X_n) \tag{2}$$

où F est une forme fermée, X_n la solution du système à l'itération n et X_{n+1} la solution à l'itération n+1 ; la solution de départ étant pour chaque planète une ellipse képlerienne. X_n et X_{n+1} sont des vecteurs à 48 composantes : les éléments des 8 planètes. Un inconvénient de la méthode itérative est qu'elle nécessite le calcul global des seconds membres des équations de Lagrange (2) ce qui, dans le cas des planètes intérieures, entraîne, pour atteindre la précision recherchée, une prolifération d'arguments dès la deuxième itération. Cela s'explique par l'importance des petits paramètres excentricités et inclinaisons de Mercure, et de Mars également. A titre d'illustration signalons, par exemple, dans la théorie de Mars, au premier ordre, la présence de l'argument $17\lambda_T - 32\lambda_M$ d'ordre 15 en excentricités-inclinaisons qui donne une contribution de l'ordre de 0"002 dans la longitude. Cela s'explique aussi par un rapport élevé des demi-grands axes entre Vénus et la Terre, par la présence d'un grand nombre d'arguments quasi-résonants qui nécessitent une grande précision dans le calcul, et, en outre, par le fait que Mars surtout est perturbée d'une manière importante par presque toutes les planètes du système solaire. Ainsi, le calcul n'est plus limité par la précision souhaitée mais la taille de l'ordinateur. Travaillant actuellement dans des tableaux de 8192 arguments Φ_i, notre précision s'est révélée très insuffisante pour déterminer correctement les arguments quasi-résonants tels que, par exemple, l'argument $4\lambda_T - 8\lambda_M + 3\lambda_J$ dans la longitude de Mars, argument de période 1783 ans soit 1000 fois environ la période de la planète.

Nous avons donc, pour les planètes intérieures, abandonné la méthode itérative pour une détermination ordre par ordre par rapport aux masses.
Ecrivons l'équation de Lagrange d'une variable quelconque x d'une planète perturbée par une autre :

$$\frac{dx}{dt} = f_x(y_i) \; ; \; i = 1, 2, \ldots, 12 \tag{3}$$

où les y_i sont les éléments des deux planètes et x est l'un quelconque d'entre eux. La solution s'exprime sous la forme :

$$x = x^{(0)} + \Delta^{(1)}x + \Delta^{(2)}x + \Delta^{(3)}x + \ldots \tag{4}$$

où $x^{(0)}$ est la solution képlerienne, $\Delta^{(1)}x$ les perturbations au premier ordre des masses, $\Delta^{(2)}x$ celles au deuxième ordre ...
Nous avons donc :

$$\frac{d\Delta^{(1)}x}{dt} = f_x\left(y_i^{(0)}\right) \tag{5}$$

$$\frac{d\Delta^{(2)}x}{dt} = \sum_{i=1}^{12} \frac{\partial f_x\left(y_i^{(0)}\right)}{\partial y_i} \Delta^{(1)}y_i \tag{6}$$

On voit sur l'équation (6) que, contrairement à ce qui se passe dans la méthode itérative, il est aisé de déterminer avec la précision souhaitée un argument quelconque fixé a priori dans le produit de la dérivée $\frac{\partial f_x\left(y_i^{(0)}\right)}{\partial y_i}$ par le premier ordre $\Delta^{(1)}y_i$. Nous sommes en train de construire actuellement ce deuxième ordre pour les planètes Mercure, Vénus, la Terre et Mars. Il faudra bien sûr construire le troisième ordre et probablement le quatrième pour atteindre la précision recherchée.

La méthode itérative a toutefois donné, avec trois itérations, une détermination assez bonne pour les arguments de période pas trop grande, c'est-à-dire de l'ordre de quelques dizaines d'années. La solution obtenue sous la forme (1) contient, pour chaque élément, entre 500 et 1000 arguments suivant la planète et l'élément. Nous avons fait des comparaisons à l'intégration numérique sur des périodes courtes : 10 ans. Les écarts sont de quelques $0\overset{''}{.}01$ pour les longitudes et de quelques $0\overset{''}{.}001$ pour les autres éléments, mais il est bien évident que sur un intervalle de temps si court on ne peut voir apparaître les erreurs dues aux termes quasi-résonants.

LES GROSSES PLANETES

Pour les grosses planètes, la méthode itérative donne des résultats plus satisfaisants. Il y a en effet une prolifération d'arguments moindre et, toujours dans des tableaux de 8192 arguments Φ_i, on obtient une précision acceptable : $0\overset{''}{.}01$ sauf pour quelques arguments quasi-résonants pour lesquels on est limité à quelques $0\overset{''}{.}01$. Ces arguments devront être améliorés, pour la partie périodique et la partie en t sin t de la solution, par la méthode d'accroissement par rapport aux masses. Par ailleurs, pour les grosses planètes, la méthode itérative permettra d'atteindre dans la formule (1) des puissances élevées du temps. Notons quelques problèmes de convergence pour l'argument $\lambda_U - 2\lambda_N$ de période 4200 ans dans les longitudes d'Uranus et de Neptune.

Actuellement nous avons effectué 5 itérations sur des seconds membres des équations de Lagrange limités à leur partie périodique et leur partie en t sin t. Mis à part l'argument $\lambda_U - 2\lambda_N$ dans les longitudes d'Uranus et de Neptune, nous avons obtenu la convergence de notre solution dans les composantes que nous avons conservées, c'est-à-dire les termes périodiques et les termes de Poisson. Nous avons donc une solution de la forme (1) tronquée :

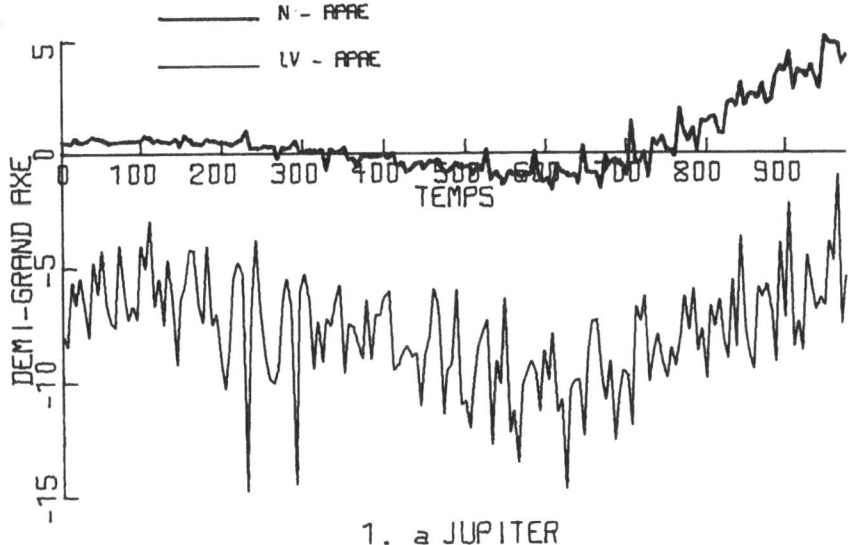

1. a JUPITER

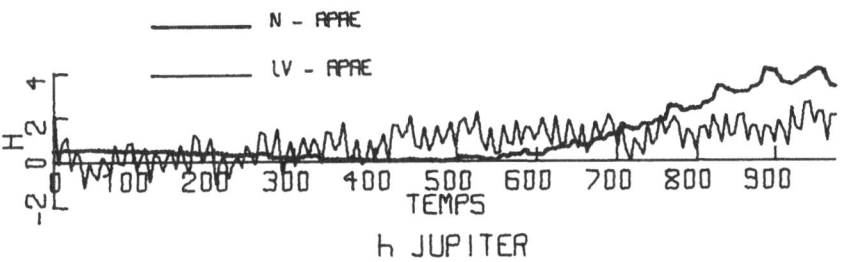

h JUPITER

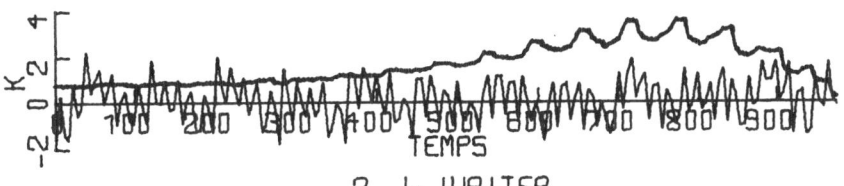

2. k JUPITER

$$x = x_0 + x_1 t + x_2 t^2 + \sum_i (s_0^i + s_1^i t) \sin \Phi_i$$
$$+ \sum_i (c_0^i + c_1^i t) \cos \Phi_i \tag{7}$$

Des comparaisons de cette solution à l'intégration numérique ou aux APAE donnent sur 1000 ans des écarts de quelques dizaines de secondes d'arc pour les longitudes et de quelques secondes pour les autres variables.

Nous avons représenté sur les figures 1 à 4 les comparaisons à l'intégration numérique concernant Jupiter. Les temps sont portés en années. La courbe épaisse représente les différences entre notre théorie et les APAE ou l'intégration numérique (N - APAE ou N - IN). La courbe fine représente les différences entre la théorie de Le Verrier et les APAE (LV - APAE). Il s'agit en fait de la théorie de Le Verrier-Gaillot d'où sont tirées les éphémérides publiées dans la Connaissance des Temps.

La figure 1 représente les différences exprimées en UA×secondes d'arc, concernant le demi-grand axe. On constate, comme pour les autres éléments, une différence pour t = 0 entre la constante utilisée dans la théorie et celle des APAE. Ces différences de constantes d'intégration expliquent une partie des oscillations à courtes périodes que comporte notre courbe. Le reste des écarts est dû à l'insuffisance actuelle de la théorie ; en particulier, on voit très bien apparaître la grande inégalité entre Jupiter et Saturne de période 883 ans et d'amplitude croissant avec le temps. Ce terme est de la forme :

$$s_2 t^2 \sin (2\lambda_J - 5\lambda_S) + c_2 t^2 \cos (2\lambda_J - 5\lambda_S)$$

composante que nous n'avons pas encore déterminée dans notre solution (7). La solution de Le Verrier-Gaillot est ici nettement moins bonne que pour les autres éléments.

La figure 2 représente les différences exprimées en secondes d'arc concernant les variables h = e sin $\tilde{\omega}$ et k = e cos $\tilde{\omega}$. On peut faire les mêmes remarques que pour le demi-grand axe. On constate que la théorie de Le Verrier-Gaillot est meilleure pour ces variables que pour le demi-grand axe. Elle ne se dégrade pas avec le temps mais comporte, sur tout l'intervalle, des écarts de quelques secondes. Les écarts entre notre théorie et les APAE, à une différence de constante d'intégration près, restent inférieurs à la seconde pendant plusieurs siècles.

La figure 3 représente les différences exprimées en secondes d'arc concernant la longitude. Dans la longitude, notre solution se dégrade assez rapidement avec le temps car les termes en $t^2 \sin t$, actuellement négligés, atteignent une centaine de secondes au bout de 1000 ans. Mais on voit que cette courbe traduit presque uniquement les effets de la grande inégalité entre Jupiter et Saturne. En effet, la figure 4 représente, à une autre échelle, les deux mêmes courbes et la courbe tiretée (F. INT.) représente une analyse de Fourier des résidus de notre théorie. Dans cette analyse, nous n'avons recherché que la grande inégalité c'est-à-dire que nous avons ajouté à la théorie la fonction :

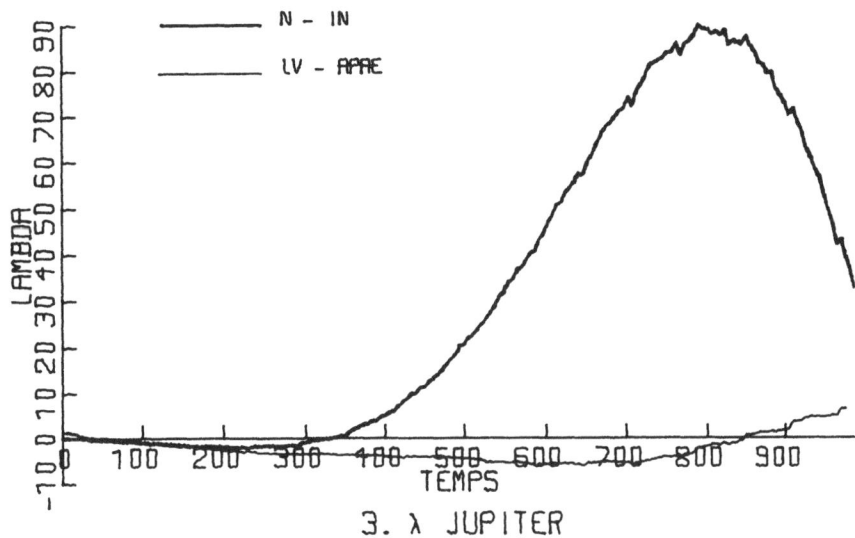

3. λ JUPITER

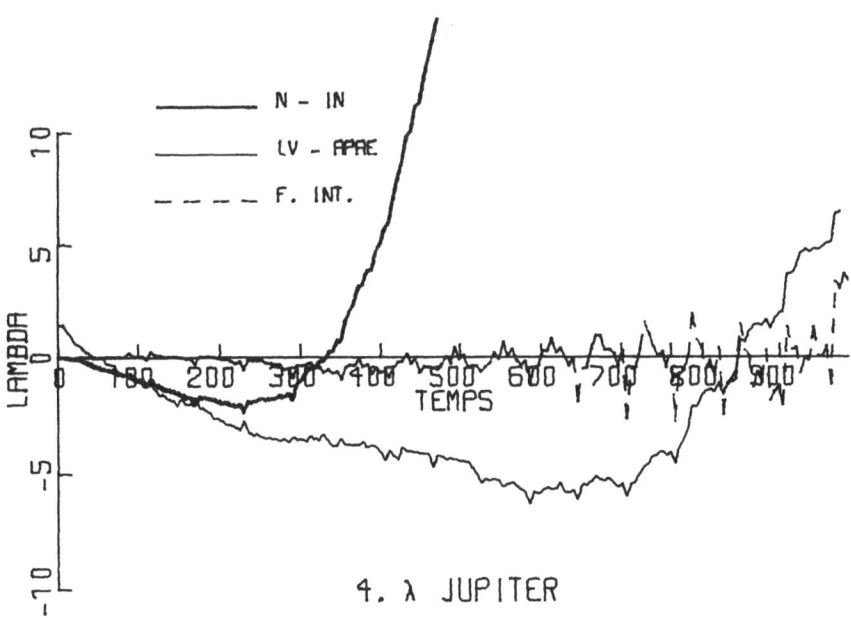

4. λ JUPITER

$$f = x_0 + x_1 t + (s_0 + s_1 t + s_2 t^2 + s_3 t^3) \sin (2\lambda_J - 5\lambda_S)$$
$$+ (c_0 + c_1 t + c_2 t^2 + c_3 t^3) \cos (2\lambda_J - 5\lambda_S)$$

avec :

$$x_0 = -12{,}96''$$

$$x_1 = -2{,}54 \times 10^{-3} \; ''/_{an}$$

$$s_0 = -4{,}07''$$

$$c_0 = -12{,}18''$$

$$s_1 = +55{,}43 \times 10^{-3} \; ''/_{an}$$

$$c_1 = -31{,}75 \times 10^{-3} \; ''/_{an}$$

$$s_2 = +55{,}38 \times 10^{-6} \; ''/_{an^2}$$

$$c_2 = +86{,}26 \times 10^{-6} \; ''/_{an^2}$$

$$s_3 = -5{,}04 \times 10^{-9} \; ''/_{an^3}$$

$$c_3 = +12{,}56 \times 10^{-9} \; ''/_{an^3}$$

Cette seule amélioration de la grande inégalité réduit les écarts à moins d'une seconde pendant plus de 500 ans. Nous n'avons pas cherché à analyser les courtes périodes restant, notre but étant de déterminer analytiquement ces termes, ainsi que la grande inégalité. L'étude de la solution actuelle et des résidus montre qu'il sera nécessaire d'atteindre, pour Jupiter et Saturne, des puissances élevées du temps dans les développements de la formule (1), probablement 5 ou 6 pour obtenir la précision de $0{,}''01$ pendant 1000 ans.

CONCLUSION

Les progrès restant à faire sont de nature différente pour les planètes intérieures et pour les grosses planètes. Pour les planètes intérieures il nous faut améliorer la précision des termes quasi-résonants afin d'élargir l'intervalle de validité de notre théorie. Pour les grosses planètes il nous faut développer notre solution jusqu'à une puissance élevée du temps et résoudre les problèmes de convergence dans les longitudes d'Uranus et de Neptune.

Ensuite, se posera le problème de l'ajustement des constantes d'intégration. On pourra, pour cela, ajuster notre solution à l'intégration numérique des APAE ou à des intégrations numériques plus récentes comme celle de C. Oesterwinter et C. J. Cohen (1972).

REMERCIEMENTS

Je remercie vivement Monsieur J. Chapront pour les conseils qu'il m'a prodigués tout au long de ce travail et Monsieur J.L. Simon dont les recherches m'ont permis de nombreuses vérifications. Je remercie également Monsieur J. Schubart qui nous a donné, par son programme d'intégration numérique, un outil puissant de comparaison. Je remercie enfin Monsieur R. Dvorak qui m'a initié à l'utilisation de ce programme.

'Discussion on the Results of Planetary Theories' by P. Bretagnon

ABSTRACT. Various aspects of the construction of the developments of the solutions are discussed: problem of convergence in iterative methods, comparisons with numerical integrations and ephemerides, special differences between major and minor planets, and precision of the solutions.

REFERENCES

Chapront, J., Bretagnon, P., Mehl, M. 1975, Celes. Mech. Vol 11, 379
Eckert, W.J., Brouwer, D., Clemence, G.M., 1950, APAE Vol 12
Le Verrier, U.J.J. 1855, Ann. Obs. Paris, Vol 1
Oesterwinter, C., Cohen, C.J. 1972, Celes. Mech. Vol 5, 317-395
Schubart, J., Stumpff, P. 1966, Veröffentlichungen des Astronomischen
                                  Rechen-Instituts. Heidelberg. Nr 18
Simon, J.L., Bretagnon, P. 1975, Astron. & Astrophys. 42,259
Simon, J.L., Bretagnon, P. 1975, Astron. & Astrophys. Suppl. 22,107

RELATION OF A CONTRACTING EARTH TO THE APPARENT ACCELERATIONS OF THE
SUN AND MOON

R.A. Lyttleton
Institute of Astronomy, Cambridge, England and Jet Propulsion
Laboratory, Pasadena, California, U.S.A.

ABSTRACT

   The tidal theory of the evolution of the lunar orbit has remained
inconsistent with the observational values of the apparent secular
accelerations of the Sun and Moon since it was first developed by
Jeffreys in 1920. Allowance for a changing moment of inertia of the
Earth enables the discrepancy to be completely removed if a decrease
is occurring at a rate of just about the amount already required by
the phase-change theory of the nature of the terrestrial core. The
agreement of the resulting theory with the latest determinations of
the lunar acceleration increases confidence in the phase-change
hypothesis. On the other hand the theory renders it most unlikely
that a changing constant of gravitation will prove necessary to
account for the observations. On the present theory of itself the
Moon would have been extremely close to the Earth only about 10^9 years
ago which suggests that some additional process may at times have influenced the lunar orbit.

The full text of this paper will appear in "The Moon"

THE ASTEROIDAL PLANET AS THE ORIGIN OF COMETS

THOMAS C. VAN FLANDERN
U.S. Naval Observatory
Washington, D.C. 20390

ABSTRACT. Recently, M. W. Ovenden has raised seemingly plausible
dynamical arguments which suggest that a 90-earth-mass planet existed in
the present location of the asteroid belt until 16×10^6 years ago, and
then rapidly disintegrated. He mentions supporting evidence from the
cosmic ray exposure ages of chondritic meteorites. If the long-period
comets originated from the recent disintegration of such a planet,
several otherwise improbable characteristics of their orbits would be
predicted, including a tendency for those orbits which are least
perturbed to return to the site of the original break-up. In this
investigation, we compare observed characteristics of long-period comet
orbits with expected characteristics, based on the missing planet hypothesis. The conclusion is that long-period comet orbits are wholly
consistent with the hypothesis; indeed, certain of their characteristics
are difficult to explain in any other way.

OVENDEN'S HYPOTHESIS

Ovenden (1972,1973) has presented seemingly plausible evidence for
the existence of a planet of 90 earth masses revolving in the present
location of the asteroid belt until 16×10^6 years ago, at which time it
rapidly dissipated. In essence, he notes that the solar system is not
currently in an equilibrium configuration (the time average of the disturbing function is not a minimum, despite an estimated relaxation time
of considerably less than 10^9 years). If Ovenden's calculations are
correct, they constitute a proof that the planets have been disturbed in
the relatively recent past, and that they may now be seeking a new
equilibrium configuration.

Since Ovenden's method leads to a dynamical derivation of the
Titius-Bode law, the results themselves suggested the possibility that
the disturbance consisted of the disintegration of a former planet in
the present location of the asteroid belt. With the aid of the assumptions that the planets were in equilibrium and in orbits close to their
present ones before the break-up of the hypothetical planet, he found

V. Szebehely (ed.), Dynamics of Planets and Satellites and Theories of Their Motion, 89-99.

the mass of the planet to have been 90 earth masses; and by invoking
the Meffroy (1955) formula for the secular variations in semi-major
axis, he estimated the epoch of disintegration at about 16×10^6 years
ago. Under these conditions, the present planetary semi-major axes
correspond to former equilibrium values to within about one per cent.

The dynamical principles behind Ovenden's arguments seem sound, and
are in no way less so even if recent criticism of his time scale proves
correct (Ovenden, 1976). He cites additional supportive evidence from
the fact that the cosmic ray exposure ages of chondritic meteorites have
a sharp cutoff at about 22×10^6 years, whereas, by contrast, some iron
meteorites have ages of 10^8 to 10^9 years. This, too, suggests a relatively recent break-up event in the solar system.

BREAK-UP SCENARIO

We have posed the following question: If Ovenden's hypothesis is
correct, are there any other observable consequences? It seems clear
that there is at present only about 0.001 of an Earth-mass left in the
asteroid belt. Therefore, to dissipate the original mass so completely,
and to leave no single fragment of significant size, the break-up would
seemingly have involved very great energy. Let us assume a normal distribution of post-break-up velocities, with small masses sent out in all
directions from the site of the break-up. Since the circular orbital
velocity of the hypothetical asteroidal planet would have been about
18 km/s, an additional forward velocity of just 7 km/s would be sufficient
for escape from the solar system (since escape velocity at 2.8 AU from
the Sun is about 25 km/s); whereas in the opposite direction, a total
increment of about 43 km/s relative to the planet is required to reach
escape velocity. Despite these asymmetries, if the break-up is sufficiently energetic, most of the mass will escape from the solar system.
Masses near the tail end of the velocity distribution curve would continue in direct orbits around the Sun. Between these extremes there will
be many masses in a great variety of orbits. Most retrograde orbits and
those of high inclination are unstable - these would be drastically
altered by planetary perturbations. Almost all would eventually be swept
up by the Sun or planets, especially Jupiter, or be ejected from the
solar system. A few will find stable orbits, while others might be cap-
tured and become planetary satellites. The only salient point here is
that it is quite difficult to predict the evolution of the orbits of the
surviving objects after about 10^7 years. However a few objects with
velocities very close to escape velocity will enter very long-period
orbits which will have considerably less mixing with the planets. These
we may hope to be able to trace for 10^7 years. Such objects are, of
course, observed - the very long-period comets. We will now examine to
what extent the known characteristics of such comets are consistent with
the Ovenden hypothesis. The accepted division between short- and long-
period comets is at a period of 100 years. For present purposes we need
to introduce another division - the "very-long-period" comets, with the
periods all on the order of 10^6-10^7 years.

VERY-LONG-PERIOD COMETS

Much of our knowledge about characteristics of comet orbits is summarized by Marsden (1974). There is no certain case of any comet having entered the domain of the planets on a hyperbolic trajectory, although something approaching half of the long-period comets leave on hyperbolic trajectories, never to return. They are clearly solar system members before encounter, and therefore could not come from other stars. Yet they come in from all directions on the sky, and their orbit planes have all different orientations.

Any theory attempting to explain the origin of very long-period comets must deal with one striking fact - although they are solar system members, they are apparently approaching the Sun and planets for the first time! One proof of this was recently reviewed by Marsden and Sekanina (1973). There is a tendency for the pre-encounter aphelion distances to cluster near 50,000 astronomical units; whereas there is no such tendency for the post-encounter aphelia. If these long-period comets had passed through a previous perihelion, the clustering of aphelia would have been widely dispersed. Members of this group of first-appearance comets are called "new" comets.

EVOLUTION OF VERY-LONG-PERIOD PLANETOIDS

Let us now examine the evolution of very-long-period planetoids resulting from the break-up up the hypothetical planet, assuming for the moment that the break-up epoch was 16×10^6 years ago. Initially, let us ignore interstellar perturbations. Then those planetoids with aphelia greater than 125,000 au corresponding to a period of 16×10^6 years will not yet have returned, and those with aphelia less than 125,000 AU will have already returned at least once in the past and have had a pseudo-random perturbation of their periods. As a result, about half of these will be given additional energy, resulting in escape. The other half will lose energy, resulting in a second encounter with the solar system after a much shorter time - typically, about 50,000 years. In later encounters these will eventually escape or evolve into short-period planetoids.

Since all of the non-escaping orbits are fixed ellipses until perturbed, they must all originally have had a common intersection point at the site of the original break-up. Moreover, although all periods will be represented by the various orbits, only those ellipses with periods exactly equal to the time since break-up will be returning for the first time at the present epoch. We therefore see that the clustering of aphelion distances of comets within a small range toward all directions on the celestial sphere follows in a natural way from Ovenden's hypothesis.

In order to agree with Ovenden's time scale, the period of these first-return comets must be nearly 16×10^6 years, corresponding to

aphelia of about 125,000 AU. The clustering noted by Marsden and
Sekanina (1973) peaks at about 60,000 AU. The quantity they determine,
1/a, the reciprocal of the pre-encounter semi-major axis, has a mean
value of about 0.000034 ± 0.000007 for the first-return comets. This
corresponds to periods of between $4x10^6$ and $7x10^6$ years, and is therefore somewhat shorter than Ovenden's time scale. Ovenden (1976) has,
however, recently criticized his own time scale calculation; and there
is clearly no conflict between his results and the above estimate.

SELECTION CRITERIA

    In order to further test the connection, if any, that comets may
have with Ovenden's hypothetical planet, the next step is to select all
of the "new" comets from the Marsden catalog (1975), and to compare
the distribution of their orbital elements with the predicted distributions. In order to minimize the number of multi-return comets in the
sample, stringent selection criteria were established. If the perihelion
distance was less than 0.5 AU, it was assumed that the non-gravitational
forces would preclude estimation of the period with sufficient precision.
And, the determined values of the reciprocal semi-major axes, 1/a, were
corrected to their pre-planetary-encounter values by removing the effects
of planetary perturbations, using the method of Everhart and Raghavan
(1970). Then only those comets with pre-encounter values of $1/a < +2x10^{-4}$
were kept. After application of these criteria, 60 comets remained in
the sample; and these 60 'very-long-period' (VLP) comets were used in
the subsequent analysis.

TESTING THE HYPOTHESIS

    If these 60 comets originated in the energetic break-up of a planet,
certain specific characteristics in the distributions of their elements
will result, most of which are unlikely to occur with any other type of
origin. A clustering in the periods of first-return comets is one such
characteristic. Another is a common point of intersection of all of the
orbits at the original epoch. Beyond that there are certain asymmetries
in the distributions which should result, even for a symmetric break-up.

    Most importantly, we must see the 'Sun-selecting influence' in the
distribution of aphelion directions. The Sun-selecting influence results
from the fact that the planetoids resulting from a break-up go immediately into a solar orbit; those energetic enough to move in a straight line
or other hyperbolic trajectory are of no interest here, because they
would escape from the solar system directly. Since the Sun is in a
certain specific direction from the point of break-up, the distribution
of perihelion and aphelion directions of VLP planetoids will be symmetric
about that direction, but will differ markedly in the solar and anti-
solar directions. This characteristic is classified as most important
because it is the least likely to be altered by either of the two important perturbation sources for VLP comets - passing stars and the galactic

field. If VLP comets are connected with the Ovenden hypothesis, the
Sun-selecting influence must be present.

SUN-SELECTING INFLUENCE

     For spherical symmetry the principal effect of the presence of the
Sun is to place approximately 71% of the perihelia in the break-up
hemisphere, with only 29% in the opposite one. Including the effect of
an original circular velocity for the hypothetical planet changes these
percentages to 82% and 18%, respectively.

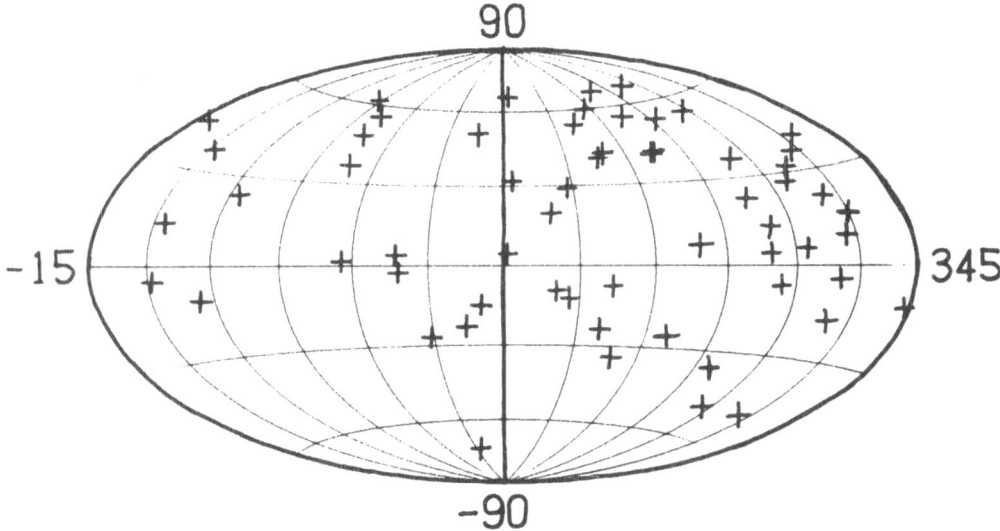

Figure 1. Perihelion directions of 60 VLP comets in ecliptic coordinates
          centered at longitude 165°, with celestial equator shown.

Figure 1 is a plot of the perihelion directions in ecliptic coordinates
for the sample of 60 VLP comets. It is centered on ecliptic longitude
165°. The hemisphere from 165° to 345° contains roughly twice as many
perihelia as the opposite one (42 vs. 18, which is 70% vs. 30%). However, observational selection effects have operated to reduce the ratio
greatly. It is well known that there are more cometary perihelia
observed north of the celestial equator than south, because of the bias
in the locations of observatories on the Earth; but the break-up direction (presumably at longitude 255°, latitude 0°) lies well south of the
equator. Hence we can discover relatively fewer comets toward the break-
up direction, and relatively more toward the opposite direction (which
lies well north of the equator). By integrating the distribution function
over the northern hemisphere only, and dividing the result into the
number of observed perihelia from Figure 1 in each part of the northern

hemisphere, we arrive at estimates of the total number of perihelia
which would have been observed over the whole sky if there were no
selection effects - 145 and 27, which is 84% and 16%. The agreement
between prediction and observation is excellent - the cometary perihelia
clearly exhibit a strong Sun-selecting influence, as well as the effects
of an initial circular orbital velocity around the Sun, and there is
negligible probability of such a strong directional bias occurring by
chance.

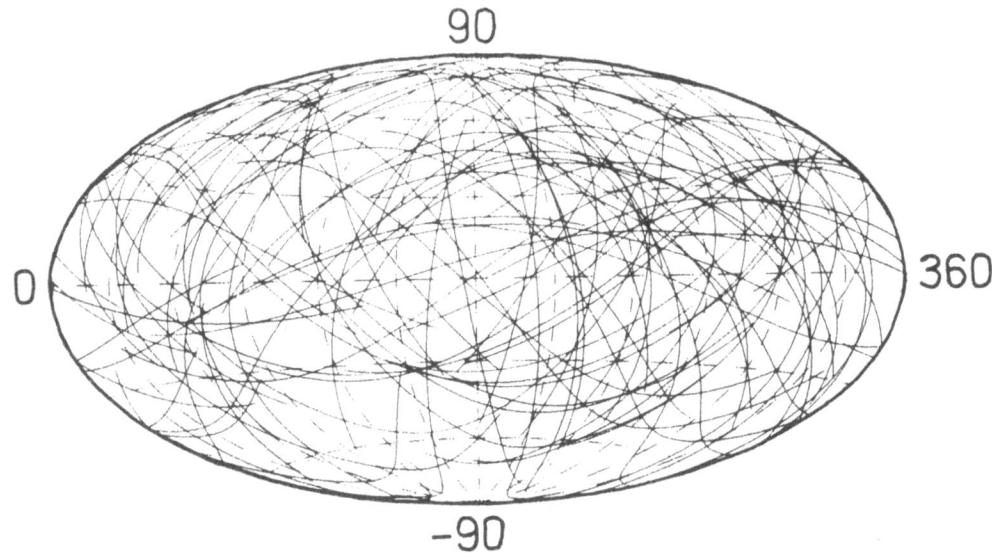

Figure 2. Heliocentric orbit planes for 60 VLP comets at present epoch,
          in ecliptic coordinates centered at longitude 180°.

DISTRIBUTIONS OF OTHER ELEMENTS

Figure 2 is a plot of the portions of the orbit planes within 30 AU
of the Sun, or twice the perihelion distance, whichever is greater, at
the present epoch as seen from the Sun. If all of these comets originated
simultaneously from a single point, then the initital orbits will all
intersect at that point. In Figure 2 there is indeed a statistically
significant intersection point at ecliptic longitude 249°, latitude +17°.
Since all of these orbits are great circles, there is a second intersection point diametrically opposite on the celestial sphere, at (69°,
-17°). The former is a denser clustering only because more comets are
near perihelion there. Near the center of this clustering, four orbits

with widely different aphelion directions and inclinations intersect
within one one-hundredth of a square degree. A second independent
clustering at (312°,+20°) is also statistically significant. However,
many orbits do not come close to either of these points. If the planetary break-up hypothesis is correct, then perturbations on the orbital
planes over some 10^7 years must be able to account for the present
dispersion of most of the orbits. Only two possible sources of such
perturbations are known – encounters with passing stars, and the tide-
like force of the gravitational field of the galaxy.

STELLAR ENCOUNTERS

     To those not already familiar with the statistics of stellar
encounters, it may come as somewhat of a surprise that stars frequently
pass closer to the Sun than the aphelion distances of VLP comets. A
study of stars in the solar neighborhood has led to the result that in
5×10^6 years there will probably be half-a-dozen stars passing closer to
the Sun than the aphelion distance of a comet with that period, including one encounter perhaps within 25,000 au. Although somewhat
disruptive, such close encounters do not 'rip off' the cloud of comets
completely because they occur so rapidly. Typically a star will recede
to more normal distances again within a few times 10^4 years. Hence the
total integrated perturbing impulse to the comets is quite small, except
for individual cases of very close approaches. The often-quoted analogy
is that such a stellar encounter with the comet cloud is like 'a bullet
shot through a swarm of bees'.

     The effect of chance encounters with passing stars will be a
tendency to randomize the velocities, and hence elements, of the long-
period comets. Oort (1950) argued that the randomization must be complete if the process has continued for 3×10^9 years. For a time as short
as 5×10^6 years, however, the randomization is far less complete; and we
may still expect to discover traces of the original distribution of
elements. But we are forced to treat the program statistically.

     To order of magnitude accuracy, we can estimate the mean size of
stellar perturbations on typical comets. For the orbital plane, the
mean perturbation in 5×10^6 years would be ± 2.1 radians. The perihelion
direction would be disturbed only ± 0.01 radians. For $1/a$, we predict
$\pm 0.5 \times 10^{-6}$ AU^{-1}; and for q, we estimate a mean error of $\pm 4.2q$. Although
these might seem somewhat pessimistic for being able to test the Ovenden
hypothesis, we can consider that many comets will be perturbed by much
less than these mean amounts, and that in general, the comets we see
are selected from among the least perturbed, since their perihelion
distances must not have been too greatly altered.

GALACTIC FIELD

     The perturbing influence of the galactic field on VLP comets has

been investigated by other authors (e.g. Chebotarev, 1966), but
primarily for the case of long-term, nearly circular motion at the
distance of the 'comet cloud'. Since our VLP comets presumably never
had anything like circular motion around the Sun, but instead almost
cease moving near their aphelia (typical aphelion velocity for a = 30,000
AU and q=1 AU is 0.3 m/s), the galactic influence is relatively quite a
bit larger than it is usually considered to be. At the Sun's distance
from the galactic center, 10,000 pc = 2.06×10^9 AU, the galactic field is
close to a simple inverse square force field directed toward the galactic
center.

Our interest is to numerically integrate the comet orbits backwards
to their previous perihelion passage, taking account of the galactic
field perturbations. As we increase the adopted period of the comet the
perturbations increase from several independent causes - greater exposure
time to the perturbations, greater semi-major axes resulting in greater
galactic shear, and slower aphelion velocities resulting in less inertial
resistance to perturbation. For periods over 10^7 years (a = 46,000 AU),
the galactic perturbations are drastic.

As we go backwards in time, the two clusterings near longitude
249° and 312° draw closer together until they finally merge; and at the
same time they approach closer to the ecliptic plane - a very important
attribute, if the origin point was indeed a planet in the solar system.
At 6×10^6 years (Figure 3) the clusterings have essentially merged and
are quite close to the ecliptic, and the integration is still fully
reversible - i.e. the present configurations (Figure 2) result inevitably
from the configurations in Figure 3 after 6×10^6 years of galactic perturbations. The clustering in Figure 3 is obviously quite pronounced; and
there is no longer any question of its arising by chance. The degree of
scatter of orbits not participating in the clustering is somewhat less
than the ±2.1 radians predicted from stellar encounters, as we had
reason to hope it would be.

Regarding the orbital inclinations, the initital distribution is
close to being proportional to sin i, reflecting random orbit pole
locations. At earlier epochs it evolves closer to a flat distribution
which is independent of i, as would tend to occur if all of the orbits
intersected at one point on the ecliptic. Moreover, an initial bias
toward retrograde orbits slowly reverses,and becomes a bias toward
direct orbits earlier than 5×10^6 years ago. It is consistent with what
might be expected from a normal curve for the initital break-up velocity
distribution with a peak at about 26 km/s. The amount of energy involved
is comparable with the kinetic energy of rotation of the original planet
(assuming Ovenden's characteristics), since rotational velocities on the
equators of Jupiter and Saturn are 13 and 10 km/s, respectively.

In connection with the energy of the original event, it may be
relevant to note two points here. Although the disappearance of most of
the original mass and the absence of a core or single large fragment led
us to suspect an event of great energy, ejecting most of the mass from

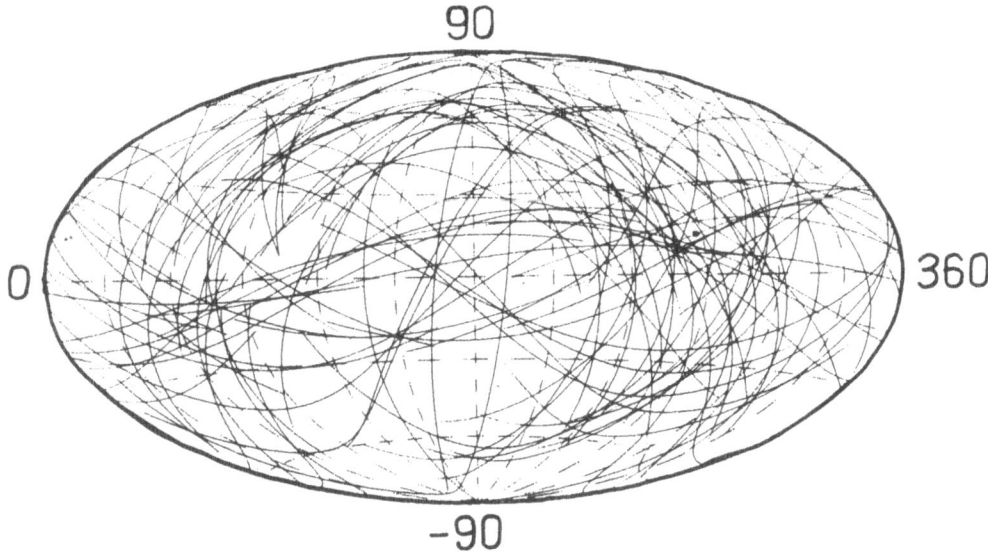

Figure 3. Heliocentric orbital planes for 60 VLP comets 5,800,000 years
         ago.

the solar system, there are other possibilities. Almost all of the
debris from the break-up which is injected into Jupiter-crossing orbits
(any part of the orbit exceeding 5 AU from the Sun) will be swept up by
Jupiter within about 10^5 years (Wetherill, 1974). This might include a
core or substantial fraction of the original mass (See, e.g., Byl and
Ovenden, 1975). Also, there are reasons for believing that most of the
planet was liquid, which makes it easier to dispose of a great deal of
mass without a trace than for a solid planet.

SUMMARY

     Ovenden has previously presented dynamical arguments which indicate
that a large planet existed in the present location of the asteroid belt
until the astronomically-recent past, and than broke up. The cosmic ray
exposure ages of chondritic meteorites also argue for a relatively
recent break-up event involving asteroidal material. The kinetic energy
of rotation alone for such a planet would likely have been sufficient to
propel some material with velocities equal to and exceeding escape
velocity from the solar system.

     In the course of searching for logical consequences of such an
event which might lead to a test of Ovenden's conclusions, we have

examined long-period comet orbits to determine whether they could have
originated in such a way. It has already been shown that short-period
comets would have originated from planetary perturbations of long-period
comets. We find that the properties of long-period comet elements are
completely consistent with the Ovenden hypothesis; indeed many of their
characteristics would be difficult to explain in any other way. The
Ovenden hypothesis predicts in a natural way the single most remarkable
property of long-period comet elements - a clustering in their energy
distributions close to the energy of escape from the solar system - a
property discovered by Oort, and which could not have survived a single
previous perihelion passage. Moreover the cometary perihelion directions
show a strong 'Sun-selecting influence' - a property almost immune to
perturbations, which is most unlikely to have resulted from any other
proposed type of cometary origin.

A break-up type of origin demands that, to the extent that the
orbits are unperturbed, they must all pass through the point of break-up
upon return to the vicinity of the planets. Such a property is observed;
and the clustering of orbital planes, which lies close to the ecliptic
plane, increases in intensity as the galactic perturbation effects are
taken out of the element distributions. The degree and kind of scatter
in the elements at an epoch of 6×10^6 years ago is consistent with that
expectable from encounters with other stars in the solar neighborhood.
This model for cometary origin is also highly successful at predicting
some statistical properties of the orbital element distributions,
including some not previously known. The epoch of the break-up event can
be bounded in several ways, and must lie close to $(6 \pm 1.5) \times 10^6$ years ago.

In short, there are now several independent arguments for the
validity of the Ovenden hypothesis, and for a common origin of comets
and asteroids from the break-up of a former massive planet of the solar
system. As remarkable as this hypothesis is, it would require no major
contortion of existing knowledge and theories to accept, but for the
single fact that the event is datable to an epoch so recent that primates
already walked the Earth!

BIBLIOGRAPHY

Byl, J. and Ovenden, M. W. 1975, 'On the Satellite Capture Problem'.
    Monthly Notices Roy. Astron. Soc. 173, 579-584.
Chebotarev, G. A.: 1966, 'Cometary Motion in the Outer Solar System',
    Soviet Astron. - AJ 10, 341-344.
Everhart, E. and Raghaven, N.: 1970, 'Changes in Total Energy for 392
    Long-Period Comets, 1800-1970', *Astron. J.* 75, 225-272.
Marsden, B. G. and Sekanina, Z.: 1973, 'On the Distribution of 'Original"
    Orbits of Comets of Large Perihelion Distance', *Astron. J.* 78,
    1118-1124.
Marsden, B. G.: 1974, 'Comets', *Ann. Rev. Astron. Astrophys.* 12, 1-12.
Marsden, B. G.: 1975, *Catalogue of Cometary Orbits*, Smithsonian Astro-
    physical Observatory, Cambridge, pp. 1-83.

Meffroy, J.: 1955, 'Contribution à l'étude de la stabilité du système
    solaire', *Bull. Astron.*, Ser. 2, 19, 1-224.
Oort, J. H.: 1950, 'The Structure of the Cloud of Comets Surrounding
    the Solar System, and a Hypothesis Concerning its Origin', *Bull.
    Astron. Inst. Neth.* 11, 91-110.
Ovenden, M. W.: (1972), 'Bode's Law and the Missing Planet', *Nature* 239,
    508-509.
Ovenden, M. W.: 1973, 'Planetary Distances and the Missing Planet',
    Recent Advances in Dynamical Astronomy ed. by B. D. Tapley and
    V. Szebehely), Reidel, Dordrecht, Holland, 319-332.
Ovenden, M. W.: 1976, 'The Principle of Least Interaction Action' (In
    process).
Sekanina, Z.: 1968, 'On the Perturbations of Comets by Nearby Stars'
    Bull. Astron. Inst. Czech. 19, 291-301.
Wetherill, G. W.: 1974, 'Solar System Sources of Meteorites and Large
    Meteoroids', *Ann. Rev. Earth Planet. Sci.* 2, 303-331.

COMETS AND THE MISSING PLANET

Michael W. Ovenden and John Byl
Department of Geophysics and Astronomy,
University of British Columbia,
Vancouver, B.C., Canada

ABSTRACT. Integrating backwards in time in the circular restricted three-body problem Galaxy-Sun-Comet, for both the real long-period comets and fictitious random sets of orbital elements,we have confirmed van Flandern's conclusion that there is a statistically-significant clustering of the orbits of real long-period comets, in heliocentric direction, some 5×10^6 years ago. The clustering is also significant in heliocentric distance, and is more marked if it is assumed that the comets have gone round the Sun more than once since the epoch of maximum clustering. We suggest that the "event" discovered by van Flandern is not the explosive disruption of a planet formerly in the asteroid belt, but the latest in a series of minor catastrophies, such as the collisional break-up of a pair of large asteroids.

...

In a series of papers (1,2,3) one of us has developed dynamical arguments leading to the conclusion that there once existed, in the region of the asteroids, a massive planet which subsequently disrupted. T.C. van Flandern (4,5) has claimed that the statistics of the orbital elements of long-period comets gives direct evidence that such a planetary disruption took place about 6 million years ago. His conclusion was based on (a) the distribution of orbital elements at the present time, and (b) the clustering of long-period comet orbits at a critical time in the past, as discovered by numerical integrations of the comets' orbits backwards in time, allowing for the perturbation by the non-uniform field of the Galaxy.

In this paper, we are concerned only with the evidence under (b). We have repeated van Flandern's calculations, using a different integration technique. We have also carried out integrations for some sets of fictitious comets, whose orbital elements were chosen at random. We wish to decide in what respects, if any, the real long-period comets differ, in their orbital characteristics, from the fictitious (random) sets.

The present orbital elements of 60 long-period comets, corrected for planetary perturbations during the apparition of observation to give pre-encounter values, have been listed by van Flandern (4). We adopted the same

101

V. Szebehely (ed.), Dynamics of Planets and Satellites and Theories of Their Motion, 101-107.

orbital elements for our "Real" comets. Our "Random" sets of elements were
generated using a pseudo-random-number generating program. The only restriction placed upon the elements was that the perihelion distance, l ,
should lie between 0.5 a.u. and 4.5 a.u. These limits correspond roughly
to the range found in van Flandern's sample, since he rejected all real
comets whose pre-encounter value of l was less than 0.5 a.u. on the grounds
that for such comets non-gravitational perturbations would be significant.
In this paper, for reasons of economy, we display the results of only one
of our random sets.

Integrations (backwards in time) were performed using a library program
for the solution of the circular restricted three-body problem. The program uses adjusted time-steps, and integrations were found to be reversible over a time of 10^6 years with a precision *better than* $0\overset{.}{.}1$ in angle
and 0.001 in l . The mass and distance of the galactic center were taken
to be 2.32×10^{11} solar masses and 11.7 kpc respectively. With these values,
in the restricted problem Galaxy-Sun-Comet, the correct galactocentric
solar angular velocity $\hat{\omega}$ = 25 km s^{-1}kpc^{-1}(6) and the correct gradient
$\partial\hat{\omega}/\partial R$ = - 3.2 km s^{-1}kpc^{-2} (7) are obtained.

If the hypothetical planetary disruption occurred T years ago,the periods
of those long-period comets actually observed within the last few centuries must be very close to T/n , where n is an integer, since $T{\sim}10^7$
years ago (2; *vide* 3). On the grounds that the pre-encounter aphelion
distances of long-period comets tend to cluster about 50,000 a.u.,whereas
the post-encounter aphelion distances do. not, van Flandern [following
Marsden and Sekaninan (8)] argues that for the selected comets n=1. It is
therefore of interest to note that the period of any comet observable near
perihelion, with an aphelion distance of 50,000 a.u., is ${\sim}10^7$ years. However, the actual aphelion distances are poorly determined for such long-
period comets. van Flandern therefore integrates his orbits backwards for
various assumed values of T, taking n = 1 for each value of T.

In the plots of our integrations, given in Figure 1, the upper left-hand
number is the value of T in years, and the upper right-hand number the
value of n .· All the plots show the comets' heliocentric coordinates on
Mollweide's Equal Area Projection [which is not the same projection as
that used by van Flandern (4)]. The zero of ecliptic longitude is taken
to be the direction of the vernal equinox at the present time, taken to
be fixed in an inertial frame. The zero of galactic longitude is taken
to be the direction of the center of the Galaxy, as seen from the Sun, at
the time T years ago, allowing for a constant angular velocity of the Sun
around the center of the Galaxy, in an inertial frame, of $\hat{\omega}$=25 km s^{-1}kpc^{-1}.

When the integration backwards of a comet, for a time T, is begun, it is
assumed that the osculating period of the comet at the present epoch is
T. This will not be precisely true, so that if the times of integration
were taken to be exactly T for all comets, the various comets would, at
that time, be in different phases of their orbits. Since the orbits are
nearly rectilinear, this means that the orbital elements of the different
comets would not be strictly comparable. In a few cases, we actually found

the sidereal period T' of the comet, and then adjusted the initial assumed
osculating period by trial and error until $T' = T$. However, we found that
it was quite sufficiently accurate to carry out our integrations for a
time slightly in excess of T , but to evaluate the orbital elements *at
perihelion passage*. The plots are produced from these perihelion orbital
elements, and are restricted to those phases of the comets' motions for
which the heliocentric distance is less than 30 a.u.

In each figure, the left-hand diagrams show the results for the real
comets - i.e. the comets in van Flandern's list. The right-hand diagrams
show the corresponding results for the fictitious random set of elements.

Figure 1 shows that the present pre-encounter orbital elements of van
Flandern's list show no more preference for clustering than does the
random set. Figure 2 shows that, for $T \sim 10^7$ years, $n = 1$, both the real
set and the random set show extreme clustering. The plots in terms of
galactic longitude show that the clustering is strongly related to the
direction to the galactic center. This clustering is called by van Flandern "galactic polarization". It arises from the fact that for a comet
to have had $l < 30$ a.u. one period $T \sim 10^7$ years ago, and *still retain*
a value of $l < 5$ a.u., it must have had an original orbit that was in-
sensitive to galactic perturbation. This "galactic polarization" has
the consequence that if the hypothetical explosive event had occurred
more than 10^7 years ago, the evidence for it in the comets' orbits would
have been lost.

Figure 1 also shows that for $n = 1$, $T \simeq 5 \times 10^6$ years, the clustering of
orbits is more marked for the "real" set than for the "random" set. Of
course, it might be argued that if the clustering is purely accidental,
the value of T has no significance, and that we could have found as
strong a clustering of the fictitious set for some other value of T.
This seems not to be the case for the random sets which we have investigated, with T restricted to past time. We find that the strongest clus-
tering for the real set occurs for $T \simeq 5 \times 10^6$ years, as compared with 6×10^6 years found by van Flandern. The difference is probably not significant. In these direction-coordinate plots, two centers of clustering,
in opposite directions, are to be expected simply as a result of the
fact that any orbit has two nodes on any given plane.

van Flandern did not investigate the consequences of assuming values of
n different from 1 . Figure 3 shows an important distinction between the
real comets and the fictitious comets. The plots show the results of integrations assuming $T \simeq 5 \times 10^6$ years, but $n = 3$. The clustering is markedly
enhanced for the real comets, whereas this is clearly not true for the
random set.

The plots of heliocentric directions give no indication of whether the
clustering orbits also intersect in space, in the region of the belt of
asteroids. van Flandern argues that the effects of stellar encounters
will have a greater effect on l than on the direction elements. This is
undoubtedly true. Nevertheless we might hope to find some residual evid-

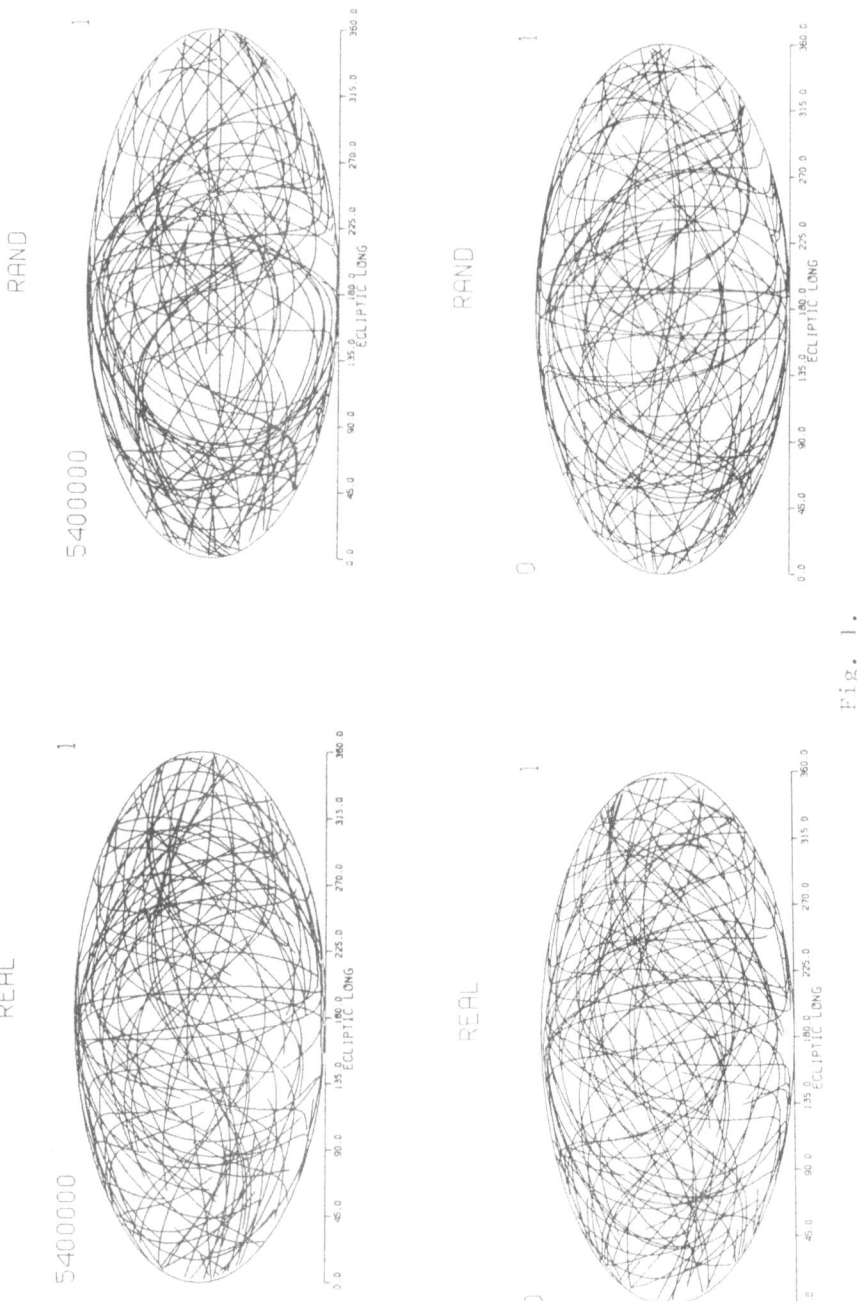

Fig. 1.

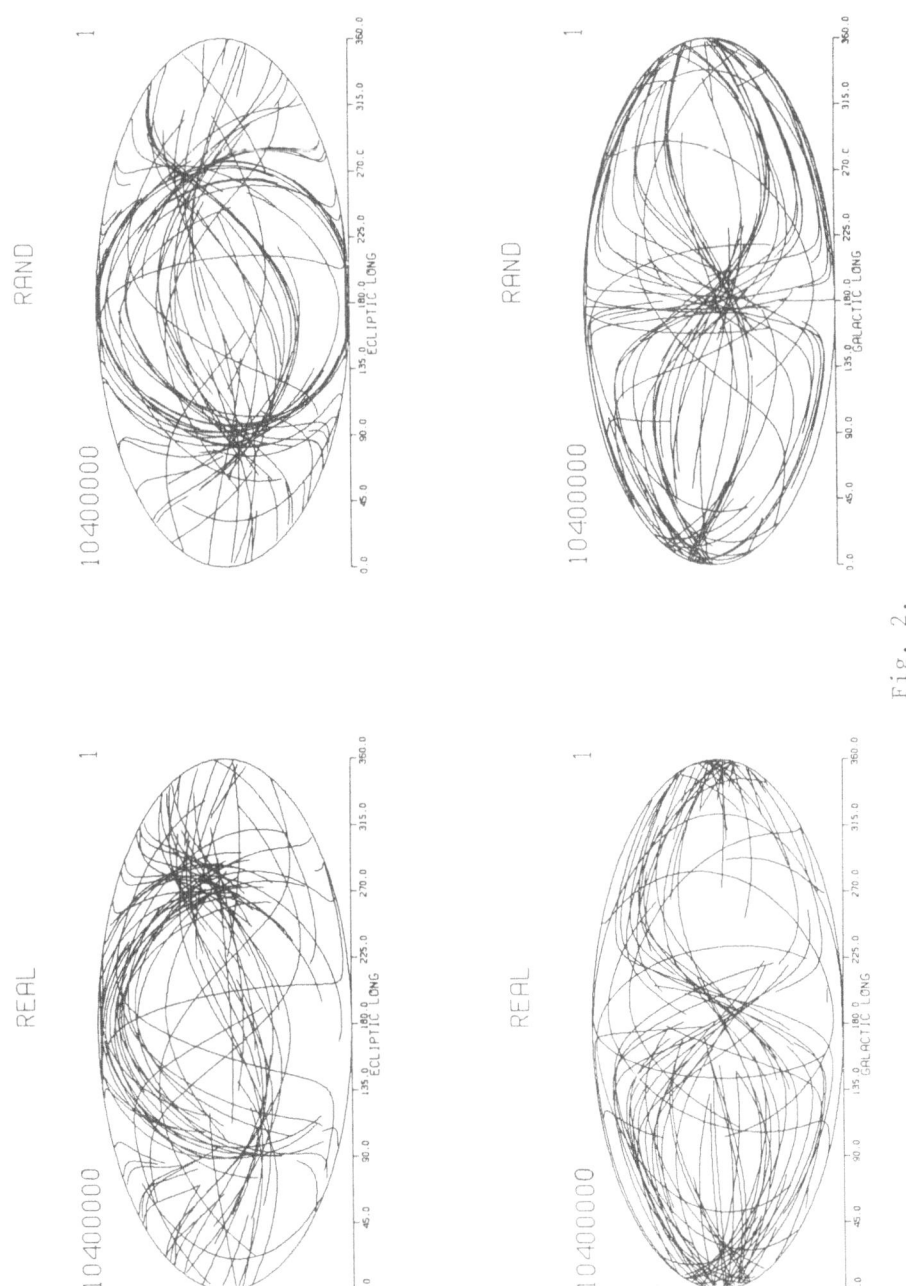

Fig. 2.

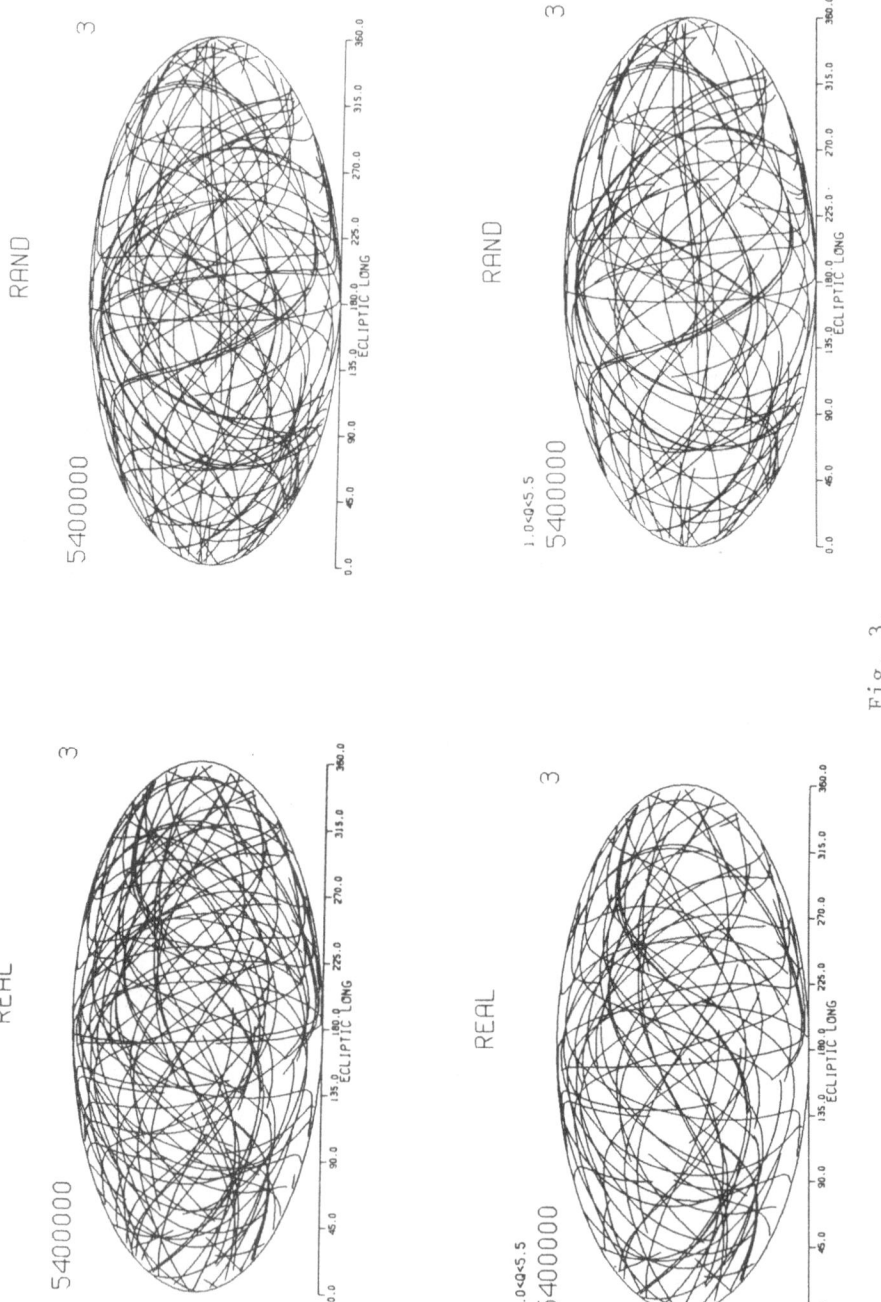

Fig. 3.

ence for the explosive event also in the clustering of Q, the heliocentric distance at the node of the comet's orbit on the ecliptic. The bottom plots in Figure 3 show the same calculations as the upper plots, except that they have been restricted to those orbits for which $1.0 < Q < 5.5$ a.u. A clear distinction now emerges between the real and the random sets. There is a correlation between those real orbits which contribute to the clustering, and those for which Q lies in the restricted range about the asteroidal distance. No such correlation can be seen for the residual clustering in the random set.

We conclude that the long-period comets of van Flandern's list do show a significant clustering in space in the region of the asteroid belt some 5×10^6 years ago (when allowance is made for the perturbing field of the Galaxy) as compared with randomly-chosen sets of "fictitious" comets. However, the clustering is more marked if it is assumed that the real comets have made more than one revolution about the Sun since the epoch of greatest clustering.

The time of 5×10^6 years is uncomfortably short for the dynamical arguments for a missing planet (3). Since "galactic polarization" would preclude discovery of an "event" prior to 10^7 years ago, and since the total cometary mass involved is much less than that of a planet, we suggest that the "event" discovered by van Flandern, and confirmed by our work, is not the catastrophic disruption of a major planet, but the latest in a series of minor catastrophies, such as (for example) the collisional break-up of two large asteroids.

REFERENCES

1. Ovenden,M.W., *Vistas in Astronomy* 18, 473, 1975.
2. Ovenden,M.W., *Recent Advances in Dynamical Astronomy*, ed.B.D.Tapley and V.Szebehely (Dordrecht: Reidel) p.319, 1973.
3. Ovenden,M.W., *Long-time Predictions in Dynamics*, ed.V.Szebehely and B.D.Tapley (Dordrecht: Reidel) p.295, 1976.
4. van Flandern, T.C. Private Communication, 1976.
5. van Flandern, T.C. Communication to I.A.U.Colloquium #41 .
6. Schmidt, M., in *Galactic Structure*, ed.G.P.Kuiper and B.M.Middlehurst (Chicago: University of Chicago Press) p.513, 1965.
7. Ovenden,M.W. and Byl,J., *Astrophys.J.* 206, 57, 1976 .
8. Marsden,B.G. and Sekaninan, Z., *Astron.J.* 78, 1118, 1973 .

MAC REVISITED: MECHANISED ALGEBRAIC OPERATIONS ON FOURTH GENERATION COMPUTERS

A. Deprit
University of Cincinnati

ABSTRACT

Via the facilities provided by PL/I, the operating system is extended to allow automatic manipulations of Poisson series in literal form.

PART II

LUNAR THEORY AND MINOR PLANET MOTIONS

CONTRIBUTION A L'ETUDE DES PERTURBATIONS PLANETAIRES DE LA LUNE

N. Abu el Ata
Bureau des Longitudes, Paris, France

I. INTRODUCTION

L'étude des perturbations planétaires de la Lune soulève un problème fondamental provenant de la différence de caractère entre la
théorie de la Lune et la théorie des planètes. Si, les deux théories
sont des cas particuliers du problème de trois corps, la distinction
entre elles résulte d'une analyse classique basée sur les résultats
de l'observation et de plusieurs études théoriques.

Il est utile de citer quelques aspects de cette distinction:

1. Dans le développement de la fonction perturbatrice en fonction
du rapport des rayons vecteurs corps perturbé/corps perturbateur, il
apparait, pour atteindre une précision relative analogue, un nombre
de termes beaucoup plus grand dans le cas des planètes que celui de
la Lune. Il en découle des méthodes de développement de l'inverse de
la distance, et de ses puissances, différentes dans les deux cas.
2. Dans la théorie des planètes les perturbations sont proportionn-
elles aux rapports des masses des planètes au Soleil et les théories
les plus élaborées négligent les puissances quatrièmes de ce rapport.
Dans la théorie de la Lune, le petit paramètre est le rapport n'/n des
moyens mouvements du Soleil à celui de la Lune, et les développements
dans ce paramètre doivent être poussés très loins pour obtenir en
valeur relative une précision analogue à celle d'un cas planétaire.
3. Les mouvements séculaires des nœuds et des périhélies ne sont
pas comparables, car dans le cas des planètes les périodes sont très
grandes (Kovalevsky, 1967) et l'étude du mouvement de la Lune par suite
d'une période de révolution très courte, nécessite une durée de validité
de la théorie sur un intervalle de temps très long comparativement à
sa révolution mensuelle.

Ces différences citées, parmi les plus fondamentales, montrent
que les deux théories doivent être traitées différemment.

Paradoxalement, le "mélange" des deux théories, sous leurs formes

V. Szebehely (ed.), Dynamics of Planets and Satellites and Theories of Their Motion, 113-124.

les plus élaborées, est indispensable pour étudier les perturbations
planétaires de la Lune. Il est donc avantageux de conserver une séparation entre planètes et Lune jusqu'à l'étape la plus avancée dans la
construction analytique de la solution.

D'après Kovalevsky (1976), la plus grande amélioration à apporter
à la théorie de Brown, qui sert actuellement de base pour la construction des éphémerides conventionelles, avec quelques corrections minimes,
porte sur la partie concernant le calcul des perturbations planétaires.
Aussi la conservation par Brown d'un trop petit nombre d'inégalités a
restreint la généralité de ses résultats. Pour Henrard (1973) l'imprécision du travail de Brown concernant les corrections qu'il a apportées
au problème central (perturbations planétaires et autres perturbations)
est due aux erreurs de troncature (précision de l'ordre 0'',01) et d'autre
part à l'utilisation d'une solution du problème central où n'/n a pris
une valeur numérique, fixée une fois pour toutes, empêchant toute généralisation de ses résultats.

Il est nécessaire actuellement de construire une meilleure solution
du problème pour aboutir à l'établissement d'éphémerides plus précises
en vue des applications aux mesures des distances Laser – Lune et des
missions spatiales, afin de rattraper le décalage entre "théorie" et
"observation". Il faut, d'autre part, assurer la cohérence entre les
différents types de perturbations de la Lune (planétaires,aplatissement,
marées,..) avec les précisions obtenues actuellement dans les solutions
les plus élaborées pour le problème central.

II. UNE APPROCHE DU PROBLEME

Dans l'élaboration de la solution, nous avons procédé par étapes
et considéré qutre niveaux:
 1. Construction formelle de la fonction perturbatrice et des équations
    du mouvement.
 2. Séparation des quantités "Lune" et "planète" au sens de Brown.
 3. Elaboration d'une technique de tamis pour obtenir les arguments qui
    donnent des termes sensibles (dits "sensible terms" par Brown).
 4. Calcul pratique des inégalités.

La première étape du travail a consisté à développer des méthodes
de calcul analytique des puissances de l'inverse de la distance D entre
deux planètes (D^{-s} s=1,3,5,..).

Hill (1879) s'était borné pour le calcul de ces quantités à s$\leqslant$5.
Nous nous sommes efforcés (Abu el Ata, Chapront 1975) d'effectuer un
calcul précis des fonctions de α = a/a', rapport des demi – grands axes
des deux planètes, dans ces développements, afin de réaliser une amélioration par rapport aux résultats de Le Verrier utilisés par Brown (1908),
qui sont, en partie, la cause d'une imprécision dans son travail. Si ces
méthodes apportent une grande précision interne, elles ne permettent pas
d'atteindre des arguments avec des multiples très élevés (fréquences)pour

les longitudes moyennes, par suite du volume considérable des calculs
que cela entrainerait, sous une forme analytique, pour construire de
tels arguments. Ici D^{-s} est représentée par une formule du type:

$$D^{-s} = \sum_{j,n_k} A_{j,n_1,\ldots,n_8} (\alpha)\, x'^{n_1} \bar{x}'^{n_2} x_p^{n_3} \bar{x}_p^{n_4}$$
$$y'^{n_5} \bar{y}'^{n_6} y_p^{n_7} \bar{y}_p^{n_8}\, \exp\sqrt{-1}\, j\,(T-P) \tag{1}$$

où, avec des notations classiques, $x' = e'\exp\sqrt{-1}\,l$; $y' = \gamma'\exp\sqrt{-1}(l+\tilde{\omega})$
$\alpha = a/a'$ et $\gamma = \sin i/2$, x désigne la quantité conjugée de x, les variables primées sont relatives à la Terre et les variables d'indice p sont
relatives à la planète. Les angles T et P sont respectivement les longitudes moyennes de la Terre et de la planète.

On a pu obtenir les multiples élevés qui sont indispensables au
calcul des termes "sensibles" en utilisant les résultats des calculs
de P. Bretagnon (1976): D^{-s} est construit avec une manipulation de séries
de Fourier à coefficients semi-numériques à partir de la valeur du D^{-2}.

D^{-s} est alors écrit sous la forme:

$$D^{-s} = \sum_{p,q} C_{p,q}^{(s)}\, \cos\,(pT + qP) + S_{p,q}^{(s)}\, \sin\,(pT + qP) \tag{2}$$

où $C_{p,q}^{(s)}$ et $S_{p,q}^{(s)}$ sont deux coefficients numériques.

Dans une étape suivante nous avons défini le sens donné aux perturbations "directes" et "indirectes" de la Lune (Chapront,Abu el Ata 1977).
On rappelle ici les formules :

- Perturbations planétaires directes (d pour directe):

$$m_p \frac{d\vec{\sigma}^{(d)}}{dt} = \sum_{j=1}^{6} \frac{\partial \vec{X}^{(c)}}{\partial \sigma_j} (\, \sigma_k^{(c)},\, \sigma_k^{(0)}\,) \cdot m_p\, \sigma_j^{(d)}$$
$$+ m_p\, \vec{X}^{(d)}\, (\, \sigma_k^{(c)}, \sigma_k^{(0)}, \sigma_p^{(0)}) \tag{3}$$

- Perturbations planétaires indirectes (i pour indirecte) :

$$m_p \frac{d\vec{\sigma}^{(i)}}{dt} = \sum_{j=1}^{6} \frac{\partial \vec{X}^{(c)}}{\partial \sigma_j} (\sigma_k^{(c)}, \sigma'^{(0)}_k) \cdot m_p\, \sigma_j^{(i)}$$
$$\sum_{j=1}^{6} \frac{\partial \vec{X}^{(c)}}{\partial \sigma'^{(0)}_j} \cdot m_p\, \sigma'^{(1)}_j \tag{4}$$

σ, σ', σ_p sont les vecteurs respectifs de la Lune, la Terre et de la
planète , $\sigma(c)$ est la solution du problème central. L'indice (0) indique
qu'il s'agit d'un mouvement képlerien, $\sigma'^{(1)}$ représente les perturbations

du premier ordre de la Terre.

Nous nous sommes efforcé d' assurer à nos développements:
- Une grande généralité (validité pour un choix de paramètres quel-
  conques, et sur une échelle de temps longue).
- Une forme aussi fermée que possible afin d'éviter les problèmes de
  troncature.
- La maniabilité dans le cadre de la solution du problème central dont
  nous disposons (choix des variables, calculs analogues,...etc).

Nous avons procédé ensuite à la mise en équations du problème, en
exprimant les dérivés de la fonction perturbatrice en polynômes de
Legendre, et en utilisant un système d'équations de Lagrange avec les
variables $\vec{\sigma} = (a, \lambda, z, \zeta)$ où a et λ sont respectivement le demi - grand
axe de la Lune, et sa longitude moyenne, $z = e \exp\sqrt{-1}\, \omega$ et $\zeta = \gamma \exp\sqrt{-1}\,\Omega$
($\gamma = \sin i/2$).

Dans la suite de ce travail, nous utiliserons les notations de
l'article (Chapront, Abu el Ata, 1977) et nous rappellerons la numéro-
tation des formules avec le symbole "n", où n est le numéro de la formule.

Les équations de Lagrange pour la variable σ_k ($\vec{\sigma} = (\sigma_k)$) ont la
forme:

$$\frac{d\sigma_k}{dt} = \frac{1}{n\,a^2} \, D_{\sigma_k} \vec{V} \cdot \vec{\text{grad}}\, R^{(c)}$$

(5)

"19"

ou bien une formule analogue pour $R^{(d)}$ pour les perturbations directes.
Par simplicité d'écriture, nous travaillerons ici avec $R^{(c)}$ seulement.
Alors:

$$\vec{\text{grad}}\, R^{(c)} = k\, m' \, (R_1^{(c)} \vec{V} + R_2^{(c)} \vec{V}')$$

(6)

et

$$R_k^{(c)} = \frac{1}{r^3} \sum_{j=1} C_{j-1}^k (\frac{r}{r'})^{j-1} \frac{\partial P_j}{\partial \Theta} \quad (k=1,2)$$

(7)

Les $C_{j-1}^{(k)}$ sont des fonctions du rapport des masses Terre – Lune, et les
P_j les polynômes de Legendre avec:

$$\Theta = \frac{1}{r\,r'} \, \vec{V} \cdot \vec{V}'$$

Ecrivons (7) sous une forme générale, en explicitant les polynômes
de Legendre, on trouve:

$$R_k^{(c)} = \sum_{j-1} \sum_{n=0}^{m} P_n^j \cdot \alpha_{j-1}^{(k)} (\frac{r}{a})^{2n} (\frac{a'}{r'})^{2j-2n+1} \Theta^{j-2n+1}$$

(8)

où $m = E (\frac{2j-1}{2}) + \text{Mod} (j,2)$

$$P_n^j = (-1)^n \frac{(2j - 2n)!}{2^j \, n! \, (j-n)! \, (j-2n-1)!}$$

$\alpha_{i-1}^{(k)}$ est une fonction des $C_{j-1}^{(k)}$ et de $\alpha = a/a'$. Dans le cas de $R^{(d)}$, il faudra remplacer r' par D et Θ par Θ_p , où:

$$\Theta_p = \frac{1}{r.D} \; \vec{V} \; . \; (\; \vec{V}' + \vec{V}_p \;)$$

$\vec{V}_p$ est le vecteur de position de P dans un repère héliocentrique. Nous avons realisé la transformation de l'équation (5) sous la forme:

$$\frac{d\sigma_k}{dt} = m \; X_{\sigma_k} = m \; \sum_i \; L_i^{(k)} \; P_i^{(k)} \tag{9}$$

m représente soit m' soit m . Les $L_i^{(k)}$ ne dépendent que des élements de la Lune et s'expriment à l'aide des crochets de la forme:

$$\{m , n\} = (\frac{r}{a})^m \; \exp n \sqrt{-1} \; (w - \lambda) \tag{10}$$

w est la longitude vraie . Les $P_i^{(k)}$ sont des fonctions des planètes et de la Terre, où seule l'inclinaison de la Lune apparait à travers son carré. Elles sont exprimées à l'aide d'un jeu de notations q_k, k =0,1, 2,3,.. définies par "42" pour les termes principaux (indépendants de i'inclinaison) et i_m, m = 0,2 en "50" pour les termes en inclinaisons.

Dans une première approche du problème afin d'étudier quelques aspects et difficultés numériques nous avons exprimé les seconds membres sous la forme:

$$X_\sigma = \sum_k \; \sum_{1_1,1_2} \; a_{1_1,1_2,k}^{(\sigma)} \; x^{1_1} \; \overline{x}^{1_2} \; \exp - k\sqrt{-1} \, \lambda \; . q_k \tag{11}$$

les x ont la signification donnée en (1). Les $a_{1_1,1_2,k}$ sont des coefficients numériques dont l'ordre en α est lié à $1_1,1_2,k$ l'indice k.

III. FORMULAIRE DU PROBLEME (suite)

a. Développements fermés des seconds membres

Dans les équations "41" nous nous sommes limités aux termes parallactiques en α_1 , où:

$$\alpha_1 = \frac{m_T - m_L}{m_T + m_L} \; . \; \frac{a}{a'}$$

dans les dérivées de la fonction perturbatrice dont la forme générale est donnée par (8). Nous pousserons maintenant les développements jusqu' aux termes de l'ordre $(a/a')^2$.
D^{-s} apparait avec des puissances s $\leqslant$9. On peut alors étudier l'effet de la troncature en a/a' sur les développements.

Avec l'utilisation de nos variables et de nos notations, en respect-tant la séparation symbolisée par (9), on peut réécrire les seconds membres de nos équations de façon à mettre en évidence un grand nombre de quantités semblables dans les trois équations "44". Les calculs sont réduits en exprimant ces quantités sous une forme fermée.

Rappelons que nous avons séparé les seconds membres de nos équations en deux parties:

$$X_\sigma = X_\sigma^1 + X_\sigma^2 \tag{12}$$

L'indice 1 représente la "partie principale". L'indice 2 représente les "termes complémentaires" en inclinaison. Nous nous bornerons à l'évaluation des termes principaux. Aux expressions "42", qui ne dépendent que des coordonnées de la Terre et de la planète, il faut ajouter:

$$q_4 = \frac{3}{2}\delta^5 - \frac{15}{2} u\,\bar{u}\,\delta^7 + \frac{105}{16} u^2\,\bar{u}^2\,\delta^9$$

$$q_5 = \left(- \frac{15}{8} u^2\,\delta^7 + \frac{35}{16} u^3\,\bar{u}\,\delta^9 \right) . \exp 2\sqrt{-1}\,\lambda \tag{13}$$

$$q_6 = \frac{36}{16} u^4\,\delta^9\,\exp 4\sqrt{-1}\,\lambda$$

$\delta = \dfrac{a'}{r'}$, dans le cas des perturbations indirectes. $\delta = \dfrac{a'}{r_p}$ pour les perturbations directes, la signification de u est donnée en "43", avec les définitions "25" et "26". Notons que q_4, q_5 et q_6 sont multipliés par $(a/a')^2$.

La nouvelle forme des équations "44" peut alors s'écrire:

On pose

$$S_0 = q_0 + \alpha_2 \frac{\rho}{a} \frac{\bar{\rho}}{a} q_4$$

$$S_1 = \alpha_1 \frac{\bar{\rho}}{a} q_1 + \alpha_2 \frac{\bar{\rho}^2}{a^2} q_5$$

$$S_2 = q_2 + \alpha_1 \frac{\rho}{a} q_1 + \alpha_1 \frac{\bar{\rho}}{a} q_3 + \alpha_2 \frac{\rho}{a}\frac{\bar{\rho}}{a} q_5 + \alpha_2 \frac{\bar{\rho}^2}{a^2} q_6$$

on trouve

$$\frac{X^1}{a} = \frac{n'^2}{n}\,\beta_\phi\,2a\,\mathrm{Im}\,\{\,\bar{z}\,\frac{\rho}{a}\,S_0 + (\,\bar{z}\,\frac{\rho}{a} - z\,\frac{\bar{\rho}}{a}\,)\,S_1$$
$$(\,\bar{z} + \bar{\tau}\,)\,\frac{\bar{\rho}}{a}\,S_2\,\}$$

$$\frac{X^1}{z} = -\sqrt{-1}\,\frac{n'^2}{n}\,\beta\,\{\,\phi\,\frac{\rho}{a}\,(\,S_0 + S_1 + \bar{S}_1\,) + \phi\,\frac{\bar{\rho}}{a}\,S_2\,\} \tag{14}$$
$$\frac{n'^2}{n}\,\beta\,\frac{1}{\phi}\,(\,z + \tau\,)\,\mathrm{Im}\,(\,\frac{\bar{\rho}^2}{a^2}\,S_2\,)$$

$$X_\lambda^1 = -2\,\frac{n'^2}{n}\,\beta\,\mathrm{Re}\,\{\,\frac{\rho}{a}\,(\,\frac{\bar{\rho}}{a} + \frac{1}{2}\,\psi\phi\,\bar{z}\,)\,S_0 + \frac{\bar{\rho}}{a}\,(\,\frac{\bar{\rho}}{a} + \frac{1}{2}\,\psi\phi\,\bar{z}$$
$$+ \frac{\psi\,\bar{\rho}}{4\phi a}\,(\,\bar{z}\,\tau - z\bar{\tau}\,)\,)S_2 + 2\,\mathrm{Re}\,(\,\frac{\rho}{a}\,(\,\frac{\bar{\rho}}{a} + \frac{1}{2}\,\psi\phi\bar{z}\,)\,S_1\,)\}$$

où

$$\alpha_2 = \{ (\frac{m_T - m_L}{m_T + m_L}) + \frac{m_T m_L}{(m_T + m_L)^2} 2 \} \ (\frac{a}{a'})^2 \qquad (15)$$

et

$$\rho = r \ \exp\sqrt{-1} \ w$$

$$\phi = \sqrt{1- e^2}$$

$$\psi = \frac{1}{1 + \phi}$$

Ces formules sont sans doute plus compactes que "44" et peuvent être généralisées pour des puissances supérieures de a/a' .

b. Les perturbations indirectes

Nous avons défini en (4) le système d'équations pour les perturbations planétaires indirectes. Ici $\vec{\sigma}'^{(1)}$ traduit les perturbations planétaires du premier ordre des masses du système Terre - Lune par la planète P. Aux termes en inclinaisons près, les variables osculatrices définissant l'orbite lunaire interviennent toujours dans les coefficients des q_k dans les équations (14). Ceux-ci sont calculés une fois pour toute et utilisés pour toutes les perturbations, les longitudes des planètes apparaissent à travers les quantités $\vec{\sigma}'^{(1)}$

Adoptons le symbole $\vec{Y}_{\sigma'}$ pour définir les seconds membres de (4),soit:

$$\vec{Y}_{\sigma'} = \sum_{j=1}^{6} \frac{\partial \vec{X}^{(c)}}{\partial \sigma'_j}(0) \cdot \sigma'^{(1)}_j \qquad (16)$$

où la partie principale de $\vec{X}^{(c)}$ est donnée en (14) ou "44". Tout en respectant la séparation de Brown on peut exprimer facilement les perturbations planétaires indirectes sous la forme:

$$Y^{(i)}_{\sigma'} = \sum_{j=1}^{6} \sum_{k} A^{(i)}_k (a, \lambda, z, \zeta) \ \text{grad}_{\vec{G}} \ q_k \ \frac{\partial \vec{G}}{\partial \sigma'_j}(0) \cdot \sigma'^{(1)}_j \qquad (17)$$

$Y^{(i)}_{\sigma'}$ est une composante du vecteur $\vec{Y}_{\sigma'}$. Les $A_k(a, \lambda, z, \zeta)$ sont les éléments dépendants de la Lune seulement. Le vecteur $\vec{G}$ est un vecteur à trois composantes:

$$\vec{G} = (r', u', u')$$

où ici :

$$u = \frac{\pi'}{a'} \ \exp - \sqrt{-1} \ \lambda$$

la signification de π' étant donnée par "24".

Nous rassemblons le résultat des calculs des composantes de $\text{grad}_{\vec{G}} \ q_k$ dans le tableau 1 et celui des $\partial\vec{G}/\partial\sigma'$ dans le tableau 2, les deux dernières lignes du tableau 1 représentent les dérivées de i_0 et i_2 qui remplacent les q_k dans l'équation en ζ (voir "50" et "51")

IV. APPLICATION DU FORMULAIRE

Nous nous intéressons dans cette partie à la construction pratique des séries formant les seconds membres pour les perturbations planétaires "directes" ou "indirectes", ainsi qu'au problème du tamis des termes sensibles dans un second membre donné. Nous nous proposons d'analyser les résultats de ces calculs dans une prochaine publication.

a. Les étapes du calcul pour la formulation des seconds membres

Il a été construit plusieurs programmes de calcul que nous utilisons basé sur des techniques présentées par ailleurs (Chapront et al. 1974) correspondant aux différents étapes:
- Calcul de D^{-s} ($s \leqslant 9$, ici) sous une forme analytique ou semi-numérique.
- Transformation des crochets $\{m,n\}$ de la formule (10) en séries de puissances des variables x et $\overline{x}$.
- Calcul des séries du type Lune, soient $L_i^{(k)}$ dans la formule (9) en utilisant la solution semi - numérique du problème central construite par M. Chapront- Touzé (1974).
- Calcul des quantités q_k ,k=0,1,2,.. et i_m, m=0,2 du type "planète", avec les séries de P. Bretagnon et J.L. Simon (1975) dans le cas des perturbations "directes".
- Calcul des derivées partielles (tableaux 1 et 2) dans le cas des perturbations "indirectes".
- Calcul du produit des deux séries Lune x (planète et Soleil) de la formule (9).
- Détermination des "termes sensibles" avec une technique qui est discutée plus bas.

b. Tamis des termes sensibles

Les termes sensibles sont ceux qui donnent naissance, après l'inté-gration à des inégalités dont les coefficients sont plus grands qu'une borne inférieure fixée, à priori, en tenant compte de la précision du calcul et de la nature des théories utilisées. Afin d'éviter une trop grande prolifération d'arguments dans les combinaisons des coordonnées de la Lune avec celles des planètes, nous avons établi une méthode de tamis qui permet à la fois de calculer systématiquement tous les termes sensibles et d'étudier certaines combinaisons qui apportent des arguments des petits diviseurs. Cette idée a été suggérée par Radau (1895) et Brown (1908).

En suivant les idées de Brown, on développe une formule qui permet de calculer la période d'un argument choisi, en variables de Delaunay, et d'autre part une formule qui donne (en considerant les caractéris-tiques associés aux arguments, d'après la propriété de d'Alembert) un ordre de grandeur du coefficient correspondant à l'argument. En conser-vant toujours la séparation entre argument "Lune" et argument "planète" nous avons calculé toutes les périodes provenant de toutes les combin-aisons possibles Lune x planète, en nous limitant à une période maximum de 3500 ans. Brown estime que les coefficients des inégalités ayant des

Tableau 1 $\overrightarrow{grad}_G\, q_k$ et $\overrightarrow{grad}_G\, i_m$

$k = 0, 1, 2, 3\; ;\; m = 0, 2$

composante du gradient / fonction	$\dfrac{\partial}{\partial r'}$	$\dfrac{\partial}{\partial \bar{u}'}$	$\dfrac{\partial}{\partial u'}$
q_0	$3\, r_1'^{-4} - \dfrac{15}{2}\, r_1'^{-6}\, u'\bar{u}'$	$\dfrac{3}{2}\, \bar{u}'\, r_1'^{-5}$	$\dfrac{3}{2}\, u'\, r_1'^{-5}$
q_1	$\dfrac{15}{2}\, u'\, r_1'^{-6} - \dfrac{105}{8}\, u'^2\bar{u}'\, r_1'^{-8}$	$-\dfrac{3}{2}\, r_1'^{-5} + \dfrac{15}{4}\, u'\bar{u}'\, r_1'^{-7}$	$\dfrac{15}{8}\, r_1'^{-7}\, u'^2$
q_2	$-\dfrac{15}{2} u'^2\, r_1'^{-6}$	$3\, u'\, r_1'^{-5}$	0
q_3	$\dfrac{105}{8}\, u'^3\, r_1'^{-8}$	$\dfrac{45}{8}\, u'^2\, r_1'^{-7}$	0
i_0	$\dfrac{15}{2\chi}\, u'\, H'\, r_1'^{-6}$	0	$\dfrac{3}{2\chi}\, H'\, r_1'^{-5}$
i_2	$-\dfrac{15}{2\chi}\, u'\, H'\, r_1'^{-6}$	$\dfrac{3}{2\chi}\, H'\, r_1'^{-5}$	0

$$H' = y\,(\, \bar{u}'\, \Xi'\, \rho_1' - \mu'\, \bar{\Xi}'\, \rho_1' \,)$$

$$= 2\,\sqrt{-1}\;\bar{y}\; \text{Im}\,(\, \bar{u}'\, \Xi'\, \rho_1' \,)$$

μ' et Ξ' sont des fonctions des inclinaisons

Tableau 2 Dérivées par rapport à $\sigma'_j(0)$

fonction $\big/ \dfrac{\partial}{\partial \sigma'_j(0)}$	r'	ρ'	ρ̄'	μ'_j	ν'_j
$\dfrac{\partial e}{\partial \gamma'}$	$-\dfrac{1}{\phi'}\,\mathrm{Im}\,(\,\bar{z}',\tau'\,)$	$\dfrac{\sqrt{-1}}{\phi'}\,(\ )\,(\,z'+\tau'\,)$	$-\dfrac{\sqrt{-1}}{\phi'}\,(\bar{\tau}'+\bar{z}')$	0	0
$\dfrac{\partial e}{\partial z'}$	$-\dfrac{1}{2}(\bar{\tau}'+\sqrt{-1}\,\bar{z}')\,z'\,\dfrac{\psi'}{\phi'}\,\mathrm{Im}(z'\bar{\tau}')$	$-a'-\dfrac{1}{2\phi'^2}z'\,(\bar{z}'+\bar{\tau}')(z'+\tau')\rho'-\dfrac{\psi'}{2\phi'^2}\bar{z}'(z'+\tau')\,a'$	$\overline{\dfrac{\partial \rho'}{\partial z'}}$	0	0
$\dfrac{\partial e}{\partial \bar{z}'}$	$-\dfrac{1}{2}(\tau'+\sqrt{-1}\,z')\,\bar{z}'\,\dfrac{\psi'}{\phi'}\,\mathrm{Im}(\bar{z}'\tau')$	$\dfrac{1}{2\phi'}\Big(\dfrac{1}{\phi'}\,\rho'+a'\,\psi'\,z'\Big)(z'+\tau')$	$\overline{\dfrac{\partial \rho'}{\partial z'}}$	0	0
$\dfrac{\partial e}{\partial \bar{z}'}$	0	0	0	0	$-X'$
$\dfrac{\partial e}{\partial \bar{z}'}$	0	0	0	0	0

$$w' = v' + g' + h'$$
$$w' = \sqrt{-1}\,\exp r',\quad r' = \tau',\quad r' = \rho$$

périodes supérieures à 3500 ans ont peu de chance d'être sensibles

Nous calculs sont obtenus à partir des séries réelles constituant les seconds membres de nos équations, alors que Brown s'est basé sur une formule empirique.

V. CONCLUSION

Afin de rechercher à améliorer les résultats de Brown nous avons pour l'instant considéré plusieurs aspects:
- Effectuer un développement de la fonction perturbatrice à un degré plus élevé en $\alpha = a/a'$.
- Calculer les expressions "Lune" et "planète" sous une forme aussi dense que possible afin d'éviter les erreurs de troncature.
- Utiliser, soit pour l'expression du D^{-s}, soit pour les séries $\sigma'^{(1)}$, des expressions plus précises que celles de Le Verrier utilisées par Brown.
- Réaliser un tamis aussi systématique que possible pour être sûr d'atteindre tous les termes de petits diviseurs essentiellement.
- Utiliser une solution du problème central aussi précise que possible.

Cette liste n'est sans doute pas exhaustive, et il faudra attendre la comparaison avec plusieurs méthodes pour garantir l'exactitude des résultats .

REMERCIEMENTS

l'Auteur remercie vivement M.J. Kovalevsky pour ses conseils et ses encouragements et M.J. Chapront qui lui a fait bénéficier de son expérience tout au long de ce travail.

'Planetary Perturbations of the Motion of the Moon' by N. Abu el Ata

ABSTRACT. The need for accurate ephemerides describing the motion of the Moon warrants a new determination of the inequalities in the Moon's coordinates due to the action of the planets (direct and indirect). Here we expose the different aspects of the problem and the methods to treat them. The difficulties that arose during the work and the elimination of their effects are explained.

REFERENC

Abu el Ata,N., Chapront,J.   1975   Astron. & Astrophys.38,57
Brown,E. W.   1908a   Adams Prize Ess.,Cambridge Uni.P.
Brown,E. W.   1908b   Mem. Roy. Astro. Soc. ,59

Bretagnon,P.	1976	Communication privée
Chapront,J.,Abu el Ata,N.	1977	Astron. & Astrophys.55,83
Chapront,J.,Bretagnon,P., Mehl,M.	1975	Celes. Mech. 16,379
Chapront,J.,Chapront,M., Simon,J.L.	1974	Astron. & Astrophys. 31,151
Chapront-Touzé,M.	1974	Astron. & Astrophys. 36,5
Henrard,J.	1973	Ciel et Terre 89, 1
Hill,G.W.	1878	Am. J. Math. 1,5
Kovalevsky,J.	1967	Celestial Mechanics, Rei.Publ.
Kovalevsky,J.	1977	Phil.Trans.R.Soc.Lond.A.284,565
Radau,R.	1895	Mem. Ann. de l'Obs. 21
Simon,J.L.,Bretagnon,P.	1975	Astron. & Astrophys. 42,259

HAMILTONIAN THEORY OF THE LIBRATION OF THE MOON

Jacques Henrard and Michèle Moons
Department of Mathematics, Facultés Universitaires
de Namur,B-5000 Namur, Belgium

ABSTRACT

The feasibility of applying the Lie transform method to the problem of
the physical libration of the Moon is investigated. By a succession of
canonical transformations, the Hamiltonian of the problem is brought
under a form suitable for perturbation technique. The mean value of
the inclination of the angular momentum upon the ecliptic and the frequencies of the free libration are computed.

1. INTRODUCTION

One of the main tools needed to analyze the data of the laser ranging of
the Moon is a precise theory of the libration of the Moon.

Several authors (Eckhardt 1973, Cook 1976, Migus 1976) have worked and
are still working on the improvement of such a theory. All of them use
a technique of successive approximations to the solution of a set of
differential equations.

We felt it would be interesting to investigate whether the problem could
be treated by an Hamiltonian perturbation method such as the Lie transform method. This method has been used successfully in several problems
of celestial mechanics and presents some advantages. One of them is
that it enables (or forces) the scientist to take one difficulty of the
system at a time and thus gives him a better understanding of it, especially in case of resonance. On the other hand, this technique is
often more difficult to implement and requires more care in choosing
coordinate systems.

Whatever the advantages or drawbacks of these two methods, it seems to
us very interesting to compare their results. Indeed their philosophy
and especially the way they treat resonance is quite different. Thus,
if the results of both of them agree, one can feel confident that the
problem has no hidden traps and that the solution is valid.

125

V. Szebehely (ed.), Dynamics of Planets and Satellites and Theories of Their Motion, 125-135.
All Rights Reserved. Copyright © 1978 by D. Reidel Publishing Company, Dordrecht, Holland.

In this paper, we study the feasibility of applying the Lie transform method to the problem of the physical libration of the Moon. This task is not a trivial one as the libration of the Moon is not obviously the perturbation of an integrable problem or, if it is (one can think of the constant rotation around the axis of inertia), the integrable problem is too degenerate to be of any use.

2. THE PHASE SPACE AND THE FREE ROTATION PROBLEM

To describe the problem of the libration of the Moon as an Hamiltonian system, we choose the Tisserand's canonical variables (Deprit 1967). They are the three angles μ_1 , μ_2 , μ_3 (see figure 1) and their conjugate momenta :

$M_1 = M_2 \cos I$

M_2 = norme of the angular momentum of the Moon

$M_3 = M_2 \cos b$

where the angles I and b are defined in figure 1. The frame of reference X , Y , Z is an inertial frame (the plane X , Y being the ecliptic) and the frame x , y , z is the frame of the principal axis of inertia of the Moon.

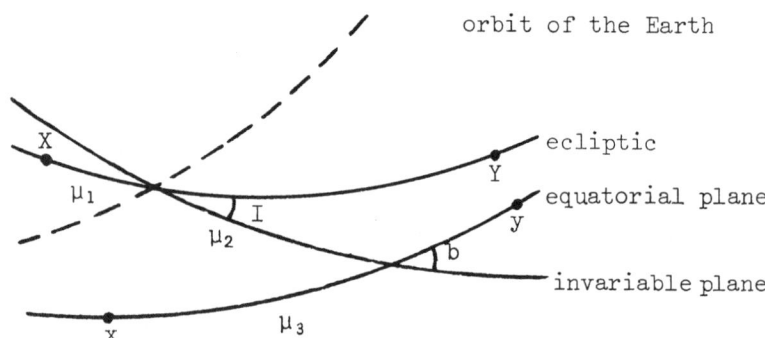

Figure 1.   Geometry of the libration of the Moon.

The invariable plane is the plane perpendicular to the angular momentum of the Moon. Note that we have taken the angle I as negative so that the mean value of the longitude μ_1 , will be equal to the mean value of the node of the orbit of the Earth as seen from the Moon.

The Hamiltonian of the free rotation is then written as (Deprit 1967) :

$$H = \frac{1}{2} \frac{M_3^2}{C} + \frac{1}{2} (M_2^2 - M_3^2) [\frac{\sin^2 \mu_3}{A} + \frac{\cos^2 \mu_3}{B}] \qquad (1)$$

where A , B , C (A $\leqslant$ B $\leqslant$ C) are the moments of inertia of the Moon with respect to the axis x , y , z .

The Tisserand's canonical variables are singular when I and b are equal to zero (or to π). When $I = 0$ (resp. $b = 0$) the angles μ_1 and μ_2 (resp. μ_2 and μ_3) are undefined (although their sum is well defined). This is a situation very similar to the one appearing when one uses the Delaunay's variables for small eccentricities or inclinations in the two body problem.

As usual, this situation can be dealt with easily by using the properties of d'Alembert characteristic if the singularity can be shown to be of the polar-coordinates type (Henrard 1974). To do this, we introduce the modified Tisserand's elements :

$$\Lambda_1 = M_2 \qquad\qquad\qquad \lambda_1 = \mu_1 + \mu_2 + \mu_3 \qquad\qquad (2.1)$$

$$\Lambda_2 = M_2 - M_3 \qquad\qquad \lambda_2 = -\mu_3 \qquad\qquad\qquad (2.2)$$

$$\Lambda_3 = M_2 - M_1 \qquad\qquad \lambda = -\mu_1 \qquad\qquad\qquad (2.3)$$

The angle λ_2 (resp. λ_3) is undefined when $\Lambda_2 = 0$ (resp. $\Lambda_3 = 0$) but the Poincaré-type variables associated with (2.2) and (2.3) :

$$\xi = \sqrt{2\,\Lambda_2}\,\sin\,\lambda_2 \qquad\qquad \Xi = \sqrt{2\,\Lambda_2}\,\cos\,\lambda_2 \qquad\qquad (3.1)$$

$$\eta = \sqrt{2\,\Lambda_3}\,\sin\,\lambda_3 \qquad\qquad H = \sqrt{2\,\Lambda_3}\,\cos\,\lambda_3 \qquad\qquad (3.2)$$

are non singular. Thus, the virtual singularities of the modified Tisserand's variables (2) will not present any difficulty if the functions we shall be dealing with present the d'Alembert characteristic with respect to the couple $(\lambda_2, \Lambda_2), (\lambda_3, \Lambda_3)$.

The Hamiltonian of the free rotation now reads :

$$H = \frac{1}{2}\frac{(\Lambda_1 - \Lambda_2)^2}{C} + \frac{1}{2}\Lambda_2\,(2\,\Lambda_1 - \Lambda_2)\,[\frac{\sin^2\,\lambda_2}{A} + \frac{\cos^2\,\lambda_2}{B}] \qquad (4)$$

or

$$H = \frac{1}{2}\frac{\Lambda_1^2}{C} + 2\frac{\Lambda_1^2}{C}\sin^2\frac{b}{2}\cos^2\frac{b}{2}[\frac{C - A}{A}\sin^2\,\lambda_2 + \frac{C - B}{B}\cos^2\,\lambda_2](5)$$

From (4), it is obvious that this Hamiltonian function presents the d'Alembert characteristic with respect to the couples (λ_2, Λ_2) and (λ_3, Λ_3). The expression (5) with the auxiliary quantities :

$$\sin^2\frac{b}{2} = \frac{\Lambda_2}{2\,\Lambda_1} \qquad\qquad\qquad \sin^2\frac{I}{2} = \frac{\Lambda_3}{2\,\Lambda_1} \qquad\qquad (6)$$

which are geometrically meaningfull and undimensional, will be prefered. It makes obvious that, when $A = B = C$ (which is almost the case for the Moon), the problem is trivial.

3. THE PERTURBATION FROM THE FREE ROTATION PROBLEM

The function which has to be added to the Hamiltonian of the free

rotation to take into account the attraction of the Earth on the rigid
Moon is :

$$V = - G \int_{Earth} \int_{Moon} \frac{d\mu \, d\mu'}{r'}$$

where r' is the distance between an element of mass $d\mu'$ in the
Earth and an element of mass $d\mu$ in the Moon. We assume that the
Earth is a mass-point, thus neglecting terms of the order of :

$$[J_2 \, (\frac{R}{r})^2 \,]_{Earth} \simeq 10^{-7}$$

with respect to the mean terms of the perturbation (r being the distance of the centers of mass and R the equatorial radius of the body).

Furthermore, in this preliminary computation, we shall neglect the terms
of the third order in the expansion of the potential of the Moon, thus
neglecting terms of the order of :

$$[\frac{J_3}{J_2} \, (\frac{R}{r}) \,]_{Moon} \simeq 5.10^{-4}$$

with respect to the mean terms of the perturbation.

Neglecting terms independant of our phase space, we are thus considering
as perturbing potential :

$$V = - \frac{3}{2} \frac{G \, E}{r^3} [\, (C - A) \, \xi_1^2 + (C - B) \, \xi_2^2 \,] \tag{7}$$

where ξ_1 and ξ_2 are the first two coordinates of the unit vector
pointing to the Earth in the frame of the principal axis of inertia of
the Moon.

Defining the mean semi-major axis of the orbit of the Earth by :

$$a^3 = G \, (E + M) \, / \, n^2 \tag{8}$$

where n is the mean notion in longitude of this orbit, we write the
equation (7) under the form :

$$V = - \frac{3}{4} \frac{n^2 \, C}{1 + \kappa} (\frac{a}{r})^3 \, \{\delta \, (\xi_1^2 + \xi_2^2) + \gamma \, (\xi_1^2 - \xi_2^2)\} \tag{9}$$

where γ and κ are defined as usual by $\kappa = M / E$ and $\gamma = (B - A) / C$.
The quantity δ is defined by :

$$\delta = (2 \, C - A - B) / C \tag{10}$$

and is related to the usual quantity $\beta = (C - A) / B$ by :

$$\delta = \frac{2 \, \beta - \gamma \, (1 - \beta)}{1 + \beta} = 2 \, \beta - \gamma - 2 \, \beta^2 + 2 \, \gamma \, \beta + \vartheta(\beta^3) \tag{11}$$

The expressions of ξ_1 and ξ_2 in (9) are obtained by a succession of
rotations (of angles μ_1 , I , μ_2 , b , μ_3) from the expressions of X_1 ,
X_2 , X_3 , the components of the unit vector pointing to the Earth in the

ecliptical frame. In our program, the quantities :

$$s = \sin(b/2) \qquad k = \cos(b/2)$$
$$S = \sin(I/2) \qquad K = \cos(I/2) \qquad (12)$$

are used to express the rotation of angles I and b .

The expressions of (a/r), X_1, X_2, X_3 in turn are obtained from the Theory of the Moon; they are multiple Fourier series in the arguments λ, D, F, l, l' (resp. the mean longitude of the Moon, the difference of longitude between Moon and Sun, the arguments of latitude, the mean anomaly of the Moon and the mean anomaly of the Sun). In this computation, we have used the solution of the main problem called ALE (Deprit et al. 1971a, 1971b, Henrard 1972) truncated at 50" for longitude and latitude and at 0"5 for the sine-parallax. This corresponds to a relative accuracy of 2.10^{-4} in the computation of V .

Note that if we use in the expression (a/r) the definition (8) for the mean semi-major axis and not the inverse of the constant term of sine parallax, we do not have to take into account the correcting factor λ = 1.0027 of Jeffreys (Jeffreys 1961). Actually, our computation corresponds to λ = 1.002726 as the value given by Jeffreys is truncated.

The system we are considering is a system with three degrees of freedom but it depends explicitly on the time through the frequencies n (of the longitude of the Moon), n' (of the longitude of the Sun), n_g (of the perigee of the Moon) and n_h (of the longitude of the node of the Moon). We transform it into an autonomous system with seven degrees of freedom by introducing artificial momenta L, L', G, H , conjugated respectively to the angles λ, l', g, h (note that L, G, H are not the Delaunay's momenta).

Eventually, taking into account (5) and (9), we obtain as Hamiltonian function for the problem of the libration of the Moon :

$$H = n L + n' L' + n_g G + n_h H + \Lambda_1^2 / 2 C +$$

$$+ \frac{\Lambda_1^2}{C} s^2 k^2 (\delta - \gamma \cos 2\lambda_2) - \qquad (13)$$

$$- \frac{3}{4} \frac{n^2 C}{1+K} \delta P_1 - \frac{3}{4} \frac{n^2 C}{1+K} \gamma P_2$$

where P_1 and P_2 are the expansions respectively of $(a/r)^3 (\xi_1^2 + \xi_2^2)$ and $(a/r)^3 (\xi_1^2 - \xi_2^2)$.

In writing (13), we have neglected a term :

$$\frac{\Lambda_1^2}{C} s^2 k^2 [\frac{\delta^2 + \gamma^2}{2} - \gamma \delta \cos 2\lambda_2] + \vartheta(\delta^3) \qquad (14)$$

of the order of δ^2 in the expression of the Hamiltonian of the free rotation.

The principal terms in P_1 and P_2 are approximatively :

$$P_1 = (K^4 + S^4)(k^4 + s^4) + 8 K^2 S^2 k^2 s^2 +$$
$$+ 2 K^4 k^2 s^2 \cos (2 \lambda - 2 \lambda_1 - 2 \lambda_2) +$$
$$+ 2 \sin i \, KS (K^2 - S^2)(k^4 + s^4 - 4 k^2 s^2) \cos (\lambda_3 + h) - \qquad (15)$$
$$- 0.015 K^4 k \, s (k^2 - s^2) \cos (\lambda - \lambda_1 - \lambda_2 - g) +$$
$$+ \ldots$$

$$P_2 = 2 k^2 s^2 (K^4 + S^4 - 4 K^2 S^2) \cos (2 \lambda_2) +$$
$$+ K^4 k^4 \cos (2 \lambda - 2 \lambda_1) - \qquad (16)$$
$$- 0.015 K^4 k^3 s \cos (\lambda - \lambda_1 + \lambda_2 - g) +$$
$$+ \ldots$$

4. CASSINI'S LAWS AND THE FREQUENCIES OF THE SYSTEM

In our notations, Cassini's laws which provide an approximation for the physical libration of the Moon, can be written :

$$<\lambda_1> = \lambda$$
$$<\lambda_3> = -h \qquad (17)$$
$$<2S> \simeq -\sin (1° \, 30')$$

where $<\ldots>$ stands for "the mean value of".

To take into account those laws and bring forward the librations around these mean values, we propose the following canonical transformation :

$$x_1 = \lambda_1 - \lambda \qquad\qquad y_1 = \frac{\Lambda_1}{n \, C} - \nu \qquad (18.1)$$

$$x_2 = \sqrt{\frac{2 \, \Lambda_2}{n \, C}} \sin \lambda_2 \qquad\qquad y_2 = \sqrt{\frac{2 \, \Lambda_2}{n \, C}} \cos \lambda_2 \qquad (18.2)$$

$$x_3 = \sqrt{\frac{2 \, \Lambda_3}{n \, C}} \sin (\lambda_3 + h) \qquad y_3 = \sqrt{\frac{2 \, \Lambda_3}{n \, C}} \cos (\lambda_3 + h) - 2 \mu \quad (18.3)$$

$$\lambda^+ = \lambda \qquad\qquad L^+ = (L + \Lambda_1) / n \, C \qquad (18.4)$$

$$h^+ = h \qquad\qquad H^+ = (H - \Lambda_3) / n \, C \qquad (18.5)$$

$$g^+ = g \qquad\qquad G^+ = G / n \, C \qquad (18.6)$$

$$l'^+ = l' \qquad\qquad L'^+ = L' / n \, C \qquad (18.7)$$

The constant ν (close to one) which appears in (18.1) comes from the fact that the mean value of Λ_1 is close but not quite equal to n C and the constant μ which appears in (18.3) is close to the constant

<-S> . These constants will be computed so that the coordinates (x_i,y_i) will be cartesian-like coordinates centered at a mean equili-brium. The multiplier of the transformation has been made equal to $1/nC$ in order to obtain undimensional (x_i,y_i) .

The Hamiltonian function (13) now reads :

$$H = n L + n' L' + n_g G + n_h H +$$

$$+ n (\nu - 1) y_1 + 2 \mu n_h y_3 +$$

$$+ \frac{n}{2} y_1^2 + \frac{n}{4} [\delta (x_2^2 + y_2^2) + \gamma (x_2^2 - y_2^2)] \nu + \qquad (19)$$

$$+ \frac{n_h}{2} (x_3^2 + y_3^2) + \ldots -$$

$$- \frac{3}{4} \frac{n \delta}{1+K} P_1 - \frac{3}{4} \frac{n \gamma}{1+K} P_2$$

where we have dropped the stars. The principal terms in P_1 and P_2 are approximatively :

$$P_1 = [-2 \sqrt{\mu^2/\nu} - \sin i + \vartheta(\mu^2)] y_3 +$$

$$+ [\sqrt{\mu^2/\nu} \sin i + \vartheta(\mu^2)] y_1 +$$

$$+ \frac{1}{2} [-(x_3^2 + y_3^2) - 2 x_2^2 + 0.15 y_1 y_3 + \qquad (20)$$

$$+ 0.16 \cos g (y_2 y_3 - x_2 x_3) -$$

$$- 0.16 \sin g (y_2 x_3 + x_2 y_3)] + \vartheta(\mu) +$$

$$+ \text{cubic terms in} (x_i,y_i)$$

$$P_2 = [-2 \sqrt{\mu^2/\nu} - \sin i + \vartheta(\mu^2)] y_3 +$$

$$+ [\sqrt{\mu^2/\nu} \sin i + \vartheta(\mu^2)] y_1 +$$

$$+ \frac{1}{2} [-(x_3^2 + y_3^2) - 2 x_2^2 - 4 x_1^2 +$$

$$+ 0.15 y_1 y_3 + 0.32 y_1 x_3 + \qquad (21.)$$

$$+ 0.16 \sin g (y_2 x_3 - x_2 y_3) -$$

$$- 0.16 \cos g (y_2 y_3 + x_2 x_3)] + \vartheta(\mu) +$$

$$+ \text{cubic terms in} (x_i,y_i)$$

where $\sin i$ is the sine of the inclination of the orbit of the Earth.

The origin of the phase space will be a mean equilibrium for the system if μ and ν are such that the mean values of the coefficients of the linear terms in (x_i,y_i) are zero.

Then, in order to compute the basic frequencies of the system around this mean equilibrium, we should, by a linear canonical transformation, bring the quadratic part in (x_i, y_i) of the Hamiltonian under the form of three uncoupled harmonic oscillators.

$$\frac{n_1}{2} (x_1^{!2} + y_1^{!2}) + \frac{n_2}{2} (x_2^{!2} + y_2^{!2}) + \frac{n_3}{2} (x_3^{!2} + y_3^{!2}) \tag{22}$$

As we have defined only a mean equilibrium in the first step, the frequencies n_i obtained would be only approximations of the true frequencies around the true equilibrium. In view of this, we shall simplify this second step by asking only that the quadratic part of the Hamiltonian becomes approximatively (22) and restrict ourselves to linear canonical transformations of the scaling-type :

$$x_i = \alpha_i \, x_i^{!} \qquad\qquad\qquad y_i = \alpha_i^{-1} \, y_i^{!} \tag{23}$$

Eventually, the transformation :

$$\begin{aligned} x_1 &= \sqrt{2 \, P} \sin p & y_1 &= \sqrt{2 \, P} \cos p \\ x_2 &= \sqrt{2 \, Q} \sin q & y_2 &= \sqrt{2 \, Q} \cos q \\ x_3 &= \sqrt{2 \, R} \sin r & y_3 &= \sqrt{2 \, R} \cos r \end{aligned} \tag{24}$$

will introduce the action-angles coordinates used in the following step of the theory.

5. NUMERICAL VALUES

The constants μ, ν, α_i and n_i $(1 \leqslant i \leqslant 3)$ of the preceding paragraph depend upon the values of $\kappa, n_h, \delta, \gamma$. We have taken :

$$\begin{aligned} \delta &= 0.00103 & \kappa &= 1 / 81.30 \\ \gamma &= 0.00023 & n_h / n &= -0.00402133375326 \end{aligned} \tag{25}$$

Of course, the constants δ and γ are not well determined and we should allow for their variation by computing the derivative with respect to them of the final solution.

With the above values, we find :

$$\mu = 0.013499866212 \tag{26}$$

$$\nu = 1.000001465746$$

which, as we check, can be compared with the values given for the mean inclination of the axis of rotation of the Moon on the ecliptic. Assuming (x_i, y_i) to be zero, we find :

$$I = -2 \sin^{-1} (\sqrt{\mu^2 / \nu}) = -1° \, 32' \, 49'' \tag{27}$$

which is to be compared to $-1° \, 32' \, 28''$ given by Eckhardt (Eckhardt 1965) or to $-1° \, 32' \, 57''$ given by Migus (Migus 1976). Not too much emphasis should be placed upon this comparison at this stage. Indeed, if both the quoted authors give as data the final mean value of I, it should be compared to the mean value of :

$$2 \sin^{-1} \left(\frac{1}{2} \left[\frac{(2 \mu + y_3)^2 + x_3^2}{\nu + y_1} \right]^{1/2} \right) \tag{28}$$

which is close to (because y_1, x_3, y_3 are small and of zero mean value) but not quite equal to (27).

Figure 2 shows the variation of this last constant when δ and γ varie. The level lines are lines of constant offset from the value $- 1° 32' 49"$.

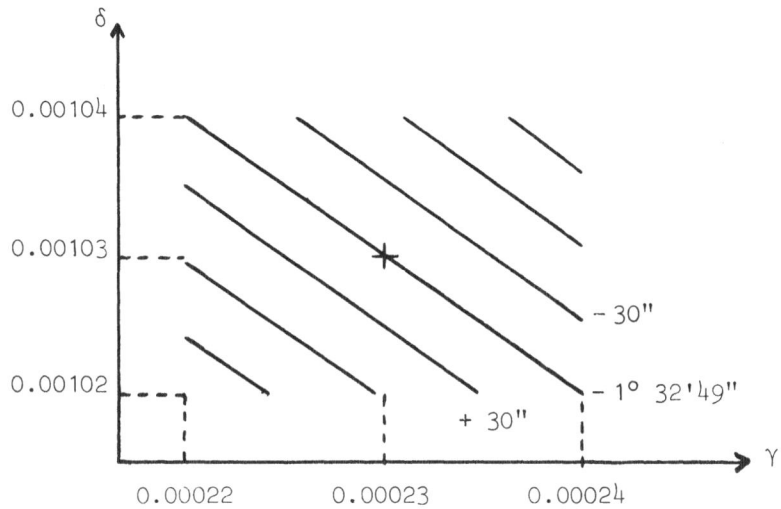

Figure 2. Variation of I with δ and γ .

For the values of α_i $(1 \leqslant i \leqslant 3)$ (see equation (23)), we find :

$$\alpha_1 = 6.207191342059$$
$$\alpha_2 = 0.633750505567 \tag{29}$$
$$\alpha_3 = 1.001132806819 \;.$$

The approximations of the frequencies at the equilibrium are thus :

$$n_1 / n = 0.025954386109$$
$$n_2 / n = 0.000993516224 \tag{30}$$
$$n_3 / n = -0.003098977293$$

The corresponding periods are respectively : 2.88 years, 75.23 years and 24.14 years.

6. CONCLUSIONS

The Hamiltonian function of the problem of the libration of the Moon has now been transformed into a form suitable for perturbation theory :

$$H = n\,L +$$
$$+ n'\,L' + n_g\,G + n_1\,P +$$
$$+ n_2\,Q + n_3\,R + n_h\,H + \tag{31}$$
$$+ n\,H_3\,(P,Q,R,p,q,r,\lambda,l',l,F,D)$$

where the principal terms in H_3 are approximatively :

$$H_3 = -0.00020\,\sqrt{P\,R}\,\cos\,(p - r) +$$
$$+ 0.00018\,\sqrt{P\,R}\,\cos\,(p + r) -$$
$$- 0.00013\,\sqrt{Q\,R}\,\cos\,(q + r + g) -$$
$$- 0.00007\,\sqrt{Q\,R}\,\cos\,(q - r - g) + \tag{32}$$
$$+ 0.00004\,\sqrt{P\,Q}\,\cos\,(p - q - g) -$$
$$- 0.00004\,\sqrt{P\,Q}\,\cos\,(p + q + g) -$$
$$- 0.000004\,\sqrt{P}\,\cos\,(p - 2\,g) +$$
$$+ \ldots$$

As suggested by the way we have written equation (31), the elimination of the periodic terms could be done in three steps, according to their frequencies. First, one could eliminate the monthly terms, then the terms of a period of a few years. They are the terms in l' , g , p with the exception of the resonant term in $p - 2\,g$. In the last step, one could eliminate the terms of a period of twenty years and more, i.e. the terms in q , r and the resonant term in $p - 2\,g$.

In the near future, we plan to implement this elimination of periodic terms and thus compute the generator of the canonical transformation which brings the Hamiltonian (31) into an integrable one. This transformation known, the series describing the libration of the Moon can be obtained easily.

Before obtaining what we hope will be a usefull theory, we shall have to include several neglected terms in the Hamiltonian and compute the derivatives of the series with respect to δ and γ .

REFERENCES

Deprit, A. : 1967, Free Rotation of a Rigid Body Studied in the Phase Plane, *Amer. J. of Physics*, 35, 424.

Deprit, A. and Henrard, J. : 1971a, Analytical Lunar Ephemeris : the Variational Orbit, *Astron. J.*, **76**, 273.

Deprit, A., Henrard, J. and Rom, A. : 1971b, Analytical Lunar Ephemeris : Delaunay's Theory, *Astron. J.*, **76**, 269.

Cook, A.H. : 1976, Theories of Lunar Libration, in Methods and Applications of Ranging to Artificial Satellites and the Moon, (print).

Eckhardt, D.H. : 1965, Computer Solutions of the Forced Physical Librations of the Moon, *Astron. J.*, **70**, 466.

Eckhardt, D.H. : 1973, *The Moon*, **6**, 127.

Henrard, J. : 1972, Analytical Lunar Ephemeris : a report, *Publication du Département de Mathématique F.U.N.*, Namur, Belgium.

Henrard, J. : 1974, Virtual Singularities in the Artificial Satellite Theory, *Celestial Mechanics*, **10**, 437.

Jeffreys, H. : 1961, On the Figure of the Moon, *Monthly N. of the R.A.S.*, **122**, 421.

Migus, A. : 1976, Théorie Analytique Programmée de la Libration Physique de la Lune, *The Moon*, **15**, 165.

NEW RESULTS ON THE COMMENSURABILITY
CASES OF THE PROBLEM SUN-JUPITER-ASTEROID

Joachim Schubart
Astronomisches Rechen-Institut, Heidelberg, Germany

ABSTRACT

The short-period terms are removed by averaging from special
equations of motion for commensurability cases of the three-dimensional, elliptic, restricted three-body problem. Some examples of
retrograde motion corresponding to the -1/1 commensurability, and
an application to Hilda-type motion demonstrate the possibilities
given by the method.

1. LONG-PERIOD EFFECTS STUDIED BY AVERAGING OF THE
    EQUATIONS OF MOTION

In 1963 the late celestial mechanician Imre G. Izsak invited me
to work at the Smithsonian Astrophysical Observatory on long-period
effects in commensurability cases according to the ideas of Poincaré
(1902). I did so on the basis of the circular, restricted three-body
problem given by the sun, Jupiter, and a small body. However, I
replaced Poincaré's way of removing the short-period terms from the
basic equations of the problem, by an averaging procedure (Schubart,
1964, 1966). The work done by Message (1966) is closer to the way
proposed by Poincaré, but I had the advantage of including the
treatment of very eccentric orbits in my work.

Later on D. Brouwer asked me to work on the 3/2 commensurability
case that is represented by an asteroid of Hilda-type and Jupiter.
For this I generalized my method of averaging to the elliptic,
restricted three-body problem, but to the planar case only. The inclination of real orbits had to be neglected (Schubart, 1968). The
inclination of (153) Hilda, for instance, is not very large, but when
Giffen (1973) applied the same method of averaging to the 2/1 case, he
had to neglect a much larger inclination to obtain a model for the
motion of (1362) Griqua. As a consequence, the model gave only a rough
approximation to real motion. Recently, I have dropped the restriction
to the planar case, and I present some first results obtained in this

137

V. Szebehely (ed.), Dynamics of Planets of Satellites and Theories of Their Motion, 137-143.

way in the present paper.

In the mean time, Scholl and Froeschlé (1974, 1975) used the method of averaging for the planar case in a treatment of the 3/1, 5/2, 7/3, and 2/1 resonances with respect to Jupiter. They tested a collision hypothesis for the formation of the Kirkwood gaps in the asteroid belt at these resonances. As an application of their results, they will present a paper "The Kirkwood Gaps as an Asteroidal Source for Meteorites" at IAU Colloquium No. 39 at Lyon. In a recent paper, Froeschlé and Scholl (1976) confirmed former results obtained by Scholl and Giffen (1974) with respect to a conjecture by Giffen (1973), which is evidently not true.

R. Bien, working on a dissertation at the Rechen-Institut, is treating the 1/1 commensurability of the planar, elliptic restricted problem by the method of averaging. He made a search in a wide range of phase space for orbits with interesting long-period effects in the orbital elements, especially in e and $\tilde{\omega}$. Orbits of Trojan-type appear in his material, but also other orbits that represent a kind of very remote, retrograde satellites of Jupiter.

2.   THE METHOD OF AVERAGING FOR THE THREE-DIMENSIONAL,
     ELLIPTIC, RESTRICTED THREE-BODY PROBLEM

In generalizing the computer program for the planar, elliptic problem (Schubart, 1968) to the three-dimensional case, I retained the basic definitions, units and constants. The reader is referred to the former paper (Schubart, 1968) for details, and especially, for the definition of the averaging process applied to the differential equations of the problem, and for the way of numerical integration. An IBM/360-44 computer was available for the recent computations.

As before, a commensurability case is described by the approximate ratio of the mean motions of an asteroid and Jupiter, given by (p+q)/p, where p and q are relative prime integers, and p > 0. The six variables to be determined from averaged differential equations by integration, are now :

$$G = a^{1/2} (1-e^2)^{1/2}$$

$$\mu = 1-1_J (p+q)/p$$

$$\psi_1 = e \cos\tilde{\omega}$$

$$\psi_2 = e \sin\tilde{\omega}$$

$$\psi_3 = \text{tg} (i/2) \cos \Omega$$

$$\psi_4 = \text{tg} (i/2) \sin \Omega$$

Here, l and l_J are the mean longitudes of the asteroid and Jupiter. a, e, $\tilde{\omega} = \omega + \Omega$, i, and Ω are the usual designations of the osculating elements of the asteroid, but the orbital plane of Jupiter is the plane of reference, and the longitude of the node, Ω , is counted in this plane from the fixed direction of the perihelion of Jupiter. This direction is optional, if the eccentricity of Jupiter, e_J, is neglected. e_J, $a_J = 1$, and $\tilde{\omega}_J = 0^\circ$ describe the orbit of Jupiter.

I omit a listing of the six differential equations that follow from the corresponding equations of the orbital elements, but I want to remark, that comparatively simple equations result for the derivatives of ψ_3 and ψ_4 with respect to t, compare my former treatment of Hill's Problem (Schubart, 1963).

The set of the six variables is not suitable for cases of retrograde motion, if such cases are described by $i > 90^\circ$, and if i is close to 180°. However, if a decreasing mean longitude, or a negative mean motion, $n < 0$, is introduced, such a case can be described by $i < 90^\circ$, and $i = 0^\circ$ is not an exceptional case then. If the sign of $(p+q)$ is changed together with the sign of n, μ will vary slowly as before. The new computer program can integrate many retrograde cases in two ways, either by $i > 90^\circ$, or by a negative mean motion. In the latter case, a negative starting value of G causes $a^{1/2} < 0$ and thus $n = a^{-3/2} < 0$.

Following Poincaré (1902), I had used quantities σ and τ in my studies of the circular problem (Schubart, 1964). They are given by :

$$\sigma = 1 - \tilde{\omega} - (1-l_J) \cdot (p+q)/q = - \tilde{\omega} - \mu \cdot p/q$$

$$\tau = 1 - \Omega - (1-l_J) \cdot (p+q)/q = -\Omega - \mu \cdot p/q$$

The new program can list both $q\sigma$ and $q\tau$, as well as $\omega = \tilde{\omega} - \Omega = \tau - \sigma$ and other quantities, as functions of t. Libration of σ appears in many cases of commensurability, and these librations are an important way for many asteroids and some other objects to avoid close approaches to a disturbing body (see, for instance, Schubart, 1968).

3. SOME SPECIAL CASES OF THE CIRCULAR RESTRICTED PROBLEM

It was one of my first questions to the new program to find out, whether τ can be equally important for an object to avoid close approaches to Jupiter. It is sufficient to consider the circular restricted problem for a first answer to this question. I knew from former studies of nonplanar motion corresponding to the 3/1 case, that libration of τ is possible. If q is even, as in this case, $\psi_1 = \psi_2 = 0$ is a particular solution of the differential equations. This allows a comparatively simple study of the long-range effects in quantities corresponding to i and τ (Schubart, 1964). However, the vanishing eccentricity is sufficient to avoid close approaches, if a is small enough.

TABLE I

Starting values of eight orbits

$-$ The sign of n   equals the sign of $(p+q)/p$ $-$

No.	$(p+q)/p$	e_J	a	e	$\tilde{\omega}$	μ	i	$\mathcal{O}$
1	$-1/1$	0.0	1.0	0.0	$0°.0$	$0°.0$	$30°.0$	$90°.0$
2	$-1/1$	0.0	1.00091140	0.0	0.0	0.0	30.0	90.0
3	$-1/1$	0.0	1.0	0.4	0.0	0.0	0.0	0.0
4	$-1/1$	0.0	1.00340782	0.4	0.0	0.0	0.0	0.0
5	$3/2$	0.048	0.763806	0.14881	261.94	65.75	8.72	221.54
6	$3/2$	0.0	0.762	0.15	0.0	0.0	30.0	270.0
7	$3/2$	0.0	0.76266322	0.15	0.0	0.0	30.0	270.0
8	$3/2$	0.0	0.7625	0.44	0.0	0.0	30.0	270.0

Therefore, I increased a to 1, the value corresponding to Jupiter.
Libration of μ is important for direct motion in the range of the 1/1
commensurability, but I changed to retrograde motion in using p = 1,
q = - 2, and a negative mean motion. The first orbit in Table I has a
starting value τ = - 90°, which will give the asteroid a distance of
90° from both nodes at a moment of conjunction with Jupiter, l = l_J. The
integration of the orbit shows, that τ librates around the starting
value with an amplitude of about 5°, and that the asteroid will not come
closer to Jupiter than 2.69 a.u., on the basis of the averaged circular
problem. In this case the libration of τ prevents a close approach. I
did not study the effects caused by a variation of the starting value
of e in this case, but I varied the one of a. In this way I found
solution No. 2 (see Table I) with values of a, e = 0, i, and τ = - 90°,
which are constant as functions of t.

The next two orbits, No. 3 and 4, belong to the retrograde 1/1
case as well, but the eccentricity is different from zero instead of
the inclination. Libration of σ around 0° causes the small body to be
close to perihelion or aphelion at a conjunction with Jupiter, which
prevents close approaches again. σ librates with an amplitude of 8°.4 in
case of No. 3, while it stays at 0° in the next case. There are probably
no real objects on such retrograde orbits, but the orbits demonstrate
the possibilities given by the program.

4.   APPLICATION TO HILDA-TYPE MOTION

The remaining orbits in Table I belong to the 3/2 commensurability,
especially No. 5 represents a model for the asteroid (153) Hilda. This
model is an extension of my former model for Hilda (Schubart, 1968).
The model is based on numerical results by Akiyama (1962). The angular
elements were transformed to the orbit of Jupiter, which is the plane
of reference. Orbit No. 5 was integrated forward and backward, so that
a total period of about 12 000 yr is covered. This corresponds to more
than 180° of retrograde motion of ☊ , as it is demonstrated at the
bottom of Fig. 1. The moment t = 0 corresponds to the year 1963. Fig. 1
demonstrates the resulting variations of σ with increasing time in
analogy to the corresponding figure for the former model. Since the
interval in t is much larger now, I did not draw a curve, but I plotted
the successive maxima and minima, caused by the period of libration.
The crosses correspond to the maxima. The period of libration equals
275 yr now. The period of perihelion that equals 2650 yr, causes strong
variations in the subsequent maxima, or minima, as it is known from my
former model. The backward revolution of the node which follows a
period of about 22 300 yr, causes only small effects in the variations
of σ, as it appears from Fig. 1. The range of these variations is only
slightly larger than in case of the former model.

The period of revolution of ω, on the average, equals about 3000 yr.
This and other periods cause variations in i and in the speed of motion
of ☊ . However, i remains close to 9°.0. The deviations are less than 0°.4.

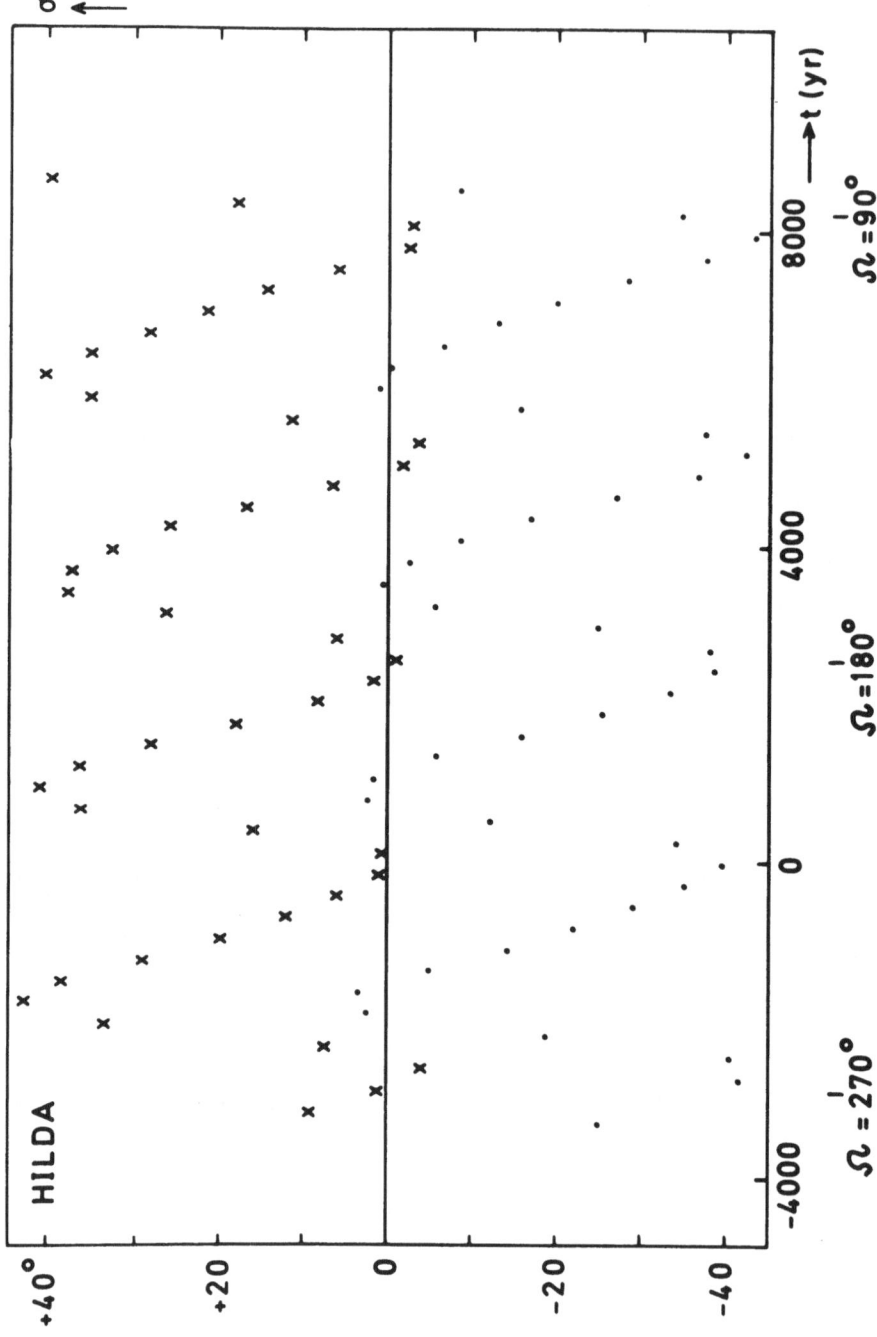

Fig. 1. Variations of σ during 12 000 yr for asteroid (153) Hilda. Instead of a curve, the figure shows the succession of the maxima (crosses) and minima (dots) of σ (t), which are caused by the period of libration. Special marks indicate the motion of Ω .

The node deviates from linear motion by not more than $2^\circ.5$ during the interval considered. As a whole, all the results obtained for this orbit show that a planar model is meaningful in a case like Hilda, but that additional effects are revealed by the three-dimensional model.

I have not yet considered real asteroids with a larger inclination, but I studied some theoretical examples with $i = 30^\circ$ that belong to the 3/2 case of the circular restricted problem. Orbit No. 6, started at $\omega = 90^\circ$, shows a libration of σ around 0°. The period of libration gives an effect in σ, but the period of the retrograde revolution of ω causes a much stronger effect. In case of orbit No. 7 there is almost no influence of the period of libration, but the revolution of ω causes an amplitude of 15° in σ. The variations of e and ω can be demonstrated in rectangular coordinates $\xi = e \cos 2\omega$ and $\eta = e \sin 2\omega$ in this case : The point ξ, η moves with a nearly constant velocity along a nearly circular curve, that has its center on the positive ξ-axis.

Finally, I selected orbit No. 8, because I suspected a libration of ω in this case, according to the information given in a paper by Jefferys and Standish (1972). A limited integration indicates indeed, that both σ and ω librate around 0° and 90°, respectively, with different main periods. The amplitudes are small in both cases. Since $\tau = \sigma + \omega$, τ is in libration as well. According to this twofold libration, an asteroid on this orbit will not come much closer to Jupiter than 3.4 a.u..

REFERENCES

Akiyama,K.: 1962, "Publ. Astron. Soc. Japan" 14, pp. 164-197.

Froeschlé,C. and Scholl,H.: 1976, "Astron. Astrophys." 48, pp. 389-393.

Giffen,R.: 1973, "Astron. Astrophys." 23, pp. 387-403.

Jefferys,W.H. and Standish,E.M.: 1972, "Astron. J." 77, pp. 394-400.

Message,P.J.: 1966, IAU Symposium No. 25, pp. 197-222.

Poincaré,H.: 1902, "Bull. Astron." 19, pp. 289-310.

Scholl,H. and Froeschlé,C.: 1974, "Astron. Astrophys." 33, pp. 455-458.

Scholl,H. and Froeschlé,C.: 1975, "Astron. Astrophys." 42, pp. 457-463.

Scholl,H. and Giffen,R.: 1974, IAU Symposium No. 62, pp. 77-79.

Schubart,J.: 1963, "Math. Annalen" 149, pp. 131-143 = Astron.
            Rechen-Inst. Heidelberg, Mitt. Ser. A Nr. 20.

Schubart,J.: 1964, Smithsonian Astrophys. Obs. Spec. Rept. No. 149.

Schubart,J.: 1966, IAU Symposium No. 25, pp. 187-193.

Schubart,J.: 1968, "Astron. J." 73, pp. 99-103.

A THEORY OF THE TROJAN ASTEROIDS

B. Garfinkel
Yale University

ABSTRACT

The paper constructs a long-periodic solution for the case of 1:1 resonance in the restricted problem of three bodies. The polar coordinates r and θ appear in the form

$$r = r(\lambda) - m|c_k| \cos(k\omega_1 t + \phi)/D$$

$$\theta = \theta(\lambda) + 2m|c_k| \sin(k_1 t + \phi)/D$$

$$\lambda = \lambda(t), \quad D = \omega_2 - k\omega_1.$$

Here λ is the mean synodic longitude, m is the small mass-parameter, k is the integer nearest to the ratio ω_2/ω_1 of the fundamental angular frequencies of the motion, and c_k is a Fourier coefficient of a certain periodic function. Only elementary functions enter $r(\lambda)$ and $\theta(\lambda)$, while the calculation of $\lambda(t)$ requires the inversion of a hyperelliptic integral $t(\lambda)$.

The _internal_ resonant terms, carrying the critical divisor D, impart to the orbit an _epicyclic_ character, in qualitative accord with the results of the numerical integration by Deprit and Henrard (1970). Our solution is valid except in the vicinity of the singularities at D = 0 and λ = 0.

The presence of the resonant terms invalidates the Brown conjecture (1911) regarding the termination of the family of the tadpole-shaped orbits at the Lagrangian point L_3. However, this conjecture holds for the mean orbits defined by $r = r(\lambda)$, $\theta = \theta(\lambda)$, and it also holds in the limit as m → 0.

145

PART III

NUMERICAL AND OTHER TECHNIQUES

STABILIZATION BY MAKING USE OF A GENERALIZED HAMILTONIAN
VARIATIONAL FORMALISM

Joachim W. Baumgarte
Mechanik-Zentrum, Technische Universität Braunschweig
Federal Republic of Germany, and Swiss Federal Institute
of Technology, Zürich, Switzerland

ABSTRACT. A generalized Hamiltonian formalism is established which is
invariant not only under canonical transformations but under arbitrary
transformations. Moreover the dependent variables, coordinates and momenta, as well as the independent variable are allowed to be transformed
This is to say that instead of the physical time t another independent
variable s is used, such that t becomes a dependent variable or, more
precisely, an additional coordinate. The formalism under consideration
permits also to include nonconservative forces.

In case of Keplerian motion we propose to use the eccentric anomaly as the independent variable. By virtue of our generalized point of
view a Lyapunov-stable differential system is obtained, such that all
coordinates, including the time t, are computed by stable procedures.
This stabilization is performed by control terms. As a new result a stabilizing control term also for the time integration is established, such
that no longer any kind of time element is needed. This holds true for
the usual coordinates as well as for the KS-coordinates.

1. INTRODUCTION

In order to obtain a flexible description of the equations of motion of
the two body problem we propose a generalized Hamiltonian variational
formalism with the following properties:

1. The formalism is invariant not only under canonical transformations
   but also under arbitrary transformations of the dependent variables.

2. In order to introduce a new independent variable s (called "fictitious
   time") instead of the time t the extended phase space is adopted.
   Hence the physical time t becomes an additional coordinate q_0 and
   therefore its canonical conjugate momentum p_0 must be introduced.
   This momentum is the negative total energy. The aim of all such transformations is to improve the stability behaviour of the differential
   system as well as to perform an appropriate step-size adaption.

149

V. Szebehely (ed.), Dynamics of Planets and Satellites and Theories of Their Motion, 149-156.

3. Nonconservative forces are also taken into account and their transformation is performed automatically in the variational formalism.

The generalized formalism proposed in this paper is more general than
the classical Hamiltonian theory in analytical dynamics.

2. THE GENERALIZED HAMILTONIAN VARIATIONAL PROBLEM

We introduce the following symbols: q_i generalized coordinates, p_i momenta, $H(p_i, q_i, t)$ Hamiltonian, $P_i(p_k, q_k, t)$ nonconservative forces. The
variational principal is then:

$$\int_{t_1}^{t_2} \left\{ \delta \left[\sum_i p_i \dot{q}_i - H \right] + \sum_i P_i \delta q_i \right\} dt = 0 \tag{1}$$

$$H = H(p_i, q_i, t) , \quad P_i = P_i(p_k, q_k, t), \quad i, k = 1, 2, \ldots, n$$

It leads to the following Euler equations:

$$\dot{q}_i = \frac{\partial H}{\partial p_i} , \quad \dot{p}_i = -\frac{\partial H}{\partial q_i} + P_i , \quad i = 1, 2, \ldots, n \tag{2}$$

The variational problem (1) is invariant under arbitrary noncanonical
transformations of the dependent variables q_i, p_i. The proof of this
statement is not too difficult if Poisson and Lagrange brackets are put
in operation.

We illustrate this statement working on the following example: We
consider an onedimensional perturbed and damped harmonic oscillator.
Equ. (1) is in this case for example:

$$\int_{t_1}^{t_2} \left\{ \delta \left[p\dot{q} - (\frac{p^2}{2m} + \frac{c}{2} q^2 + \varepsilon \frac{b}{4} q^4) \right] - \varepsilon k p \, \delta q \right\} dt = 0 \tag{3}$$

with the mass m, ε as the small (dimensionsless) perturbing parameter
and the constants c,b,k. The equations of motion are:

$$\dot{q} = \frac{p}{m} , \quad \dot{p} = -cq - \varepsilon b q^3 - \varepsilon k p . \tag{4}$$

We now perform in (3) the noncanonical transformation introducing amplitude A and angle ψ :

$$q = A \sin \psi , \quad p = \sqrt{c\,m} \; A \cos \psi. \tag{5}$$

Consequently the variational problem (3) is transformed into

$$\int_{t_1}^{t_2} \{\delta[\sqrt{c}\ m\ A\ \cos\ \psi(\dot{A}\ \sin\psi + \dot{\psi}A\ \cos\ \psi)$$

$$- (\frac{c}{2} A^2 + \varepsilon \frac{b}{4} A^4 \sin^4\ \psi)\ ] \tag{6}$$

$$- \varepsilon k\ \sqrt{c}\ m\ A\ \cos\ \psi(\sin\ \psi\delta A + A\ \cos\psi\delta\psi)\ \}dt = 0$$

The corresponding equations of Euler are

$$\dot{\psi} = \sqrt{\frac{c}{m}} + \frac{\varepsilon\ b}{\sqrt{cm}}\ A^2\ \sin^4\psi + \varepsilon k\ \cos\psi\ \sin\psi$$

$$\dot{A} = - \frac{\varepsilon\ b}{\sqrt{c}\ m}\ A^3\ \sin^3\psi\ \cos\psi - \varepsilon k\ A\ \cos^2\psi \tag{7}$$

and they are the correct equations of motion of the problem at hand.

3.   INTRODUCTION OF A NEW INDEPENDENT VARIABLE

Now we want to introduce a new independent variable s which is linked to the time t by the differential relation

$$\frac{dt}{ds} = t' = \mu > 0, \tag{8}$$

where μ is the scaling function which may depend on all depending variables. In order to incorporate this time transformation into the variational principle the formalism of the extended phase is appropriate (Stiefel and Scheifele [4]). The physical time t becomes a new coordinate q_0 and thus we are forced to introduce its conjugate momentum p_0 into the variational principle. The symbol P_0 will be explained below. Prime means differentiation with respect to the new independent variable s.

Now our variational principle has the form:

$$\int_{s_1}^{s_2} \{\delta\ [p_0q_0' + \sum_i p_iq_i' - \mu(H + P_0)]$$

$$+ \mu P_0\delta\ q_0 + \mu \sum_i P_i\delta\ q_i\}\ ds = 0 \tag{9}$$

$$H = H(p_i, q_i, q_0), \quad P_i = P_i(p_k, q_k, q_0)$$

$$\mu = \mu(p_i, q_i, P_0, q_0) > 0 \qquad i, k = 1, 2, \ldots, n$$

The variational problem (9) leads to the following set of differential equations:

$$q_i' = \mu \frac{\partial H}{\partial p_i} + \{H + p_o\} \frac{\partial \mu}{\partial p_i} \qquad (10a)$$

$$q_o' = \mu + \{H + p_o\} \frac{\partial \mu}{\partial p_o} \qquad (10b)$$

$$p_i' = - \mu \frac{\partial H}{\partial q_i} - \{H + p_o\} \frac{\partial \mu}{\partial q_i} + \mu\, P_i \qquad (10c)$$

$$p_o' = - \mu \frac{\partial H}{\partial q_o} - \{H + p_o\} \frac{\partial \mu}{\partial q_o} + \mu\, P_o \qquad (10d)$$

$$i = 1,2,\ldots,n$$

It is seen that these equations, especially the Equations (10a) and (10c) are only the correct equations of motion provided $\{H + p_o\}$ vanishes on the track. In order to achieve this we do want at first, that $\mu(H + p_o)$ is an integral of motion. This desire is satisfied by putting

$$P_o = - \sum_{j=1}^{n} P_j \frac{\partial H}{\partial p_j} \qquad (11)$$

whereby P_o is the (negative) dissipative power. Secondly we choose as initial condition for p_o : $p_o = - H$, at the instant $t = q_o = s = 0$. Then $\mu(H + p_o)$ as well as $\{H + p_o\}$ vanishes on the track. With these prescription it follows at first that p_o is the negative total energy, secondly, Equ. (10b) reduces (in the exact but not in the computed solution) to

$$q_o' = \mu ,$$

such that the definition on the time transformation

$$\frac{dt}{ds} = t' = \mu > 0 \qquad (8)$$

is included in the differential set (10). Finally Equ. (10d) reads (in the exact but not in the computed solution)

$$p_o' = - \mu \frac{\partial H}{\partial q_o} + \mu\, P_o$$

which is the wellknown equation of energy.

In practise we do not cancel the terms facterized by $\{H + p_o\}$ in the Equations (10) (called control terms) since they may modify the numerical behaviour during a computer integration in particular they may stabilize.

Remark. In the Lagrangian language of mechanics a compagnion principle was published by the author [2]. This principle can be used if the scaling function μ does not depend on the p_i, and the momenta p_i are eliminated directly in the variational problem (9) by the relation

$$q_i' = \mu \frac{\partial H}{\partial p_i} \tag{12}$$

4.   EXAMPLE: KEPLERIAN MOTION

·Let x_i be rectangular coordinates and p_i the corresponding momenta of a particle of unit mass subjected to the gravitational attraction of a central mass in distrance r. The pertinent Hamiltonian is then:

$$H = \frac{1}{2} \sum_i p_i^2 - \frac{K^2}{r} \quad , \quad r^2 = \sum x_i^2 \quad , \quad i = 1,2,3 \tag{13}$$

(K^2: gravitational parameter.)

Let us choose a scaling function

$$\mu = \frac{r}{\sqrt{2p_o}} \tag{14}$$

In this case the fictitious time s is the generalized eccentric anomaly in the sense of Stiefel and Scheifele [4]. It is wellknown that the eccentric anomaly behaves better as independent variable for numerical integration than the time or true anomaly.

Taking into account these assumptions the variational principle (9) becomes for the perturbed motion:

$$\int_{s_1}^{s_2} \{\delta [p_o x_o' + \sum_i p_i x_i' - (\frac{r}{2\sqrt{2p_o}} \sum_i p_i^2 - \frac{K^2}{\sqrt{2p_o}} + \frac{r}{\sqrt{2p_o}} \varepsilon V +$$

$$\sqrt{\frac{p_o}{2}} r)] + \frac{r}{\sqrt{2p_o}} \varepsilon P_o \delta x_o + \frac{r}{\sqrt{2p_o}} \varepsilon \sum_i P_i \delta x_i \} ds = o \tag{15}$$

$$P_o = - \sum_j P_j p_j \qquad\qquad\qquad i,j = 1,2,3$$

Remember that $x_o = q_o = t$ is the physical time, p_o the negative total energy, εP_i are forces, which may not be derivable from a potential, and ε a small (dimensionless) perturbing parameter.
The Euler equations are:

$$x_i' = \frac{r}{\sqrt{2p_o}} p_i \tag{16a}$$

$$
x_o' = \frac{r}{\sqrt{2p_o}} - \{\frac{1}{2} \sum_i p_i^2 - \frac{\kappa^2}{r} + \varepsilon V + p_o\} \frac{r}{\sqrt{2p_o}^3} =
$$

$$
= \frac{r}{2\sqrt{2p_o}} - \left[\frac{r}{2} \sum_i p_i^2 - \kappa^2 + \varepsilon r V\right] \frac{1}{\sqrt{2p_o}^3}
$$

(16b)

$$
p_i' = \frac{r}{\sqrt{2p_o}} (- \frac{\kappa^2}{r^3} x_i - \varepsilon \frac{\partial V}{\partial x_i}) - \{\frac{1}{2} \sum_i p_i^2 - \frac{\kappa^2}{r} + \varepsilon V + p_o\} \frac{x_i}{r\sqrt{2p_o}} +
$$

$$
+ \frac{r}{\sqrt{2p_o}} \varepsilon P_i = - \left(\frac{\sum_i p_i^2}{2\sqrt{2p_o}} + \sqrt{\frac{p_o}{2}}\right) \frac{x_i}{r} - \frac{\varepsilon}{\sqrt{2p_o}} \frac{\partial (rV)}{\partial x_i} + \frac{r}{\sqrt{2p_o}} \varepsilon P_i
$$

(16c)

$$
p_o' = \frac{r}{\sqrt{2p_o}} \varepsilon \left[- \frac{\partial V}{\partial x_o} + P_o\right]
$$

(16d)

The control terms in Equs. (16b) and (16c) are factorized by the same
curly brackets. Remember that these control terms (energy relation)
vanish provided the integration of the system is exact but that they
may modify the behaviour of a numerical integration. In particular we
claim that they stabilize the differential system under consideration.

More precisely we prove in the unperturbed case $\varepsilon = 0$ the following
statement:

Assumtion. The numerical value of the constant energy p_o is considered
as a a priori constant, never varied.

Statement. The system (16a-d) is Lyapunov-stable (for $\varepsilon = 0$) with re-[+)]
spect to variations of the initial conditions of x_i, x_o, p_i.

Proof: We discuss, at first, the time integration. The system (16) has
(in the unperturbed case) the following first integral, where C is an
integration constant:

$$
x_o = \frac{1}{2p_o} \left[\frac{\kappa^2}{\sqrt{2p_o}} s - \sum_i x_i p_i\right] + C
$$

(17a)

It is seen by differentiation and inserting appropriately the Equs.(16).

+) Comments: 1. Remember that the classical differential equations of
   Keplerian motion are unstable. 2. In the perturbed motion p_o is no
   longer constant thus the strict stability (of the stabilized system
   (16)) is lost, but the appearing instability is only of the order of
   magnitude of the perturbing parameter ε (compare the paper of the author [3].

Now we use the basic law of Keplerian motion

$$\sum_i x_i \, p_i = \sqrt{2p_o} \; e \sin E \tag{17b}$$

where $E = (s + \text{const})$ is the eccentric anomaly and e the eccentricity. Equ. (17b) shows that the expression $\sum_i x_i \, p_i$ is pure periodic. From Equ. (17a) it thus follows that variations of the initial conditions of x_i, p_i do influence the time x_o only in pure periodic manner. Therefore x_o is Lyapunov stable.

In order to discuss, secondly, the remaining Equs. (16a), (16c), (16d) we multiply the system (16a-d) by $\sqrt{2p_o}$ and eliminate the momenta p_i. This does not influence the stability behavior. By doing this we obtain:

$$x_i'' - \frac{\sum_j x_j \, x_j'}{r^2} \; x_i' = - \left(\frac{\sum_j x_j'^2}{2r^2} + p_o \right) x_i - r\varepsilon \frac{\partial (rV)}{\partial x_i} + r^2 \varepsilon \, p_i \tag{18a}$$

$$p_o' = - \varepsilon \left[r \frac{\partial V}{\partial x_o} + \sum_j P_j \, x_j' \right] \tag{18b}$$

$$x_o' = \frac{r}{2} - \left[\frac{\sum_j x_j'^2}{2r} - K^2 + \varepsilon r V \right] \frac{1}{2p_o} \tag{18c}$$

$$i,j = 1,2,3$$

In this system [+)] the Equs. (18a) and (18b) are identic with the equations for the x_i and p_o ($p_o = h$) discussed in the reference of the author [1] where the Lyapunov-stability of the equations was proved by the Levi-Civita-transformation.

Furthermore the time integration (18c) is stabilized by a control term. This control term is automatically produced by the choose of the scaling function $\mu = \dfrac{r}{\sqrt{2p_o}}$ instead of $\mu = r$.

We give finally a motivation for our line of approach. In the book Stiefel and Scheifele [4] as well as in the paper of the author [1] the integration of the KS-coordinates as well as usual coordinates x_i with respect to s as the independent variable was stabilized in case of the

———————————————————

+) Now the independent variable, which we call also s, is no longer the generalized eccentric anomaly, but proportional to this anomaly.

time transformation

$$\frac{dt}{ds} = t' = x_o' = r \ .$$

But the integration of t(s) was left unstable. In order to remove this instability these publications introduced a time element (or used a t"-equation). The theory above does not need this detour, but stabilizes all integrations including $t = x_o$.

We now list the corresponding stabilized KS-equation. In vector notation we obtain (with $p_o = h$, $x_o = t$):

$$\underline{u}'' + \frac{h}{2}\,\underline{u} = -\frac{\varepsilon}{4}\,\frac{\partial}{\partial\underline{u}}\,(|\underline{u}|^2 v) + \frac{\varepsilon}{2}\,|\underline{u}|^2\,L^T\,\underline{P} \tag{19a}$$

$$h' = -\varepsilon|\underline{u}|^2\,\frac{\partial V}{\partial t} - 2\varepsilon\,(\underline{u}',\,L^T\,\underline{P}) \tag{19b}$$

$$t' = \frac{1}{2}\,|\underline{u}|^2 - \left[2\,|\underline{u}'|^2 - \kappa^2 + \varepsilon|\underline{u}|^2\,v\right]\frac{1}{2h} \tag{19c}$$

In case of the canonical KS-theory we have to put

$$\overline{\underline{p}} = 4\,\underline{u}' \tag{20}$$

and obtain the canonical first order system, which corresponds to the system (16) multiplied by $\sqrt{2p_o} = \sqrt{2h}$.

ACKNOWLEDGEMENTS

The author acknowledges gratefully the support of the "Deutsche For-schungsgemeinschaft". He is indebted to Prof. Stiefel for stimulating discussions.

REFERENCES

1. Baumgarte, J.: Numerical Stabilization of the Differential Equations of Keplerian Motion, Celes. Mech. 5, 490 - 501, (1972).
2. Baumgarte, J.: Stabilization by Modification of the Lagrangian, Celes. Mech. 13 , 247 - 251, (1976).
3. Baumgarte, J.: Stabilization, Manipulation and Analytic Step Adap-tion, Long-Time Predictions in Dynamics, edited by V. Szebehely and B. D. Tapley, Nato Advanced Study Institutes Series, Series C - Mathematical and Physical Sciences, Vol. 26, D. Reidel Publishing Company, Dordrecht-Holland/Bostqn-U.S.A., (1976).
4. Stiefel, E.L. and Scheifele, G.: Linear and Regular Celestial me-chanics, Springer, Berlin-Heidelberg-New York, (1971).

A SPECIAL PERTURBATION METHOD: m-FOLD RUNGE-KUTTA

D.G. Bettis
The University of Texas at Austin

ABSTRACT

Runge-Kutta methods of order p that utilize m derivatives of the equations of motion are presented. The results of applying these methods are compared to other special perturbation techniques for several test problems.

NUMERICAL INTEGRATION OF NEARLY-HAMILTONIAN SYSTEMS

Victor R. Bond
NASA - JSC
Houston, Texas U.S.A.

ABSTRACT

Consideration is given to the solution by numerical integration of systems of differential equations that are derived from a Hamiltonian function in the extended phase space plus additional forces not included in the Hamiltonian (that is, nearly-Hamiltonian systems). An extended phase space Hamiltonian which vanishes initially will vanish on any solution of the system differential equations. Furthermore, it vanishes in spite of the additional forces, and defines a surface in the extended phase space upon which the solution is constrained.

Direct numerical comparisons are made between (1) nearly-Hamiltonian systems having vanishing Hamiltonians and (2) those having nonvanishing Hamiltonians. It is seen that for some problems, numerical solutions are more stable when computed from systems of the type (1). The problems considered are the harmonic oscillator with the van der Pol perturbation and perturbed Keplerian motion.

1. INTRODUCTION

This paper considers the solution by numerical integration of systems of differential equations that are derived from a Hamiltonian function in the extended phase space plus additional forces that are not included in the Hamiltonian. An extended phase space Hamiltonian which vanishes initially will vanish on any solution of the system differential equations. Furthermore it vanishes in spite of the additional forces, and defines a surface in the extended phase space upon which the solution is constrained.

Nacozy (1971) uses the idea of solutions of differential equations being constrained to lie on surfaces in phase space to develop a formula for corrections that, when added to the numerical solution, cancel the errors made during the preceding integration step, forcing the solution

159

V. Szebehely (ed.), Dynamics of Planets and Satellites and Theories of Their Motion, 159-173.

back to the surface. Baumgarte (1972) suggests for this purpose the use
of a control term that is theoretically zero to be added to the differential equations of Keplerian motion. The control term which is the
initial value of the energy minus the computed value at any step, vanishes on the exact solution and forces the solution back to the surface
when it departs. These techniques require that there be an integral of
the motion that defines a surface in the phase space. In case of nonconservative perturbed motion, these techniques are no longer strictly
valid but may still be useful when the instantaneous surface is only
changing slowly.

In Stiefel and Scheifele (1971) the idea of the Hamiltonian in
extended phase space such that it will vanish on any solution in the
space is discussed. This essentially says that even for nonconservative
problems where no integral exists and which might have additional forces
not included in the Hamiltonian, a surface in phase space may still be
defined provided the new Hamiltonian in the extended phase space vanishes. Several formulations of Keplerian motion have recently appeared
in the literature which are derived from Hamiltonians which vanish in
the extended phase: Scheifele and Graf (1974) introduced Keplerian elements similar to the classical Delaunay elements; and Bond (1976) developed Keplerian elements similar to the classical Poincaré elements.
These formulations use either the eccentric or the true anomaly as the
independent variable. In the last paper numerical integration results
were presented which showed that the elements similar to the Poincaré
elements with eccentric anomaly as the independent variable (henceforth
called Poincaré-Similar elements or PSu elements) showed unusually good
stability.

In order to interpret these numerical results, the approach taken
in this paper is to obtain two formulations - one with a vanishing Hamiltonian in the extended phase space and one with a nonvanishing Hamiltonian in the extended phase space - for a much simpler problem. The problem
chosen is the one-dimensional perturbed harmonic oscillator. The particular example chosen for investigation was the van der Pol equation

$$\ddot{x} + x = \epsilon(1 - x^2)\dot{x} \qquad\qquad (1.1)$$

This equation has an exact asymptotic value for its amplitude, which is
convenient in making error comparisons with numerical results.

This paper will make comparisons between a numerical solution and
a known analytical solution rather than comparing a numerical solution
computed by one method to a numerical solution computed by another method. In addition to the van der Pol problem, known solutions in the
restricted problem of three bodies will be used as examples in perturbed
Keplerian motion. One of these solutions is the stable Lagrangian
libration point in the Earth-Moon system and the other is the unstable
collinear libration point between the Earth and the Moon. The numerically integrated results from the PSu system of elements will be locally
transformed and compared with the known solutions.

2.   THE EXTENDED PHASE SPACE HAMILTONIAN

Discussions on the extended phase space Hamiltonian may be found in the book by Szebehely and in the book by Stiefel and Scheifele. This latter reference also considers the case of additional forces which are not derivable from the Hamiltonian, referring to the additional forces as "canonical forces." The paper by Murdock (1975) refers to Hamiltonian systems of differential equations which are augmented by additional forces as "nearly-Hamiltonian systems." In this paper the terms "additional forces" and "nearly-Hamiltonian systems" will be used.

In Stiefel and Scheifele (1971) the following theorem is proved: Consider the nearly-Hamiltonian system

$$\frac{d\overline{x}_k}{ds} = \frac{\partial \overline{F}}{\partial \overline{p}_k} - \overline{X}_k$$

$$k = 0, 1, 2, \dots n \qquad (2.1)$$

$$\frac{d\overline{p}_k}{ds} = - \frac{\partial \overline{F}}{\partial \overline{x}_k} + \overline{P}_k$$

where $\overline{X}_k$ and $\overline{P}_k$ are the additional forces and where $\overline{F}$ is the Hamiltonian in the extended phase space

$$\overline{F} = \mu(H + p_0)$$

H is the old Hamiltonian; and

$$p_0 = -H$$

$$x_0 = t \text{ (the time)}$$

and the new independent variable s is found from

$$\frac{dt}{ds} = \mu(x_0 \dots x_n; \ p_0 \dots p_n) \qquad (2.2)$$

On any solution of the system (2.1) satisfying the initial conditions

$$x_0(0) = 0; \ s = 0$$

$$(2.3)$$

$$p_0(0) = -H\big(x_0(0), \dots x_n(0); \ p_1(0), \dots p_n(0)\big)$$

the Hamiltonian $\overline{F}$ vanishes for any value of the independent variable s.

The new Hamiltonian $\bar{F}$ is thus equivalent to an integral of the system in the extended phase space. (It should be obvious that this integral is not always invariant under further transformation. In particular if a generating function is chosen which depends upon the independent variable s, then it is possible for the Hamiltonian to lose the property of vanishing.) Since $\bar{F} = 0$ it defines a surface in the phase space upon which the solution remains for all values of s. That is

$$\frac{d\bar{F}}{ds} = \sum_{k=0}^{n} \left[\frac{\partial \bar{F}}{\partial \bar{x}_k} \frac{d\bar{x}_k}{ds} + \frac{\partial \bar{F}}{\partial \bar{p}_k} \frac{d\bar{p}_k}{ds} \right] = 0 \qquad (2.4)$$

or in vector notation, where $z = (\bar{x}_k, \bar{p}_k)$

$$\frac{\partial \bar{F}}{\partial z} \cdot \frac{dz}{ds} = 0 \qquad (2.5)$$

The normal to the surface at any point is defined by the components of $\partial \bar{F}/\partial z$; any change in the coordinates or momenta must lie in the surface.

3. TWO CANONICAL FORMULATIONS OF THE PERTURBED HARMONIC OSCILLATOR

The Hamiltonian for the perturbed oscillator with a perturbing potential V and frequency of one may be written as

$$H = \frac{1}{2} p^2 + \frac{1}{2} x^2 + V(x,t) \qquad (3.1)$$

A perturbing force f which includes perturbations which may not be derivable from a potential may be systematically included in the equations of motion. The equations of motion are

$$\dot{x} = \frac{\partial H}{\partial p} = p \qquad [1](3.2)$$

------

[1]Throughout this paper,

$$(\dot{\ }) = \frac{d(\ )}{dt} \quad \text{and} \quad (\ )' = \frac{d(\ )}{d\tau}$$

$$\dot{p} = -\frac{\partial H}{\partial x} + f = -x - \frac{\partial V}{\partial x} + f \tag{3.3}$$

Now a new Hamiltonian in the extended phase space, H_h, will be developed that will vanish on any solution of the equations of motion. This is done by introducing a new momenta, p_0, which is the negative of the Hamiltonian (eq. 3.1) and is therefore the total energy of the system. The new coordinate x_0 that is canonically conjugate to p_0 is the time, t. The new momenta, p_0, is

$$p_0 = -H \tag{3.4}$$

and the extended phase space (x, p, x_0, p_0) Hamiltonian is

$$H_h = H + p_0 = 0 \tag{3.5}$$

The independent variable may be changed from t to τ according to

$$\mu = \frac{dt}{d\tau} = t' = 1 \qquad {}^1(3.6)$$

In the more general case, μ may be a function of all x_k and p_k. In the present case, the new independent variable τ is related to time by a constant.

Now introduce a new Hamiltonian in the extended phase space by

$$F_h = H_h\mu = \frac{1}{2} p^2 + \frac{1}{2} x^2 + p_0 + V(x,x_0) = 0 \tag{3.7}$$

The system given by the Hamiltonian (3.7) will now be transformed into two new canonical systems. The first new system corresponds to the oscillator solution of the form

$$x = \alpha \sin \beta \tag{3.8}$$

where α is a constant and β is a linear function of time. The second new system corresponds to the equivalent solution,

$$x = a \cos t + b \sin t \tag{3.9}$$

where a and b are constants. It is well known that the last two equations are equivalent solutions to the unperturbed oscillator.

The procedures for developing the two transformations are similar: generating functions will be used to transform from the old canonical variables (x, p, x_0, p_0) to the new sets $(\bar{x}, \bar{p}, \bar{x}_0, \bar{p}_0)$ and $(\tilde{x}, \tilde{p}, \tilde{x}_0, \tilde{p}_0)$. The new Hamiltonians will then be presented. The two transformations are elementary and presented without details.

(A) First New System

The generating function:

$$\bar{S}(x_0, x, \bar{p}_0, \bar{p}) = \int (2\bar{p} - x^2)^{1/2} \, dx + \bar{p}_0 \, x_0 \qquad (3.10)$$

where

$$p = \frac{\partial \bar{S}}{\partial x}, \quad p_0 = \frac{\partial \bar{S}}{\partial x_0}, \quad \bar{x} = \frac{\partial \bar{S}}{\partial \bar{p}}, \quad \bar{x}_0 = \frac{\partial \bar{S}}{\partial \bar{p}_0}$$

The transformation $(x, p, x_0, p_0 \to \bar{x}, \bar{p}, \bar{x}_0, \bar{p}_0)$:

$$x = \sqrt{2\bar{p}} \, \sin \bar{x}$$

$$p = \sqrt{2\bar{p}} \, \cos \bar{x}$$

$$x_0 = \bar{x}_0 \qquad (3.11)$$

$$p_0 = \bar{p}_0$$

The new Hamiltonian:

$$\bar{F}_h = \bar{p} + \bar{p}_0 + V(\bar{x}, \bar{p}, \bar{x}_0) = 0 \qquad (3.12)$$

The differential equations for the elements

$$\bar{x}' = 1 - \frac{x}{2\bar{p}} \left(f - \frac{\partial V}{\partial x} \right)$$

$$\bar{p}' = p \left(f - \frac{\partial V}{\partial x} \right) \qquad (3.13)$$

$$\bar{x}_0' = 1$$

$$\bar{p}_0' = - \frac{\partial V}{\partial x_0} - pf$$

(B) Second New System

The generating function:

$$\widetilde{S}(\widetilde{p}_0,\ p,\ \widetilde{x}_0,\ \widetilde{x},\ \tau) = \frac{1}{\sin \tau}\left[\frac{1}{2}(p^2 + \widetilde{x}^2)\cos \tau - p\widetilde{x}\right] - \widetilde{p}_0\,\widetilde{x}_0$$

(3.14)

where

$$x = -\frac{\partial \widetilde{S}}{\partial p},\quad x_0 = -\frac{\partial \widetilde{S}}{\partial \widetilde{p}_0},\quad \widetilde{p} = -\frac{\partial \widetilde{S}}{\partial \widetilde{x}},\quad \widetilde{p}_0 = -\frac{\partial \widetilde{S}}{\partial \widetilde{x}_0}$$

The transformation $(x,\ p,\ x_0,\ p_0 \rightarrow \widetilde{x},\ \widetilde{p},\ \widetilde{x}_0,\ \widetilde{p}_0)$:

$$x = -\widetilde{p}\cos \tau + \widetilde{x}\sin \tau$$

$$p = \widetilde{p}\sin \tau + \widetilde{x}\cos \tau$$

$$x_0 = \widetilde{x}_0$$

$$p_0 = \widetilde{p}_0$$

(3.15)

The new Hamiltonian:

$$\widetilde{F} = F_h + \frac{\partial \widetilde{S}}{\partial \tau} = \widetilde{p}_0 + V(\widetilde{x},\ \widetilde{p},\ \widetilde{x}_0) \neq 0$$

(3.16)

The differential equations for the elements:

$$\widetilde{x}' = \left(f - \frac{\partial V}{\partial x}\right)\cos \tau$$

$$\widetilde{p}' = \left(f - \frac{\partial V}{\partial x}\right)\sin \tau$$

$$\widetilde{x}_0' = 1$$

$$\widetilde{p}_0' = -\frac{\partial V}{\partial x_0} - pf$$

(3.17)

Note that equations (3.11) and (3.13) have only an implicit dependence on the independent variable τ, whereas equations (3.15) and (3.17) have an explicit dependence on τ. Also, the Hamiltonian equation (3.12) for the system A still vanishes after the transformation, whereas the Hamiltonian equation (3.16) for the system B has lost the vanishing characteristic.

The association of implicit differential equations with a vanishing Hamiltonian in extended phase space (system A) and explicit differential equations with a nonvanishing Hamiltonian in extended phase space (system B) is not entirely accidental. A generating function $\bar{S}$ which does not depend explicitly on τ generates implicit transformations between the new and old variables ($\bar{x} = \bar{x}(x,p)$, $\bar{p} = \bar{p}(x,p)$) in extended phase space and also maintains the vanishing property of the Hamiltonian. A generating function $\tilde{S}$ which does depend explicitly on τ generates explicit transformations ($\tilde{x} = \tilde{x}(x,p,\tau)$, $\tilde{p} = \tilde{p}(x,p,\tau)$) and in general does not maintain the vanishing Hamiltonian property. Since in either case the old variables are eliminated from the right hand side of the differential equations by these transformations, it follows that the implicit differential equations and the vanishing Hamiltonian stem from a generating function which does not depend upon τ. This of course does not imply that implicit differential equations must have a vanishing Hamiltonian.

4. A TEST PROBLEM (VAN DER POL EQUATION)

A van der Pol equation will be used as a test problem for the two Hamiltonian formulations of the perturbed harmonic oscillator. The van der Pol equation in its coordinate form, equation (1.1) is an oscillator with the perturbation.

$$f = \epsilon(1 - x^2)\dot{x} \tag{4.1}$$

where ϵ is a small parameter, the potential V has been set to zero.

For the system (A) where $\bar{F}_h = 0$:

The perturbation, f, is;

$$f = \epsilon p(1 + p^2 + 2\bar{p}_0)$$

The initial conditions ($x = 0$, $p = 2$, when $t = 0$) become;

$$\bar{x} = 0, \; \bar{p} = 2, \; \bar{x}_0 = 0,$$

$$\bar{p}_0 = -2$$

The amplitude, A, of the oscillation is

$$A = \sqrt{2p}$$

where $A \rightarrow 2$ asymptotically.

By averaging in the asymptotic region over one cycle of the equations of motion

Equations (3.13) yield

$$\left\langle \frac{dA}{d\tau} \right\rangle = 0$$

Has no secular increase in amplitude.

For system (B) where $\widetilde{F}_h \neq 0$:

The perturbation, f, is;

$$f = \epsilon p(1 + p^2 + 2\widetilde{p}_0)$$

The initial conditions (x = 0, p = 2, when t = 0) become;

$$\widetilde{x} = 2, \widetilde{p} = 0, \widetilde{x}_0 = 0, \widetilde{p}_0 = -2$$

The amplitude, A, of the oscillation is

$$A = \sqrt{\widetilde{x}^2 + \widetilde{p}^2}$$

where $A \rightarrow 2$ asymptotically.

By averaging in the asymptotic region over one cycle of the equations of motion

Equations (3.17) yield

$$\left\langle \frac{dA}{d\tau} \right\rangle = 0(\epsilon)$$

Has a secular increase in amplitude of the order $0(\epsilon)$.

The averaging over one cycle of equations (3.13), system A, is done with respect to the angle $\overline{x}$ which is permissible since the independent variable τ does not appear explicitly on the right hand sides of the equations. The averaging results in no secular change in amplitude.

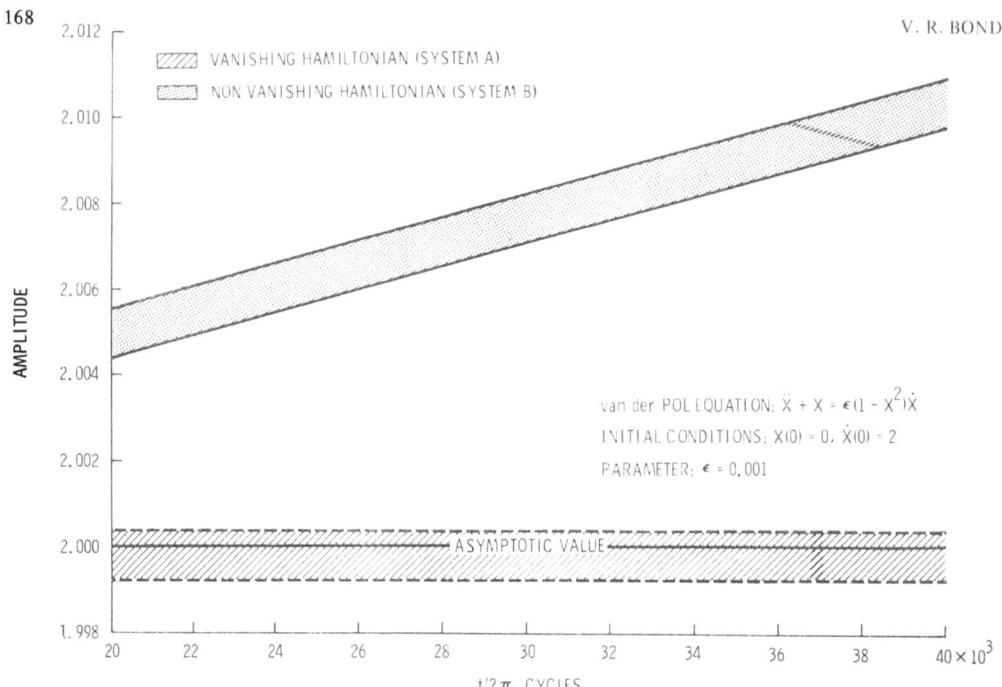

Figure 1.- Comparison of numerically computed solutions of the van der POL
equation (Runge-Kutta 2nd order).

This problem was solved by averaging using a slightly different set
of elements (non-canonical) by Cesari (1970). The averaging of equations
(3.17), system B, is slightly more difficult since the independent
variable appears on the right hand side and the average must be done
over an imprecise period, $2\pi + O(\varepsilon)$. The averaging results in a secular
change of order ε in the amplitude. This is an example of a physically
stable problem which is made mathematically unstable because of the
choice of the variables in which the problem is solved.

Both formulations of the van der Pol problem were numerically
integrated. These results are shown in figures (1) and (2). In figure
(1) the amplitudes were compared to the asymptotic value of two where
the numerical integrator was a second order Runge-Kutta method; figure
(2) shows a similar comparison using third order Runge-Kutta method.
The computations were done using approximately 15 steps per cycle. It
is obvious that the system B solution has linear deviation from the
theoretical asymptotic amplitude of two, whereas the system A solution
is quite stable oscillating slightly about two. For higher order
Runge-Kutta methods, the difference between the two solutions is less
pronounced.

5. TEST PROBLEMS IN KEPLERIAN MOTION

Two of the libration points in the restricted problem of three
bodies are used as test cases for the Keplerian motion formulations.

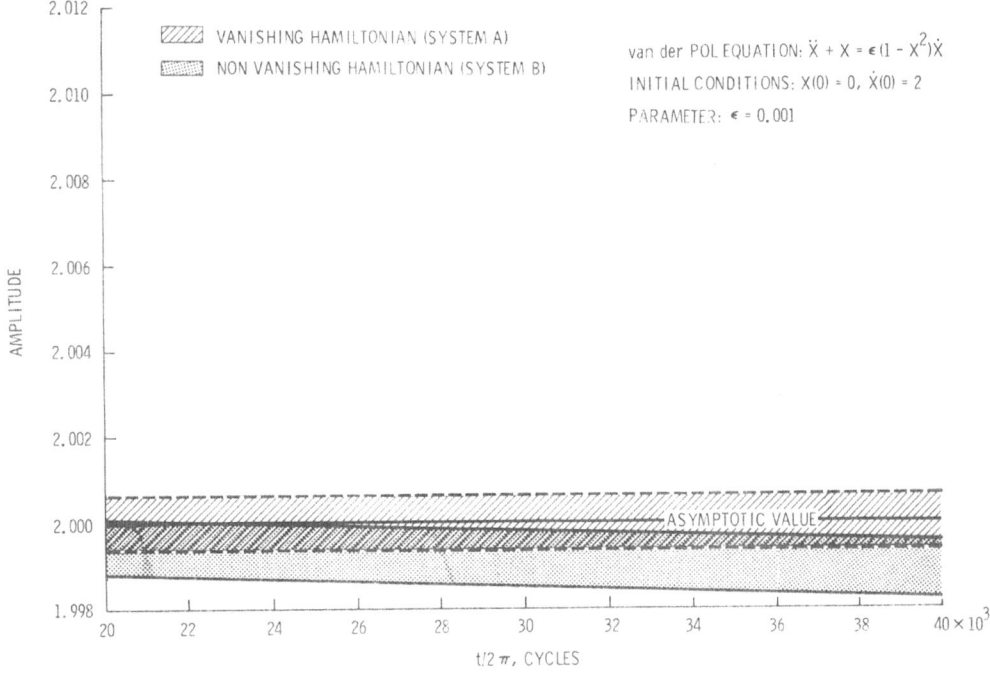

Figure 2. - Comparison of numerically computed solutions of the van der Pol
equation (Runge-Kutta 3rd order).

These solutions are known solutions in the restricted problem and
therefore make excellent test cases for perturbed Keplerian motion.
The numerically integrated results are locally transformed and compared
with the known solutions. The model for the Earth-Moon system used in
this computation was taken from the book by Stiefel and Scheifele. The
model and the initial conditions are provided for those who wish to make
their own comparisons.

Gravitational parameters:

Earth: $GM_E = 398601.0$ KM^3 SEC^{-2}

Moon: $GM_M = 4902.66$ KM^3 SEC^{-2}

Earth-Moon distance: $R_{EM} = 384400.0$ KM

Moon orbital rate: $\Omega = \sqrt{(GM_E + GM_M)/R_{EM}^3}$

The moon is considered to be initially on the x_1 axis.

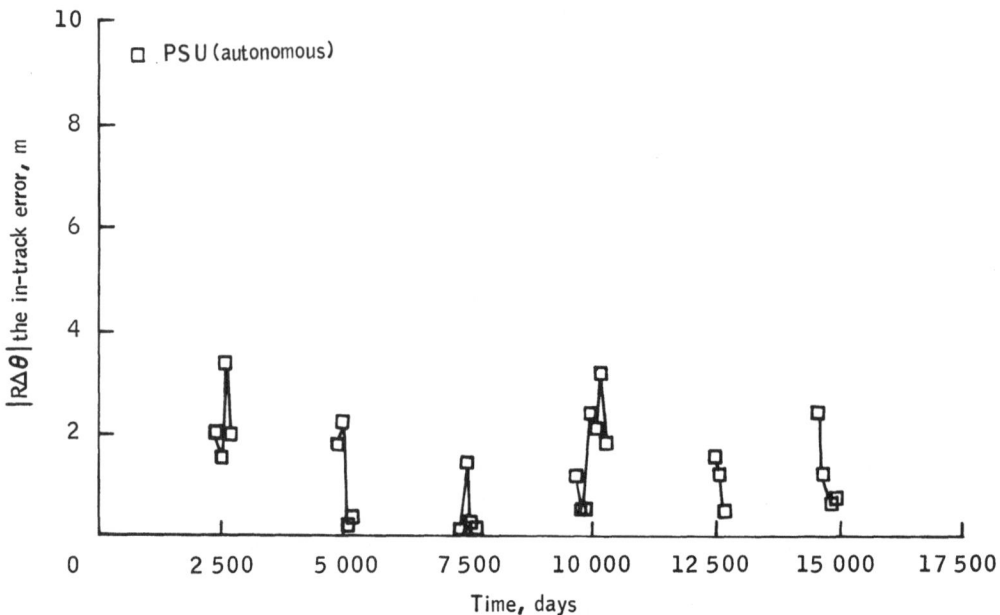

Figure 3.- The in-track error in the motion of a particle about the Lagrangian libration
point in the Earth-Moon system over a period of 15 000 days.

Figure 3 shows the intrack error of the numerical solution versus
the analytical solution for a particle at a stable libration point in
the Earth-Moon system. This test case is given by Bond (1976). The
initial conditions for the trajectory commencing at the stable point
are:

$$x_1 = \frac{R_{EM}}{2} \; ; \; x_2 = \sqrt{3} \, \frac{-R_{EM}}{2} \; ; \; x_3 = 0$$

$$\dot{x}_1 = -\Omega \, x_2 \; ; \; \dot{x}_2 = \Omega \, x_1 \; ; \; \dot{x}_3 = 0$$

The numerical solution was computed from the PSu formulation of perturbed Keplerian motion with a fifth order Runge-Kutta method at 20
steps per revolution. The solution shows no error growth over a period
of 15,000 days. The equations of motion, in PSu elements, were averaged
by quadrature for this case with the result that no secular change
occurs in the elements. This process is rather lengthy and is omitted.
But here as in the case of the system (A) formulation of the van der Pol
problem the numerical solution and the first order analytical solution
yield consistent results.

The next test case is that of problem of a particle at the collinear unstable libration point in the Earth-Moon system located between

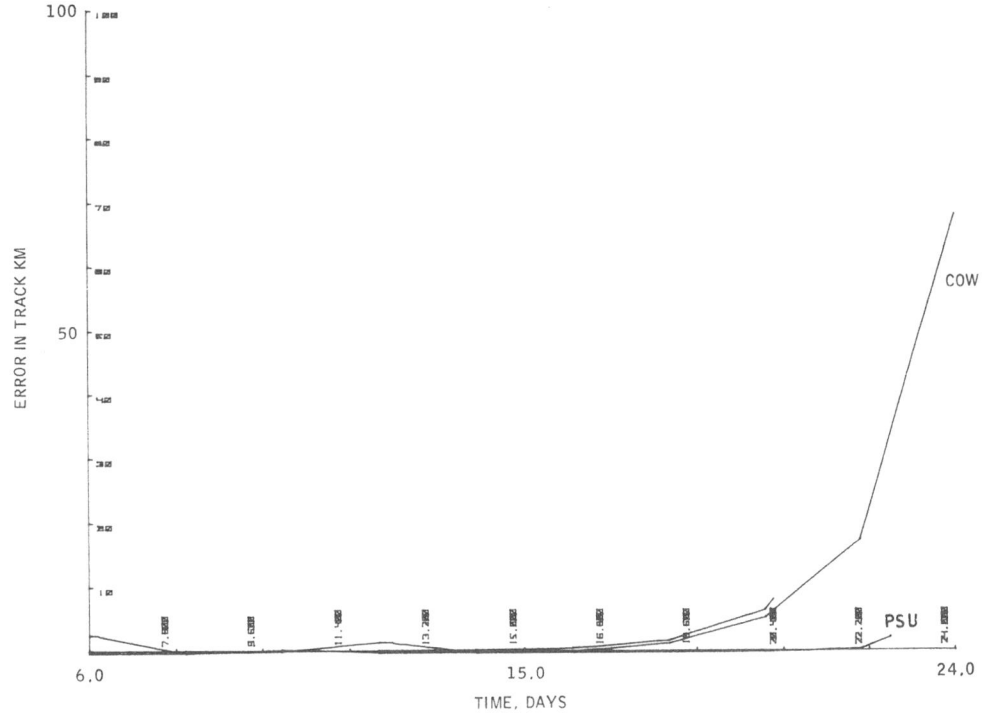

Figure 4.- Numerical solution at L2 RK45 200 SPR.

the Earth and the Moon. For example, after one lunar period (approximately 27 days) from figure 4 the error in the PSu solution was only 1 or 2 KM. The solution computed by the Cowell method had an error of about 70 KM. The integration was done with a Runge-Kutta fifth order method at 200 steps per revolution. The initial conditions for the trajectory commencing at the unstable point are:

$$x_1 = 326381.4038784118380 \text{ KM}; \quad x_2 = 0; \quad x_3 = 0$$

$$\dot{x}_1 = 0, \quad \dot{x}_2 = 0.8699095063452839935 \text{ KM SEC}^{-1}; \quad \dot{x}_3 = 0$$

This problem is extremely sensitive to the initial conditions, which were obtained by iteration using initial conditions obtained from the book by Szebehely. The full 18 digit initial conditions with the above model must be used in order to obtain results which are to compare with figure 4.

6. DISCUSSION

In the numerical solution of ordinary differential equations a distinction must be made between the stability of the physical problem

which is to be solved and the stability of the mathematical formulation
or choice of variables. The physical problem can be dynamically stable,
as in the case of the van der Pol problem, but the mathematical formulation of the problem can be unsuited for the problem and induce instability. For example from section 4.0 the solution from the formulation
of the van der Pol oscillator with a non-vanishing Hamiltonian (system
B) has a secular error growth in the average amplitude, but the solution
from the formulation with the vanishing Hamiltonian (system A) has no
error growth in the average amplitude. These results are the same for
both the analytically averaged and the numerical solutions. From section 5.0, the solution in PSu elements (vanishing Hamiltonian) of a particle at a stable libration point also shows no error growth in both the
analytically averaged and the numerical solution.

It is not surprising that numerical solutions of differential
equations should exhibit the same stability characteristics as their
analytical solutions. From the definition of stability, stable solutions are affected only slightly by small errors in initial conditions
while unstable solutions diverge for small errors in initial conditions.
The numerical integration process introduces errors in the solution at
a given step. Since the results of this step are used to initialize the
next step, the errors in initial conditions are propagated forward in
the same manner as they would be in their analytical solutions.

It is also possible that instabilities may be introduced due to
the numerical method chosen to solve the problem. This paper has not
emphasized this aspect, but the order of the integration method was
shown to have an effect. For example, increasing the order of the
Runge-Kutta method was shown to improve the stability of the numerical
solution of the van de Pol when the system with the nonvanishing
Hamiltonian was integrated, but the increase in the order made no difference in the stability of the numerical solution when the system
with the vanishing Hamiltonian was used. That is, the latter formulation gave a stable solution even for the lowest order integrator.

Some excellent discussions of particular approaches to the problem
of matching the numerical integration method to the mathematical formulation of a physical problem are given by Bettis (1970), Szebehely
and Bettis (1971), Graf and Bettis (1973), Janin (1974), Velez (1974),
Shampine and Gordon (1975), and Graf (1975).

ACKNOWLEDGMENTS

The author wishes to thank Dr. Victor Szebehely of the University
of Texas at Austin for his helpful discussions concerning stability; and
Dr. Dale Bettis of the University of Texas at Austin for providing the
numerical integration programs for these applications.

REFERENCES

Baumgarte, J.: 1972, Celestial Mechanics 5, 490-501.

Bettis, D.: 1970, Celestial Mechanics 2, 282-295.

Bond, V. R.: 1976, Celestial Mechanics 13, 287-311.

Cesari, L.: 1971, Asymptotic Behavior and Stability Problems in Ordinary Differential Equations, Springer-Verlag, New York, Heidelberg, Berlin.

Graf, O. and D. Bettis: 1975, Celestial Mechanics 11, 433-448.

Janin, G.: 1974, Celestial Mechanics 10, 451-467.

Murdock, J.: 1975, Int. J. of Non-Linear Mechanics, 10, 259-270.

Nacozy, P. E.: 1971, Astrophysics and Space Sci. 14, 40-51.

Scheifele, G. and O. Graf: 1974, AIAA Paper No. 74-838.

Shampine, L. F. and M. K. Gordon: 1975, Computer Solution of Ordinary Differential Equations, W. H. Freeman, San Francisco.

Stiefel, E. and G. Scheifele: 1971, Linear and Regular Celestial Mechanics, Springer, Berlin-Heidelberg-New York.

Szebehely, V.: 1967, Theory of Orbits, Academic Press, N. Y.

Szebehely, V. and D. Bettis: 1971, Astrophysics and Space Sci. 13, 365-376.

Velez, C.: 1974, Celestial Mechanics 10, 405-422.

Graf, O.: 1975, Johnson Space Center publication JSC Internal Note No. 75-FM-47.

ON THE SOLUTION OF THE EXTERIOR BOUNDARY VALUE PROBLEM WITH THE AID OF SERIES

M. S. Petrovskaya
Institute for Theoretical Astronomy, Leningrad

ABSTRACT

The exterior gravitational field depending on the Earth's non-sphericity is usually determined from the analysis of satellite data or by the solution of the exterior boundary value problem. In the latter case some integral equations are solved which correlate the exterior potential with the known vector of gravity and the shape of the Earth's surface (molodensky problem). In order to carry out the integration the small parameter method is applied. As a result, all the quantities which involve the equations should be expanded in powers of a certain small parameter, among these being the heights of the Earth's surface points as well as the inclination α of the Earth's physical surface. Since the angle α can be significant, especially in mountains, and in fact does not depend on any small parameter then the solution of integral equations is possible only for the Earth's surface which is smoothed enough.

Different authors expressed the desire to represent the Earth's potential V by a unified mathematical model valid both on the Earth's surface and outside the Earth. The widely accepted form of such a kind is the expansion in spherical harmonics

$$V = fMr^{-1} \sum_{n=0}^{\infty} \sum_{m=0}^{n} \left(\frac{R}{r}\right)^n P_n^{\ m}(\sin\psi) \ [C_{nm} \cos m\lambda + S_{nm} \sin m\lambda] \qquad (1)$$

where f, M, R are the gravitational constants, the Earth's mass and its mean radius and r, ψ, λ represent the spherical coordinates of the point considered. The series (1) is known to converge outside the sphere enclosing the Earth. But for the possibility of applying the terrestrial gravity data in the evaluation of the coefficients C_{nm} and S_{nm} this expansion is extended up to the Earth's surface. The problem of the convergence of the series (1) on the Earth's surface has not yet been solved but even if it does converge another problem emerges whether or not the sum of the series tends to the potential V.

V. Szebehely (ed.), Dynamics of Planets and Satellites and Theories of Their Motion, 175-176.

In the present paper, spherical harmonic expansion is developed which generalizes (1). It converges and represents the potential both on the Earth's surface and in the outer space. While the series (I) is based on the expansion

$$\Delta^{-1} = r^{-1} \sum_{n=0}^{\infty} (\frac{r_1}{r})^n P_n(\cos i) \qquad (2)$$

the new one is derived as a result of the analytical continuation of the series (2) which in the real domain corresponds to

$$\Delta^{-1} = 2 \sum_{n=0}^{\infty} \frac{(4rr_1)^n}{[r + r_1 + |r-r_1|]^{2n+1}} P_n(\cos i)$$

The exact formula in comparison with (1) has some additional terms with the coefficients of different structure than those corresponding to (1), that is

$$V = fMr^{-1} \sum_{n=0}^{\infty} \sum_{m=0}^{n} (\frac{R}{r})^n P_n^m(\sin\psi) [C_{nm}^* \cos m\lambda + S_{nm}^* \sin m\lambda] \qquad (3)$$

where

$$C_{nm}^* = C_{nm} + \Delta C_{nm}(r), \quad S_{nm}^* = S_{nm} + \Delta S_{nm}(r) ,$$

$$\Delta C_{nm}(r) = \sum_{k=0}^{\infty} C_{nm}^{(k)} T_k^*(\frac{R_i}{r}), \quad \Delta S_{nm} = \sum_{k=0}^{\infty} S_{nm}^{(k)} T_k^*(\frac{R_i}{r}) .$$

Here $T_k^*(\kappa)$ are the shifted Chebyshev polynomials and R_i means the shortest distance of the Earth's surface from the origin of coordinates.

The additional terms $\Delta C_{nm}(r)$ and $\Delta S_{nm}(r)$ represent the errors of the model (I) on the Earth's surface. They have the order of the square of the Earth's flattening and vanish outside the enveloping sphere. Thus no satellite observations can in principal provide these quantities. The values of the complements $\Delta C_{nm}(r)$ and $\Delta S_{nm}(r)$ are the larger, the lower is the point on the Earth's surface where the gravity is measured.

The series (3) may be applied for the well-grounded evaluation of the potential through a combination of terrestrial and satellite data.

All the consideration remains obviously true for the potential of any other planet in the solar system.

A NOTE ON THE DEVELOPMENT OF THE RECIPROCAL DISTANCE IN PLANETARY THEORY

Gen-ichiro Hori and Manabu Yuasa
University of Tokyo

ABSTRACT

In a development of the reciprocal distance $(1/\Delta)$ between two planets, their orbital inclinations, in a sense, improve the convergence, though the development is carried out in powers of the inclinations. This is most clearly shown in the Neptune-Pluto case: if their inclinations, 2° and 17°, are assumed to be 0°, the two orbits cross each other. The development in the actual case ought to be easier than in the assumed case with vanishing inclinations. We take advantage of this fact by introducing Δ_0 of the form

$$\Delta_0^2 = r^2 + r'^2 - 2rr'(cc' - ss')^2 \cos(v-v'),$$

such that,

$$\Delta^2 = \Delta_0^2 - 2rr' \{c^2s'^2 \cos(v+v' - 2h') + s^2c'^2 \cos(v+v'-2h)$$
$$+ s^2s'^2 \cos(v-v' -2h+2h') + 2csc's' [\cos(v-v' - h+h')$$
$$- \cos (v+v' - h-h')] + (2csc's' - s^2s'^2) \cos(v-v')\} ,$$

where $c = \cos(i/2)$ and $s = \sin(i/2)$.

On the right-hand side, Δ_0^2 is always larger than $|2rr'\{ \}|$ so far as the two orbits do not cross each other, and $1/\Delta$ is developed with $1/\Delta_0$ as the leading term. When $i' = 0$, Δ_0^2 is reduced to the form introduced by Brown & Shook in the case of asteriodal motion.

PART IV

SATELLITES OF JUPITER AND SATURN, AND ARTIFICIAL SATELLITES

AN APPLICATION OF THE STROBOSCOPIC METHOD

E. A. Roth
ESOC, Darmstadt

ABSTRACT

In this paper the motion of an orbiter of a satellite of one of the major planets is considered. The orbiter undergoes various perturbing effects. It is shown that the semi-analytical stroboscopic method is well suited to take into account all perturbations.

1. INTRODUCTION

The stroboscopic method is a semi-analytical method for orbit computation. The denomination "stroboscopic" is derived from the fact that the osculating orbital elements are only known at <u>one</u> well-defined point of the orbit, usually at perigee. The main advantages are the possibility of an easy inclusion of all types of perturbations, the speed of computation and the fair accuracy obtained even after many hundreds of revolutions. In this paper the essential steps for the application of the method will be given.

The example to be considered here is a spacecraft, the orbiter, moving around one of the large satellites (such as JI to JIV, or SVI, or NI) of a major planet (Jupiter, Saturn or Neptun). Such orbiters are technically feasible within the next decade. For various reasons the pericentre will be low and the eccentricity and the inclination can have arbitrary, large values :

$$0 < e < e^* < 1$$

$$0 < i < \pi$$

The upper limit e^* of the eccentricity is defined by the sphere of influence of the satellite (Roth, 1975).

2. PERTURBATIONS OF AN ORBITER

V. Szebehely (ed.), Dynamics of Planets and Satellites and Theories of Their Motion, 181-188.

An orbiter of a satellite S_n will undergo a number of perturbations which determine the evolution of the orbit and therefore its stability. In the case of a Galilean satellite as central body one has to consider at least the following perturbing forces.

Perturbation	Order of magnitude
Oblateness of the satellite S_n	ϵ
Third-body perturbations by the primary	ϵ
Fourth-body perturbations : Sun	ϵ^2
Satellites S_m (m ≠ n)	ϵ^2
Oblateness of the primary	ϵ^2
Atmospheric drag	ϵ^2

The orders of magnitude of the perturbations give only a rough indication. They depend considerably on the size of the orbit, the satellites and the primary planet.

In the following the perturbations by the oblateness of the primary planet is considered as an example. At the end of the paper a few remarks will be made concerning the other perturbations (section 6).

3. EXPANSION OF THE PERTURBING FUNCTION

In a first step the perturbing potential has to be expanded in an appropriate way. The potential due to the oblateness of the primary planet is, as usual, expanded in terms of Legendre polynomials[1].

$$V' = - \sum_{n=2}^{\infty} \frac{c_n}{d^{n+1}} P_n(\frac{z}{d}) \tag{3.1}$$

For Jupiter only the terms n = 2, 4 and 6 are known to a fair accuracy (Wong, 1975).

It is now necessary to expand the perturbing acceleration

$$\underline{B} = \frac{\partial V'}{\partial \underline{d}}$$

in terms of the elements of the planet and the orbiter using the wellknown expansion for d^{-1} (see fig. 1). The algebra is, however,

_____

(1) The main term (n = 0) of the potential of the primary planet gives rise to the third-body perturbation mentioned in section 2.

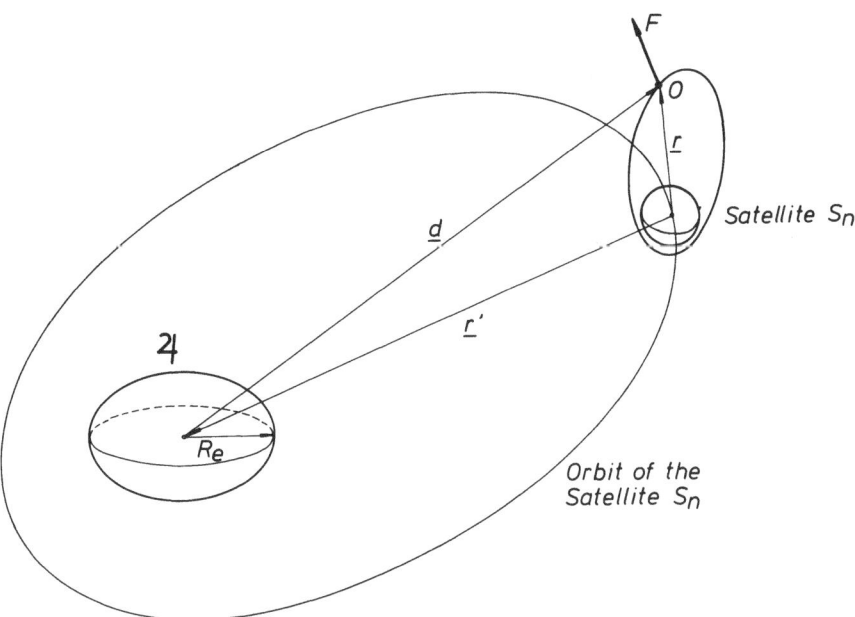

Fig. 1. Orbiter 0 of a satellite S_n.

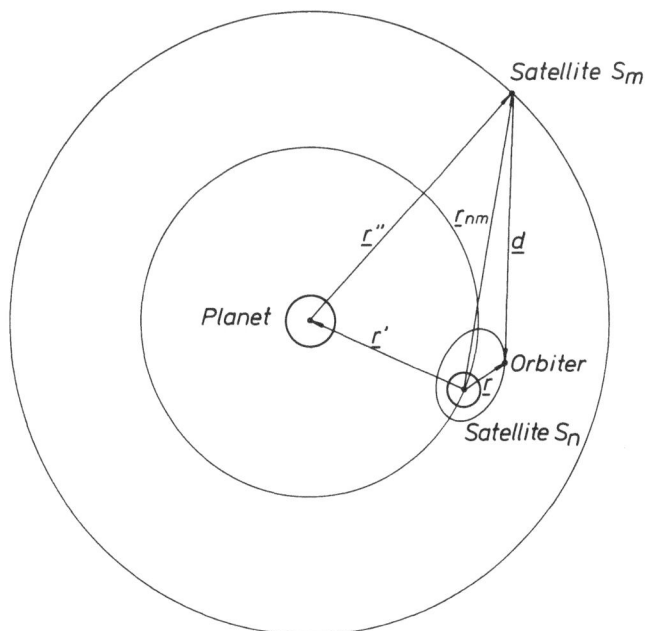

Fig. 2. Geometry of the fourth-body perturbation.

considerably more involved than in the classical case where the
orbiter revolves around the primary planet. It can be shown that the
radial, transversal and binormal component of the acceleration due
to J_n have the form (Roth, 1975)

$$F_i^{(n)} = \frac{\kappa^{(n)}}{p'^{n+2}} \sum_{j=0}^{\infty} \left(\frac{p}{p'}\right)^j \sum_h Y_{ijh}^n(u') T_{ijh}(\theta) \qquad (3.2)$$

where $p = a(1-e^2)$ is the semi-latus rectum of the orbit and u' the
longitude of the planet (primed quantities refer always to the perturbing primary). $Y_{ijh}^n(u')$ are polynomials in sin u' and cos u',

whereas $T_{ijh}(\theta)$ depends on the true anomaly θ of the orbiter. The
acceleration $F_i^{(n)}$ is therefore (in the first approximation) a function of a slow variable u' and a fast variable θ. The important step
is to consider Y(u') as a function of the time t and to expand it in
a Taylor series. This allows to introduce the mean and the eccentric
anomalies M and E of the orbiter. Y(u') becomes (omitting subscripts
for simplicity)

$$Y = \sum_{s=0}^{\infty} \frac{1}{s!} \left(\frac{T}{2\pi}\right)^s Y^{(s)}(t_0) M^s \qquad (3.3)$$

The derivates $Y^{(s)}$ are evaluated on the <u>osculating</u> orbit of the planet
around the satellite (in a system centered at the satellite) and
are therefore easily available. Introducing (3.3) into (3.2) it is
seen that the perturbing acceleration due to J_n becomes a double
sum over j and s

$$\underline{F}^n = \sum_{j=0}^{\infty} \sum_{s=0}^{\infty} \underline{F}_{js}^n \qquad (3.4)$$

with the obvious abbreviation $\underline{F}_{js}^n$.

4. INTEGRATION OF THE LAGRANGEAN EQUATIONS

The Lagrange equations are used in the Gaussian form and can be written as a matrix equation

$$\frac{d\underline{E}}{d\theta} = L \cdot \underline{F} \qquad (4.1)$$

It is essential to use orbital elements $\underline{E}$, e.g. the Keplerian elements (2) and an angle, such as the true anomaly θ, as independent

---

(2) If the time is used instead of a time-element, then a second integration becomes necessary.

variable so that the pericentre is easily defined.

Introducing now the expansion (3.4) into the equation (4.1) it is seen that for each couple (j, s) an equation of the following form is obtained

$$\frac{dE^n_{-js}}{d} = L.F^n_{-js} \tag{4.2}$$

Considering the expressions (3.2) and (3.3) it is seen that the equations (4.2) have essentially the same form for fixed values n and j and varying s. In a first-order perturbation theory the elements are kept constant on the right-hand side of (4.2). It turns out that only integrals of the following type are occuring

$$I(m,n,q,s;\theta) = \int_0^\theta M^s \frac{\sin^m\theta\cos^n\theta}{\Delta^q} d\theta \tag{4.3}$$

with Δ = 1+e cosθ and m+n<q. Numerous recursion formulae can be derived for these integrals (Sridharan, 1973). It is also possible to introduce the eccentric anomaly E as new independent variable.The integrand of (4.3) becomes in this case a finite Poisson polynomial in E which is easy to integrate. Again recursive relations exist which can be used for the evaluation of the integrals (Roth, 1971, 1974). Obviously, these relations are ideally suited for the application of formula manipulation techniques.

5. THE STROBOSCOPIC METHOD

For the application of the stroboscopic method the integrals (4.3) are evaluated between the limits E = 0 and E = 2π. The integration is therefore performed over one revolution from pericentre to pericentre. This has the advantage to lead to a considerable simplification of the final formulae since many of the terms vanish after integration.

Carrying out the integration provides the variation $\Delta\underline{E}$ of the elements after one complete revolution

$$\Delta\underline{E}^n = \sum_j \sum_s \Delta\underline{E}^n_{-js} \tag{5.1}$$

Now, the elements (at pericentre) are updated according to

$$\underline{E} + \Delta\underline{E}^n => \underline{E} \tag{5.2}$$

which provides the new initial values. With these new elements and

the new time t+Δt^n it is possible to calculate new values for the

derivates $Y^{(s)}$ and all other quantities. Performing again the integration the increments $\Delta \underline{E}$ of the next revolution are obtained. This is
repeated until the desired final time is reached.

REMARKS

1. It has some advantage to keep the indefinite integrals (4.3) since
they are needed if the short-period terms have to be recovered. Moreover, they are necessary for the derivation of a second-order perturbation theory.

2. The recursive relations mentioned in section 4 could be used to calculate directly the numerical values of the definite integrals (4.3)
with the upper limit 2π.

3. Updating the elements after each revolution (formula 5.2) implies
that the results are somewhat better than with a pure first-order perturbation theory. In addition the contributions of the long-period
terms are obtained implicitly.

Obviously the method sketched for the oblateness perturbation J_n by

the primary planet can be applied in a similar way to the other per-

turbations $\underline{F}^q$ mentioned in section 2. At each step (each revolution)

all corresponding increments $\Delta \underline{E}^q$ are separately calculated and combined to the total increment

$$\Delta \underline{E} = \sum_q \Delta \underline{E}^q \tag{5.3}$$

It will be necessary to expand the various perturbing forces in the
way explained in section 3. However, the expansion (3.3) is only
needed if the perturbation is non-conservative. In order to be consistent it is necessary to derive the perturbations to the appropriate
order. In the case of forces of the order ε it will be necessary to
use a second-order perturbation theory, at least for the main term
($j = 0$) in an expansion like (3.2).

6. SURVEY OF THE VARIOUS PERTURBATIONS

In the following we present shortly the status concerning the various
perturbations mentioned in section 2.

Oblateness of the Satellite S_n

The increments $\Delta \underline{E}$ of the first order are given by the wellknown secular terms. In some cases it will be necessary to consider also the

second-order term $J_2{}^2$ for which convenient expressions have been given

by Merson (1963).

Third-body perturbation

The first-order perturbation theory of a third body has been developed
by Lidov (1962) and later extended by Roth (1968, 1971). A second-
order perturbation theory for the main term has however not yet been
developed.

Fourth-body perturbation

The geometry of the problem is shown in figure 2. The perturbing body
(the sun, the satellites S_m ($m \neq n$) is moving around the third body

(Jupiter). The distance d depends in this case on the motion of Jupiter,
the satellite S_m and the orbiter. The expansion of the perturbing for-

ce in the satellite-centered system is considerably more involved as in
the third-body case. This expansion is presently under investigation.

Oblateness of the primary planet

The development of the corresponding perturbing function and the derivation of the first-order increments for J_2 and J_4 has been given by

Roth (1975).

Atmospheric drag

The largest satellites of the major planets are known to have an atmosphere. It is therefore necessary to consider the perturbation by
the drag. The orbits under consideration will have a relatively low
pericentre and a large eccentricity. In this case the developments
given by King-Hele (1962, 1964) are very satisfactory (see also Roth,
1970).

7. REFERENCES

King-Hele, D. G. :
-1962, The Contraction of Satellite Orbits under the Influence of
Air Drag. III. High-eccentricity Orbits (o.2$\leq$e<1). Proc. Roy. Soc. ,
A, 267, 541-557

-1964, Theory of Satellite Orbits in an Atmosphere. Butterworth,
London

Lidov, M. L. :
-1962, The Evolution of Artificial Satellites of Planets under the

Action of Gravitational Perturbations of External Bodies, Planet.
Space Science, $\underline{9}$, 719-759

Merson, R. H. :
-1963, The Perturbation of a Satellite Orbit in an axi-symmetric gravitational Field. RAE, TN Space 26,

Roth, E. A. :
-1968, Luni-solar Perturbations of a Highly Eccentric Orbit Satellite,
ESRO SR-9, Paris

-1970, A Method and a Computer Program for the Quick Determination of
a Preliminary Launch-Window for Satellites with Moderately-to-Highly
Eccentric Orbits, ESRO SR-15, Paris

-1971, The Short Periodic Luni-solar Perturbation of the Motion of a
Highly Eccentric Orbit Satellite, ESRO SR-18, Paris

-1973, Fast Computation of High Eccentricity Orbits by the Stroboscopic
Method, Cel. Mech. , $\underline{8}$, 245-249

-1974, Perturbation of a Planetary Orbiter by Radiation Pressure, ESRO
SR-24, Paris

-1975, Perturbation of a Satellite Orbiter by the Oblateness of the
Primary Planet, ESA SR-26, Paris

Sridharan, R. :
-1973, Evolution of the Orbital Elements in Geocentric Orbits of High
Eccentricity by Non-Numeric Computation, Ph. D. Thesis, Carnegie-
Mellon University

Wong, S. K. :
-1975, Gravity Field of Jupiter from Pioneer 11 Tracking Data, Science,
$\underline{188}$, 476-477

NEW FORMULATION OF DE SITTER'S THEORY OF MOTION FOR JUPITER I-IV. I. EQUATIONS OF MOTION AND THE DISTURBING FUNCTION

K. AKSNES

Tokyo Astronomical Observatory,[*] University of Tokyo, Mitaka, Tokyo and Center for Astrophysics, 60 Garden St., Cambridge, Mass. 02138 (Received 19 August, 1976)

ABSTRACT

A brief discussion is given of the basic features of de Sitter's theory. The main advantage of his theory is that it contains no small divisors, thanks to the use of elliptic rather than circular intermediate orbits in the first approximation. A 50-year extension of the satellite observations available to de Sitter makes it desirable to rederive the elements of his intermediate orbits, whose perijoves have a common retrograde motion. Furthermore, the theory suffers from a convergence problem, which can be avoided by reformulating the theory in terms of canonical variables, a task that is begun here. We adopt a formulation in Poincaré's canonical relative coordinates rather than, as customary, in ordinary relative coordinates or in the Jacobian canonical coordinates. By means of the generalized Newcomb operators devised by Izsak, the disturbing function is expanded in a form that is very convenient for use with the modified Delaunay variables, G, $L - G$, $H - G$, $\ell + \omega + \Omega$, ℓ, and Ω and their associated Poincaré variables.

1. INTRODUCTION

In a paper entitled "Outlines of a New Mathematical Theory of Jupiter's Satellites," de Sitter (1918) introduced an entirely new approach to the problem of the motions of the Galilean satellites (Io, Europa, Ganymede, and Callisto). In the many earlier treatments of this problem, whose difficulty arises from the strong mutual attractions, Wargentin, Lagrange, Laplace, Souillart, Sampson, and others adopted circular and coplanar intermediate orbits as a first approximation. The motions of the three inner satellites are characterized by an exact commensurability among their mean motions n_1, n_2, and n_3,

$$n_1 - 3n_2 + 2n_3 = 0 \ ,$$
(1)

and the near-commensurabilities

$$n_1 = 2n_2 + \kappa = 4n_3 + 3\kappa \ ,$$
(2)

[*]Visiting scientist

189

V. Szebehely (ed.), Dynamics of Planets and Satellites and Theories of Their Motion, 189-206.
All Rights Reserved. Copyright © 1978 by D. Reidel Publishing Company, Dordrecht, Holland.

with

$$\kappa = n_1 - 2n_2 = n_2 - 2n_3 \approx \frac{1}{68} n_3 \quad .$$

The first relation results in the famous Laplace libration condition on the mean longitudes λ_1, λ_2, and λ_3,

$$\lambda_1 - 3\lambda_2 + 2\lambda_3 = 180° + \theta \quad , \tag{3}$$

where the libration argument θ has a period of about 6 years and nearly zero amplitude. Equation (2) is, in a sense, more troublesome since it gives rise to small divisors and therefore slowly convergent series for the mean longitudes. de Sitter took advantage of the fact that, in a coordinate system having a prograde rotation with the angular velocity κ, the satellites will have mean motions $c_i = n_i - \kappa$, $i = 1, 2, 3, 4$, the first three of which satisfy the relations

$$c_1 - 3c_2 + 2c_3 = 0 \tag{4}$$

and

$$c_1 = 2c_2 = 4c_3 \quad . \tag{5}$$

Now, (4) corresponds exactly to (1), while in (5), the near-commensurabilities expressed by (2) are turned into exact ones. The circular orbits mentioned are periodic orbits (Poincaré's first kind) for the special initial conditions imposed by (1) and (2). de Sitter discovered that a set of elliptic periodic orbits (Poincaré's second kind) is similarly associated with (4) and (5). The prograde rotation is imparted by giving the four perijoves a common retrograde motion, $-\kappa$, in the fixed coordinate system. de Sitter derived numerical values of the elements of these so-called variation orbits by imposing the necessary periodicities on the equations of motion, with the disturbing function limited to its "secular" and "critical" parts. The resulting particular solution of the problem thus limited turns into a general solution through the addition of the purely periodic "variations," derived by the Lagrangian method of varying the arbitrary constants. Finally, the remainder of the disturbing function gives rise to periodic terms that de Sitter simply called "perturbations," also derived by means of the Lagrangian method.

The greatest advantage of de Sitter's approach is that no small divisors appear at any stage of the solution. Furthermore, the elliptic intermediaries, plus the relatively simple variations, include not only the troublesome long-period terms, which in the earlier theories contain the small divisors, but also the short-period "great inequalities" and the libration. However, de Sitter was disappointed to find that, owing to the presence of an infinite secular determinant, exponential terms of the type $\epsilon_1 \exp(\beta t) + \epsilon_2 \exp(-\beta t)$ appeared in the expressions for the perturbations in longitude, and although ϵ_1 and ϵ_2 are very small, this term will ultimately cause divergence. In his Darwin Lecture, de Sitter (1931) discussed this problem and announced the future publication of the complete expressions for the perturbations beyond the first order. Probably because of a subsequent sickness that caused his death in 1934, these expressions were not completed; at least they never appeared in print. This incompleteness of the theory and the lack of convenient tables for ephemeris calculation, such as those published by Sampson (1910), have prevented practical applications of the theory, a very unfortunate circumstance since recent results (Aksnes and Franklin, 1975) indicate that de Sitter's theory, as far as it can be applied, is at least as accurate as that due to Sampson, although both are now in need of revision (Lieske, 1975; Aksnes and Franklin, 1976).

The reasons for undertaking a new formulation of de Sitter's theory can be summarized as follows. First, it is desirable to derive new values for the elements of the variation orbits for a current epoch, on the basis of an almost 50-year extension of the satellite observations available to de Sitter. Of particular interest are the series of plates of the Galilean satellites taken by D. Pascu with the Leander-McCormick refractor during the last decade and the highly accurate photometric observations of the mutual satellite events in 1973. Even more accurate observations, in the form of range or range-rate data on the satellites, can be expected in the near future from an on-going experiment with the Arecibo radio telescope. It is vital that the orbital elements be as precise as possible since they enter the theory in numerical form and cannot be changed subsequently without redoing the theory. Second, de Sitter claimed that the afore-mentioned exponential terms can be avoided by using Delaunay's or von Zeipel's perturbation method in terms of canonical elements, instead of the Lagrangian method in terms of Kepler elements. Rather than adopting either of the two first-mentioned methods, we propose to use the more elegant canonical method due to Hori (1966), which is based on Lie series. A canonical formulation has the added advantage of simplifying the equations of motion and the construction of the theory, provided the disturbing function is expanded in an appropriate way.

Thus, the goal of our undertaking is not to revise de Sitter's theory in its original form, but to construct a new theory that will incorporate only the most essential features of the old one. In the two remaining sections, we present the first part of this work on the equations of motion and the expansion of the disturbing function, to be followed by later parts on the derivation of the variation orbits, the variations, and the perturbations, in de Sitter's terminology.

2. THE EQUATIONS OF MOTION

To apply Hori's perturbation method, we need a canonical formulation of the equations of motion with a common Hamiltonian. A formulation of this kind due to Jacobi has been widely used in investigations of the three-body problem, and Marsden (1964) adopted it in his thesis on the short-period terms in the motions of the Galilean satellites. Jacobi's method amounts to choosing a different origin for the coordinates of each body, such that the second body is referred to the first and each succeeding body is referred to the center of mass of all the preceding ones. Unfortunately, the use of the Jacobian coordinates complicates the expansion of the disturbing function considerably. There is a simpler canonical formulation[*] due to Poincaré (1897) and advocated by Charlier (1902) for use on the three-body problem, although they did not attempt to apply the method. In the following adaptation of the method to the problem at hand, we shall use a notation similar to that introduced by Marsden.

---

[*]I am indebted to Dr. Hori for pointing out the existence of this formulation.

Let m_p ($p = 0, 1, \ldots, n$) be the masses and ξ_p, η_p, and ζ_p be the cartesian coordinates, referred to the center of mass of the system, of $n + 1$ interacting bodies. We take m_0 (Jupiter) to be the central mass to which we wish to refer the coordinates x_p, y_p, z_p ($p = 1, 2, \ldots, n$) of the remaining masses (satellites plus perturbing bodies):

$$x_p = \xi_p - \xi_0 \, , \qquad y_p = \eta_p - \eta_0 \, , \qquad z_p = \zeta_p - \zeta_0 \, . \qquad (6)$$

It is well known that if $X_p = m_p \dot{x}_p$, $Y_p = m_p \dot{y}_p$, and $Z_p = m_p \dot{z}_p$ are taken as the momenta conjugate to the coordinates x_p, y_p, and z_p, only a semi-canonical formulation is achieved in which each body has its own Hamiltonian. If, instead, we define the momenta by

$$X_p = \frac{\partial T}{\partial \dot{x}_p} \, , \qquad Y_p = \frac{\partial T}{\partial \dot{y}_p} \, , \qquad Z_p = \frac{\partial T}{\partial \dot{z}_p} \, , \qquad (7)$$

where, in terms of the inertial velocities,

$$T = \frac{1}{2} \sum_{q=0}^{n} m_q (\dot{\xi}_q^2 + \dot{\eta}_q^2 + \dot{\zeta}_q^2) \, , \qquad (8)$$

we have Hamilton's canonical equations,

$$\left. \begin{array}{lll} \dfrac{dX_p}{dt} = \dfrac{\partial F}{\partial x_p} \, , & \dfrac{dY_p}{dt} = \dfrac{\partial F}{\partial y_p} \, , & \dfrac{dZ_p}{dt} = \dfrac{\partial F}{\partial z_p} \\[3mm] \dfrac{dx_p}{dt} = -\dfrac{\partial F}{\partial X_p} \, , & \dfrac{dy_p}{dt} = -\dfrac{\partial F}{\partial Y_p} \, , & \dfrac{dz_p}{dt} = -\dfrac{\partial F}{\partial Z_p} \end{array} \right\} \quad p = 1, 2, \ldots, n \, , \qquad (9)$$

where the Hamiltonian F represents (the negative of) the total energy of the system,

$$F = -T + k^2 \sum_{q=1}^{n} \sum_{r=0}^{q-1} \frac{m_q m_r}{r_{qr}} \, , \qquad (10)$$

k^2 being the constant of gravitation and $r_{qr}^2 = (x_q - x_r)^2 + (y_q - y_r)^2 + (z_q - z_r)^2$. In order to derive explicit expressions for the momenta from (7), we must express the kinetic energy T in terms of the relative velocities. In deriving the transformation from the inertial frame to the relative frame, we consider only the x components, with the understanding that the y and z components transform in the same way. By means of (6) and the relation

$$\sum_{q=0}^{n} m_q \xi_q = 0 \, ,$$

we readily deduce that

$$\xi_0 = -\frac{1}{M} \sum_{q=1}^{n} m_q x_q \quad,$$

$$\xi_p = x_p - \frac{1}{M} \sum_{q=1}^{n} m_q x_q \quad, \qquad p = 1, 2, \ldots, n \quad, \tag{11}$$

where M is the total mass of the system. If we differentiate the last two equations, and the corresponding ones for the η and ζ components, with respect to t and substitute the result in (8), we find, after some straightforward manipulation,

$$T = \frac{1}{2} \sum_{q=1}^{n} m_q (\dot{x}_q^2 + \dot{y}_q^2 + \dot{z}_q^2) - \frac{1}{2M} \left[\left(\sum_{q=1}^{n} m_q \dot{x}_q \right)^2 + \left(\sum_{q=1}^{n} m_q \dot{y}_q \right)^2 \right.$$

$$\left. + \left(\sum_{q=1}^{n} m_q \dot{z}_q \right)^2 \right] = \frac{1}{2M} \left[\sum_{q=1}^{n} m_q (M - m_q)(\dot{x}_q^2 + \dot{y}_q^2 + \dot{z}_q^2) \right.$$

$$\left. - 2 \sum_{q=1}^{n} \sum_{r=1}^{q-1} m_q m_r (\dot{x}_q \dot{x}_r + \dot{y}_q \dot{y}_r + \dot{z}_q \dot{z}_r) \right] \tag{12}$$

From (7), (11), and (12), it then follows that

$$X_p = m_p \dot{x}_p - \frac{m_p}{M} \sum_{q=1}^{n} m_q \dot{x}_q = m_p \dot{\xi}_p \quad ; \qquad p = 1, 2, \ldots, n \quad, \tag{13}$$

i.e., in Poincaré's canonical formulation, the momenta conjugate to the relative coordinates are related to the inertial velocities in the same way that they are related to the relative velocities in the semicanonical formulation. If we again make use of (6) and the relation

$$m_0 \dot{\xi}_0 = - \sum_{q=1}^{n} m_q \dot{\xi}_q = - \sum_{q=1}^{n} X_p \quad,$$

the desired form of T becomes

$$T = \frac{1}{2} \left[\sum_{q=1}^{n} \frac{1}{m_q} (X_q^2 + Y_q^2 + Z_q^2) + \frac{1}{m_0} \left\{ \left(\sum_{q=1}^{n} X_q \right)^2 + \left(\sum_{q=1}^{n} Y_q \right)^2 + \left(\sum_{q=1}^{n} Z_q \right)^2 \right\} \right]$$

$$= \frac{1}{2} \sum_{q=1}^{n} \frac{1}{m_q^*} (X_q^2 + Y_q^2 + Z_q^2) + \frac{1}{m_0} \sum_{q=2}^{n} \sum_{r=1}^{q-1} (X_q X_r + Y_q Y_r + Z_q Z_r) \quad, \tag{14}$$

where

$$\frac{1}{m_p^*} = \frac{1}{m_p} + \frac{1}{m_0} \quad, \qquad p = 1, 2, \ldots, n \quad. \tag{15}$$

If m_0 is much larger than m_p ($p = 1, 2, \ldots, n$) or if the latter masses do not come very close to each other, we can obtain an approximate solution by neglecting the last term in (14) and by limiting the potential to its main part with $r = 0$ in (10). The resulting intermediate orbits are thus obtained with a Hamiltonian,

$$F_0 = \sum_{q=1}^{n} \left[-\frac{1}{2m_q^*} (X_q^2 + Y_q^2 + Z_q^2) + \beta_q^2 \frac{m_q^*}{r_q} \right] , \tag{16}$$

where β_q^2 is a constant to be suitably chosen, and $r_q^2 = r_{q0}^2 = x_q^2 + y_q^2 + z_q^2$. With $F = F_0$, it follows from (9) that

$$\dot{x}_p = \frac{X_p}{m_p^*} \tag{17}$$

and

$$\ddot{x}_p = \frac{1}{m_p^*} \frac{\partial F_0}{\partial x_p} = -\beta_p^2 \frac{x_p}{r_p^3} ,$$

i.e., elliptic motion. Hence, there exists an energy integral,

$$\frac{1}{2} (\dot{x}_p^2 + \dot{y}_p^2 + \dot{z}_p^2) = \beta_p^2 \left(\frac{1}{r_p} - \frac{1}{2a_p} \right) , \tag{18}$$

where a_p is the semimajor axis, by means of which (16) can be written as

$$F_0 = \sum_{q=1}^{n} \frac{\beta_q^2 m_q^*}{2a_q} . \tag{19}$$

It would seem natural to put $\beta_p^2 = k^2(m_0 + m_p)$ such that, in view of (10), (15), and (16), F_0 would absorb the entire $1/r_q$ part of the potential. The mean motion n_p would then be related to a_p precisely as for two-body motion, viz.

$$n_p = \beta_p a_p^{-3/2}. \tag{20}$$

However, to satisfy the periodicity requirements of the variation orbits, it is necessary to take

$$\beta_p^2 (1 - \mu_p) = k^2(m_0 + m_p) = k^2 \frac{m_0 m_p}{m_p^*} , \tag{21}$$

where μ_p (de Sitter, 1918) is a small constant to be determined together with the elements of the variation orbits. The perturbing Hamiltonian, $F_1 = F - F_0$, then becomes

$$F_1 = -\sum_{q=1}^{n} \mu_q \beta_q^2 \frac{m_q^*}{r_q} + \sum_{q=2}^{n} \sum_{r=1}^{q-1} \left[\frac{k^2 m_q m_r}{r_{qr}} - \frac{1}{m_0} (X_q X_r + Y_q Y_r + Z_q Z_r) \right] . \tag{22}$$

It will be seen that $F/m_p^* = (F_0 + F_1)/m_p^*$ can be regarded as the pth body's Hamiltonian (we can neglect all terms not dependent on the elements of this body) to which corresponds a set of canonical variables; e.g., the Delaunay set $L_p = \beta_p \sqrt{a_p}$, $G_p = L_p \sqrt{1 - e_p^2}$, $H_p = G_p \cos I_p$, ℓ_p, ω_p, and Ω_p. Here, e_p is the eccentricity, I_p the inclination, ℓ_p the mean anomaly, ω_p the argument of the pericenter, and Ω_p the longitude of the ascending node. It follows that the combined, modified Delaunay set,

$$\left\{ \begin{matrix} L_p = m_p^* \beta_p \sqrt{a_p} & , & G_p = L_p \sqrt{1 - e_p^2} & , & H_p = G_p \cos I_p \\ \ell_p & , & \omega_p & & \Omega_p \end{matrix} \right\}_{p=1,2,\ldots,n} \tag{23}$$

obeys the canonical equations,

$$\left\{ \begin{matrix} \dfrac{dL_p}{dt} = \dfrac{\partial F}{\partial \ell_p} & , & \dfrac{dG_p}{dt} = \dfrac{\partial F}{\partial \omega_p} & , & \dfrac{dH_p}{dt} = \dfrac{\partial F}{\partial \Omega_p} \\[2ex] \dfrac{d\ell_p}{dt} = -\dfrac{\partial F}{\partial L_p} & , & \dfrac{d\omega_p}{dt} = -\dfrac{\partial F}{\partial G_p} & , & \dfrac{d\Omega_p}{dt} = -\dfrac{\partial F}{\partial H_p} \end{matrix} \right\} \begin{matrix} p = 1,2,\ldots,n \end{matrix} , \tag{24}$$

with the common Hamiltonian

$$F = F_0 + F_1 , \tag{25}$$

given by (19) and (22).

By the principle of variation of arbitrary constants, the elliptic formulas relating the Delaunay variables to the positions and velocities for unperturbed motion also hold for perturbed motion. However, in the latter case, we have to make an important distinction. Whereas the position x_p in the perturbed intermediary must be equal to the true relative position, say x_{pt},

$$x_{pt} = x_p , \qquad p = 1,2,\ldots,n , \tag{26}$$

the same will not be true of the velocities $\dot{x}_p$ and $\dot{x}_{pt}$. This can be seen as follows. From (13), in which $\dot{x}_p$ and $\dot{x}_q$ must now be replaced by $\dot{x}_{pt}$ and $\dot{x}_{qt}$, we have that $\dot{x}_{qt} = \dot{x}_{pt} - X_p/m_p + X_q/m_q$. If this expression is substituted back into (13), there results

$$\frac{M}{m_p} X_p = M\dot{x}_{pt} - (M - m_0)\left(\dot{x}_{pt} - \frac{X_p}{m_p}\right) - \sum_{q=1}^{n} X_q ,$$

or, by means of (17),

$$\dot{x}_{pt} = \frac{X_p}{m_p} + \frac{1}{m_0} \sum_{q=1}^{n} X_q = \dot{x}_p + \sum_{\substack{q=1 \\ q \neq p}}^{n} \frac{m_q^*}{m_q} \dot{x}_q , \qquad p = 1,2,\ldots,n . \tag{27}$$

The intermediaries are therefore not osculating orbits, but this matters very little. The only difference from the use of osculating orbits is that, at the very end, if we wish to compute the true relative velocities (which are usually not needed anyway), they must be obtained from (27). The velocity-dependent indirect term under the double summation sign in (22) can be expanded, as we shall see, perhaps even more easily than can the corresponding indirect term in de Sitter's semicanonical formulation, and the direct terms are identical, apart from an extra mass factor in (22). Note that, while the present formulation is closely related to de Sitter's formulation, the two do not lead to identical intermediaries.

To take advantage of the fact that the eccentricities and inclinations are very small, we shall introduce the modified Delaunay set

$$
\left\{
\begin{array}{l}
G_p = m_p^* \beta_p \sqrt{a_p(1 - e_p^2)} \ , \quad M_p = m_p^* \beta_p \sqrt{a_p}(1 - \sqrt{1 - e_p^2}) \ , \quad N_p = G_p(\cos I_p - 1) \\
\lambda_p = \ell_p + \omega_p + \Omega_p \ , \qquad\qquad\qquad \ell_p \qquad\qquad , \qquad\qquad \Omega_p
\end{array}
\right\}
$$

(28)

but since e_p and I_p may pass through zero, it is convenient also to make use of the following set, in which the conjugate pairs (M_p, ℓ_p) and (N_p, Ω_p) are replaced by the associated Poincaré variables (p_p, q_p) and (u_p, v_p) given by

$$
\left\{
\begin{array}{lll}
G_p \ , & p_p = \sqrt{2M_p} \cos \ell_p \ , & u_p = \sqrt{-2N_p} \cos \Omega_p \\
\lambda_p \ , & q_p = \sqrt{2M_p} \sin \ell_p \ , & v_p = -\sqrt{-2N_p} \sin \Omega_p
\end{array}
\right\} .
$$

(29)

Finally, closely related to the set (28) is the Hill set,

$$
\left\{
\begin{array}{lll}
G_p & , \ \dot{r}_p \ , & N_p \\
f_p + \omega_p + \Omega_p & , \ r_p \ , & \Omega_p
\end{array}
\right\} ,
$$

(30)

where f_p is the true anomaly. The Hill variables may prove useful for obtaining the perturbations in r_p and $\dot{r}_p$ which are of interest if range and range-rate observations of the satellites become available. The canonical equations for these three sets of variables can be written down immediately from (24) by replacing the conjugate pairs of variables there by the appropriate new ones. The sets (28) and (29) are particularly convenient if we expand the disturbing function in powers of the auxiliary eccentricities ϵ_p and the auxiliary inclinations γ_p, defined by,

$$
\left\{
\begin{array}{l}
\epsilon_p = \sqrt{2M_p/G_p} = \sqrt{2(1 - e_p^2)^{-1/2} - 2} = e_p + 0(e_p^3) \ , \\
\gamma_p = \sqrt{-2N_p/G_p} = 2 \sin \tfrac{1}{2} I_p = I_p + 0(I_p^3) \ ,
\end{array}
\right\}
$$

(31)

rather than, as usual, in powers of e_p and I_p (or sin I_p), whose derivatives with respect to G_p, M_p, N_p, and the Poincaré variables are rather cumbersome. de Sitter pointed out the advantages of the set (28) over the noncanonical Kepler variables, which he introduced only to be able to make easy use of the existing expansions of the disturbing function in powers of e_p and I_p. The new expansion proposed here is greatly facilitated today by utilizing an algebra program on an electronic computer. In place of (28), Marsden (1964) adopted the canonical set

$$\begin{cases} L_p = m_p^* \beta_p \sqrt{a_p} \ , \quad M_p = L_p\left(\sqrt{1-e_p^2}-1\right) \ , \quad N_p = L_p\sqrt{1-e_p^2}(\cos I_p - 1) \\ \lambda_p = \ell_p + \omega_p + \Omega_p \ , \quad \bar{\omega}_p = \omega_p + \Omega_p \ , \quad \Omega_p \end{cases} \ , \tag{32}$$

which has the advantage that it contains only one "fast" variable, λ_p. However, corresponding to (31), we now have

$$\epsilon_p = \sqrt{-2M_p/L_p} = \sqrt{2 - 2(1 - e_p^2)^{1/2}} \ ,$$

$$\tag{33}$$

$$\gamma_p = \sqrt{-2N_p/L_p} = (1 - e_p^2)^{1/4} \ 2 \sin\frac{I}{2}p \ ,$$

and e_p in the last equation introduces a considerable complication, since $\sin I_p/2$ occurs quite naturally by itself in the disturbing function. We note that with the von Zeipel method used by Marsden, it is difficult to handle more than one fast variable at once, but this is not so with Hori's method.

In the remainder of this section, we introduce a very convenient formulation due to Marsden. Since Jupiter's (m_0's) oblateness has a pronounced effect on the Galilean satellites (m_1, m_2, m_3, m_4), it is necessary to add to F_1 the potential

$$F_{1J} = -k^2 m_0 \sum_{q=1}^{4} \left[J_2 R_0^2 \ \frac{m_q}{r_q^3} \ P_2(\sin \phi_q) + J_4 R_0^4 \ \frac{m_q}{r_q^5} \ P_4(\sin \phi_q) \right] \ , \tag{34}$$

where R_0 is the equatorial radius and J_2 and J_4 are the dynamical form factors of Jupiter, ϕ_q is the latitude of m_q on Jupiter's equator, and P_2 and P_4 are the second and fourth Legendre polynomials. Note that in this potential, unlike in the corresponding one in de Sitter's formulation, there is no interaction between the oblateness and the indirect part of F_1. There is a converse effect of the satellites on the motion of Jupiter's equator. This can be described by allowing the index p to take on also the value zero in the expressions (28) for N_p and Ω_p, defining Ω_0 and I_0 to be, respectively, the longitude of the ascending node and inclination of Jupiter's equator on the fixed reference plane, e_0 to be zero, and

$$G_0 = Cn_0 = \text{constant} \tag{35}$$

to be the angular momentum of Jupiter about its polar axis, n_0 being the angular velocity of rotation and C the moment of inertia about this axis. With these definitions, putting

$$\alpha = \sin I_0 \sin \Omega_0 \ , \quad \beta = -\sin I_0 \cos \Omega_0 \ , \quad \gamma = \cos I_0 \ , \tag{36}$$

$\sin \phi_q$ in (34) can be written

$$\sin \phi_q = \frac{1}{r_q}(\alpha x_q + \beta y_q + \gamma z_q) \ . \tag{37}$$

Finally, if we define m_5 to be the Sun, its attraction can be included by taking n = 5 in (19) and (22). It is sufficient to adopt a fixed ellipse for the Sun's motion about Jupiter, such that for p = 5 in (32) all the variables are constants, except for λ_5, which is a linear function of time. (The use of the set (28) would lead to two time-dependent variables, ℓ_5 and λ_5.) It follows that we can remove the time from the Hamiltonian by including the canonical pair (L_5, λ_5) (provided that, as far as the Sun is concerned, F_1 is regarded as a function of L_5, λ_5, $\bar{\omega}_5$, and Ω_5). Our problem, then, has altogether 14 degrees of freedom — three for each of the satellites, one for Jupiter's equator, and one for the Sun. The attractions of the remaining planets could, of course, be included in the same way, but according to de Sitter, even the perturbations by Saturn are entirely negligible to the order of accuracy that he aimed for, i.e., 10^{-6} radians in the longitudes of the satellites.

3. EXPANSION OF THE DISTURBING FUNCTION

Following de Sitter, we shall take Jupiter's equator at a certain epoch, e.g. 1950.0, as our reference plane. This choice makes the inclinations of the satellite orbits less than about a tenth of that of the Sun's orbit, $I_5 \approx 3°$. For simplicity, we shall take m_0 as the unit of mass and $(n_2 - n_3)^{-1} \approx 1.1222$ ephemeris days as the unit of time. We also put k = 1 (k will denote a dummy index in what follows), which leads to a unit of length of about 0.0070854 a.u., being very close to the mean distance of the third satellite.

The approximate values (de Sitter, 1918) of the various small parameters are given in Table I, where

Table I. Collection of small parameters.

p	m_p	J_{2p}	$-J_{4p}$	e_p	I_p
1	4×10^{-5}	$\frac{2}{5} \times 10^{-3}$	$\frac{1}{2} \times 10^{-6}$	4×10^{-3}	6×10^{-4}
2	$\frac{5}{2} \times 10^{-5}$	$\frac{1}{6} \times 10^{-3}$	$\frac{2}{3} \times 10^{-7}$	9×10^{-3}	8×10^{-3}
3	8×10^{-5}	$\frac{2}{3} \times 10^{-4}$	1×10^{-8}	2×10^{-3}	3×10^{-3}
4	$\frac{9}{2} \times 10^{-5}$	2×10^{-5}	1×10^{-9}	$\frac{3}{4} \times 10^{-2}$	4×10^{-3}

$$J_{ip} = J_i R_0^i / \sigma_p^i \; ; \quad i = 2, 4 \; ; \quad p = 1, 2, 3, 4 \; , \tag{38}$$

with

$$\sigma_p = a_p (1 - e_p^2) = \frac{\nu_p G_p^2}{m_p^2} \; , \qquad \nu_p = (1 - \mu_p)(1 + m_p) \; , \tag{39}$$

which relations follow from (15), (21), and (28). The values for e_p include both the free and the forced eccentricities, and we note that the inclinations enter the disturbing function only in the combination $I_p I_q$ ($p, q = 1, 2, 3, 4$). If we regard e_p and I_p as quantities of the first order, then m_p and J_{2p} are roughly of the second order, and J_{4p}, of the third order. According to de Sitter, it is necessary to develop the principal terms of $F_1 + F_{1J}$, as given by (22) and (34), to the eighth order (sixth order in de Sitter's case, since our disturbing function contains an extra mass factor) to achieve an accuracy of 10^{-6} in the longitudes. Hence, we must include terms of the order m_p times

$$m_p e_p^4, \quad m_p e_p^2 I_p^2, \quad J_{2p} e_p^4, \quad J_{2p} e_p^2 I_p^2, \quad J_{4p} e_p^3, \quad J_{4p} e_p I_p^2. \tag{40}$$

Since $m_5/r_5^2 \approx 2.1 \times 10^{-3}$ is comparable to m_p/r_p^2 ($p = 1, 2, 3, 4$), the terms in (22) that involve the Sun and one satellite and those that involve pairs of satellites will give rise to perturbations of roughly the same order, but the former terms can be expanded much more easily on account of the smallness of $1/r_5 \approx 1.4 \times 10^{-3}$.

In the subsequent derivations, we need consider only one pair of bodies. To ease the notation, we shall drop the subscripts for the body numbers and attach primes to the symbols relating to the outer body, i.e., $a < a'$; the distance between the bodies will be denoted by Δ. For the expansion of the indirect part of F_1, we have the following formulas for elliptic motion,

$$\dot{x} = \beta\sigma^{-1/2} [-P_1 \sin f + \bar{P}_1(e + \cos f)] \quad,$$

$$\dot{y} = \beta\sigma^{-1/2} [-P_2 \sin f + \bar{P}_2(e + \cos f)] \quad, \tag{41}$$

$$\dot{z} = \beta\sigma^{-1/2} [-P_3 \sin f + \bar{P}_3(e + \cos f)] \quad,$$

where

$$P_1 = c^2 \cos \bar{\omega} + s^2 \cos (\bar{\omega} - 2\Omega) \quad,$$

$$P_2 = c^2 \sin \bar{\omega} - s^2 \sin (\bar{\omega} - 2\Omega) \quad, \tag{42}$$

$$P_3 = 2sc \sin (\bar{\omega} - \Omega) \quad,$$

and $\bar{P}_k$ ($k = 1, 2, 3$) can be obtained from the expressions for P_k by first replacing "cos" by "sin" and "sin" by "cos" and then changing the signs of the arguments. In (42), we have introduced the abbreviations

$$s = \sin \frac{I}{2} = \frac{\gamma}{2} \quad, \qquad c = \cos \frac{I}{2} \quad. \tag{43}$$

In Cayley's tables (Cayley, 1861), we find the following expansions to the fourth order in e,

$$\sin f = \sum_{i=1}^{\infty} a_i \sin i\ell = \left(1 - \frac{7}{8}e^2 + \frac{17}{192}e^4\right) \sin \ell + \left(e - \frac{7}{6}e^3\right) \sin 2\ell$$

$$+ \left(\frac{9}{8}e^2 - \frac{207}{128}e^4\right) \sin 3\ell + \frac{4}{3}e^3 \sin 4\ell + \frac{625}{384}e^4 \sin 5\ell + 0(e^5) \quad,$$

$$e + \cos f = \sum_{i=1}^{\infty} b_i \cos i\ell = \left(1 - \frac{9}{8}e^2 + \frac{25}{192}e^4\right)\cos \ell + \left(e - \frac{4}{3}e^3\right)\cos 2\ell$$

$$+ \left(\frac{9}{8}e^2 - \frac{225}{128}e^4\right)\cos 3\ell + \frac{4}{3}e^3 \cos 4\ell + \frac{625}{384}e^4 \cos 5\ell + 0(e^5) \quad ,$$

$$(44)$$

where the powers of e may easily be replaced by powers of ϵ by means of (31). By using (15), (17), (21), (39), and (41), the indirect term inside the double summation sign in (22) may now be written

$$-(XX' + YY' + ZZ') = mm'(\nu\nu'\sigma\sigma')^{-1\,2} \cdot \sum_{i=-\infty}^{\infty} \sum_{j=1}^{\infty} \sum_{k=1}^{3} \frac{1}{2}$$

$$\times [(P_k P'_k a_i a'_j - P_{\bar{k}} P'_{\bar{k}} b_i b'_j) \cos (i\ell + j\ell')$$

$$+ (P_k P'_{\bar{k}} a_i b'_j + P'_k P_{\bar{k}} a'_j b_i) \sin (i\ell + j\ell')] \quad , \qquad (45)$$

where it is understood that $a_i = -a_{-i}$ (and therefore $a_0 = 0$), $b_i = b_{-i}$, and $b_0 = 0$. Now,

$$\sum_{k=1}^{3} P_k P'_k = c^2 c'^2 \cos (\bar{\omega} - \bar{\omega}') + 2ss'cc' \cos (\bar{\omega} - \bar{\omega}' - \Omega + \Omega')$$

$$+ s^2 s'^2 \cos (\bar{\omega} - \bar{\omega}' - 2\Omega + 2\Omega') + s^2 c'^2 \cos (\bar{\omega} + \bar{\omega}' - 2\Omega)$$

$$+ s'^2 c^2 \cos (\bar{\omega} + \bar{\omega}' - 2\Omega') - 2ss'cc' \cos (\bar{\omega} + \bar{\omega}' - \Omega - \Omega') , \quad (46)$$

and by changing the signs of the last three terms, this expression turns into that for $\sum P_{\bar{k}} P'_{\bar{k}}$, while $\sum P'_k P_{\bar{k}} = \left(\sum P_k P'_{\bar{k}}\right)$, from which $\sum P_k P'_{\bar{k}}$ obtains, of course, by interchanging the primed and unprimed quantities. Substituting these results in (45), we find after a considerable amount of calculation,

$$-(XX' + YY' + ZZ') = mm'(\nu\nu'\sigma\sigma')^{-1/2} \sum_{i,\,j=-\infty}^{\infty} \frac{1}{4}(a_i + b_i)$$

$$\times \Big[(a'_j - b'_j)\{c^2 c'^2 \cos (i\ell + j\ell' + \bar{\omega} - \bar{\omega}') + 2ss'cc' \cos (i\ell + j\ell' + \bar{\omega} - \bar{\omega}' - \Omega + \Omega')$$

$$+ s^2 s'^2 \cos (i\ell + j\ell' + \bar{\omega} - \bar{\omega}' - 2\Omega + 2\Omega')\} + (a'_j + b'_j)\{s^2 c'^2 \cos (i\ell + j\ell'$$

$$+ \bar{\omega} + \bar{\omega}' - 2\Omega) + s'^2 c^2 \cos (i\ell + j\ell' + \bar{\omega} + \bar{\omega}' - 2\Omega') - 2ss'cc' \cos (i\ell + j\ell'$$

$$+ \bar{\omega} + \bar{\omega}' - \Omega - \Omega')\}\Big]$$

$$(47)$$

As a partial check, we observe that the d'Alembert rule is obeyed since the lowest powers of e, e', s, and s' occuring in the coefficients are the same as the respective multiples of ℓ, ℓ', Ω, and Ω'. As was to be expected, the indirect term has no secular part since there are not terms with i = j = 0. We shall later show that the indirect term can be conveniently combined with the direct term whose expansion we take up next.

We have that

$$\Delta^{-1} = (r^2 + r'^2 - 2rr' \cos \psi)^{-1/2} \quad , \tag{48}$$

where ψ is the angle between the radius vectors r and r'. We note that the expression (46) is the dot-product of the unit vectors directed along the apses of the two orbits, so that we may obtain $\cos \psi$ from the same expression merely by replacing ϖ and ϖ' by the true longitudes $v = \varpi + f$ and $v' = \varpi' + f'$, respectively:

$$\cos \psi = \left(\sum_{k=1}^{3} P_k P'_k \right) \begin{array}{c} \varpi \to v \\ \varpi' \to v' \end{array} \tag{49}$$

Since we are forced to treat the motions of all the bodies simultaneously, the usual expansions of Δ^{-1} in powers of the mutual inclinations of pairs of bodies cannot be used here. For the expansion in s and s', we proceed as follows, aided by a novel treatment of the planetary disturbing function by Yuasa and Hori (1975). For this expansion it is sufficient to consider circular orbits, in which case we denote Δ by Δ_0. Then, in view of (46), (48), and (49),

$$\Delta_0^{-1} = \left[a^2 + a'^2 - 2aa' \{ \cos (\lambda - \lambda') + \delta \} \right]^{-1/2} \quad , \tag{50}$$

where δ is a quantity of the order ss' given by

$$\begin{aligned}
\delta = &(c^2 c'^2 - 1) \cos (\lambda - \lambda') + 2 ss'cc' \cos (\lambda - \lambda' - \Omega + \Omega') \\
&+ s^2 s'^2 \cos (\lambda - \lambda' - 2\Omega + 2\Omega') + s^2 c'^2 \cos (\lambda + \lambda' - 2\Omega) \\
&+ s'^2 c^2 \cos (\lambda + \lambda' - 2\Omega') - 2ss'cc' \cos (\lambda + \lambda' - \Omega - \Omega') \quad .
\end{aligned} \tag{51}$$

Yuasa and Hori define δ slightly differently such that $\cos (\lambda - \lambda')$ in (50) will contain an additional factor $(cc' - ss')^2$. They found that the convergence of the expansion is thereby improved with the remarkable result that it even holds for intersecting orbits. However, since convergence is no problem in our case because s and s' are very small, and in order to introduce the familiar Laplace coefficients, we have replaced this factor by unity. By expanding the right-hand side of (50) by means of the binomial theorem, we find

$$\Delta_0^{-1} = \sum_{i=0}^{\infty} \sum_{j=-\infty}^{\infty} (2\delta)^i \, a'^{-1} b_j^{(i)} \exp [\sqrt{-1} \, j(\lambda - \lambda')] \quad , \tag{52}$$

where the $b_j^{(i)}$'s are the coefficients in the expansion

$$\binom{-1/2}{i} (-a)^i [1 + a^2 - 2a \cos (\lambda - \lambda')]^{-1/2-i} = b_0^{(i)} + \sum_{j=1}^{\infty} 2b_j^{(i)} \cos (j\lambda - j\lambda')$$

$$= \sum_{j=-\infty}^{\infty} b_j^{(i)} \exp [\sqrt{-1} \, j(\lambda - \lambda')] \tag{53}$$

with $a = a/a' < 1$ and where $b_j^{(i)} = b_{-j}^{(i)}$ depends on a only $(b_j^{(i)}$ here would be written

$$\frac{1}{2}\binom{-1/2}{i}(-a)^i \cdot b_{1/2+i}^{(j)}$$

in the usual notation for the Laplace coefficients). In our problem we need consider values of i only up to two, and we may then readily perform the expansion

$$(2\delta)^i = c_{0000} + 2\sum s \, \frac{|k_3|}{s'} \, |k_4| \, c_{k_1 k_2 k_3 k_4}^{(i)} \cos(k_1\lambda + k_2\lambda' + k_3\Omega + k_4\Omega')$$

$$= \sum s \, \frac{|k_3|}{s'} \, |k_4| \, c_{k_1 k_2 k_3 k_4}^{(i)} \exp\left[\sqrt{-1}(k_1\lambda + k_2\lambda' + k_3\Omega + k_4\Omega')\right] \quad,(54)$$

where the first expression includes a finite number of terms for all the occurring combinations of the k's with $k_1 \geq 0$. In the last sum we include for each of these terms (except when all the k's are zero) an additional term with coefficient $c_{-k_1,-k_2,-k_3,-k_4}^{(i)} = c_{k_1,k_2,k_3,k_4}^{(i)}$. We finally substitute (54) in (52) to obtain

$$\Delta_0^{-1} = \sum_{i=0}^{\infty} \sum_{j=-\infty}^{\infty} \sum_{k} a'^{-1} b_j^{(i)} c_k^{(i)} s \, \frac{|k_3|}{s'} \, |k_4| \, \exp\left[\sqrt{-1}\{(j+k_1)\lambda\right.$$

$$+ (-j+k_2)\lambda' + k_3\Omega + k_4\Omega'\}\Big] \quad, \tag{55}$$

where k is an abbreviation for k_1, k_2, k_3, and k_4.

The expansion in powers of s and s' having been completed, we may in the following regard Δ_0^{-1} as a function of only a, a', λ, and λ'. We see from (50) that Δ^{-1} is the same function of r, r', v, and v' and hence may be obtained from (55) if the former variable set is replaced by the latter. This replacement can be done most easily by introducing the complex variables

$$w_1 = \exp(\sqrt{-1}\lambda) \,, \qquad w_1' = \exp(\sqrt{-1}\lambda') \,,$$
$$w_2 = \exp(\sqrt{-1}v) \,, \qquad w_2' = \exp(\sqrt{-1}v') \,,$$
$$w_3 = \exp(\sqrt{-1}\ell) \,, \qquad w_3' = \exp(\sqrt{-1}\ell') \,, \tag{56}$$
$$w_4 = \exp(\sqrt{-1}f) \,, \qquad w_4' = \exp(\sqrt{-1}f') \,,$$

and the differential operators,

$$D = a\partial/\partial a \,, \quad D' = a'\partial/\partial a' \,, \quad D_1 = w_1\partial/\partial w_1 \,, \quad D_1' = w_1'\partial/\partial w_1' \,. \tag{57}$$

Reference is made to Izsak et al. (1964) for a detailed exposition of the following method. It is easy to show that we may write symbolically

$$\Delta^{-1} = (r/a)^D (r'/a')^{D'} (w_4/w_3)^{D_1} (w_4'/w_3')^{D_1'} \Delta_0^{-1}(a, a', w_1, w_1') \,, \tag{58}$$

where the D's may be treated formally as if they were exponents, and $(r/a)^D$ and $(w_4/w_3)^{D_1}$ are supposed to be expanded in Laurent series in positive and negative powers of w_3, and similarly for the primed quantities. The product of these series for the inner body will be of the form

$$(r/a)^D (w_4/w_3)^{D_1} = \sum_{n=-\infty}^{\infty} \sum_{q} \Pi_n^q (D, D_1)\, e^q w_3^n \,, \tag{59}$$

where q is summed over all values for which $q - |n| = 0, 2, 4, \ldots, \infty$. The corresponding result for the outer body is obtained by adding a prime on all the symbols in (59). The Newcomb operator $\Pi_n^q (D, D_1)$ is a polynomial of degree q in D and D_1.

In performing the operations indicated in (58), we observe that D affects only the factor $a'^{-1} b_j^{(i)}$ in (55) and, for any integer q, $D'^q a'^{-1} b_j^{(i)} (a/a') = (-D - 1)^q a'^{-1} b_j^{(i)} (a/a')$. Furthermore, the effect of D_1^q and $D_1'^q$ on Δ_0^{-1} is to multiply each term of Δ_0^{-1} by the factors $(j + k_1)^q$ and $(-j + k_2)^q$, respectively. We have thus succeeded in reducing the four D-operators to the single one, D, and we have the combined result

$$\Pi_{nn'}^{qq'} (D, j, k_1, k_2) = \Pi_n^q (D, j+k_1) \cdot \Pi_{n'}^{q'} (-D-1, -j+k_2) \quad . \tag{60}$$

We note that the only structural difference between this equation and the corresponding equation (26) in Izsak et al. (1964) is that we have two distinct indices k_1 and k_2 where they have only one. This is due to the fact that only one mutual inclination and one common node occur in their expansion which is equivalent to ours with $s' = 0$ and $\Omega = \Omega'$. We are otherwise led to precisely the same Newcomb operators which may be taken from an existing table or generated by means of the very convenient recursion formulas developed by those authors. Izsak and Benima (1963) have also published a computer algorithm for computation of the Laplace coefficients and their Newcomb derivatives.

By means of (55) and (58) to (60), we now get

$$\Delta^{-1} = \sum_{i=0}^{\infty} \sum_{-\infty}^{\infty} \sum_{\substack{k, q, q' \\ j, n, n'}} \Pi_{nn'}^{qq'}\, a'^{-1} b_j^{(i)} c_k^{(i)} e^q e'^{q'} s^{|k_3|} s'^{|k_4|}$$

$$\times \cos\,[n\ell + n'\ell' + (j + k_1)\lambda + (-j + k_2)\lambda' + k_3 \Omega + k_4 \Omega'] \,, \tag{61}$$

where $q - |n|$ and $q - |n'|$ take one the values $0, 2, 4, \ldots, \infty$. This expression does not yet have the desired form, since we wish to replace e and e' above by ϵ and ϵ'. We could, of course, do this by direct substitution by means of (31), but we also wish to replace a and a' by σ and σ' which involve only the canonical variables G and G', according to (39). Now, we may replace a by σ and a' by σ' in (55), (57), and (58) provided that we alter the meaning of a, as it enters through $b_j^{(i)}(a)$, to $a = \sigma/\sigma' < 1$. Furthermore, we may change a into σ and e into ϵ in (59) where $\Pi_n^q(D, D_1)$ will now be a different polynomial but of the same structure as before. This polynomial is a special case of the generalized Newcomb operators devised by Izsak et al. (1964) with $\kappa_I = -1/2$ and $a_I = \sigma$ (where the subscript I for "Izsak" has been added to avoid confusion with the meanings of κ and a in the present paper). If we in place of (28) and (31) had adopted (32) and (33), we would have $\kappa_I = 0$ and $a_I = a$. Izsak et al. have derived recursion formulas also for the generalized Newcomb operators.

Before writing down the new version of (61), we shall reformulate (47) to enable a convenient combination of the indirect and direct terms. We notice the similarity of the coefficients in (47) to those in (54) with $i = 1$. If we put $\omega = \lambda - \ell$ and $\omega' = \lambda' - \ell'$ in (47), it is easy to show that that equation may be written

$$
\begin{aligned}
-(XX' + YY' + ZZ') = mm'(\nu\nu'\sigma\sigma')^{-1/2} \sum_{n,n'=-\infty}^{\infty} \frac{1}{8}\Big[&\{(a_{n+1} + b_{n+1}) \\
\times (a'_{n'-1} - b'_{n'-1}) \cos(n\ell + n'\ell' + \lambda - \lambda') + (a_{n-1} &- b_{n-1})(a'_{n'+1} + b'_{n'+1}) \\
\times \cos(n\ell + n'\ell' - \lambda + \lambda')\} + \sum_k \frac{|k_s|}{s} \frac{|k_4|}{s'} &c_k^{(1)}(a_{n+k_1} + k_1 b_{n+k_1}) \\
\times (a'_{n'+k_2} + k_2 b'_{n'+k_2}) \cos(n\ell + n'\ell' + k_1\lambda + k_2\lambda' + k_3\Omega + k_4\Omega') \Big]&, \quad (62)
\end{aligned}
$$

where, in view of (51), k_1 and k_2 have only the values ± 1. Let p be ± 1. Then

$$
\begin{aligned}
a_{n+p} + p b_{n+p} &= p(a_{|n|+1} + b_{|n|+1}) \quad &\text{if } pn \geq 0, \\
&= -p(a_{|n|-1} - b_{|n|-1}) \quad &\text{if } pn \leq -1,
\end{aligned} \qquad (63)
$$

and because of (44) we can write,[*]

$$
a_{|n|+1} + b_{|n|+1} = \sum_q \alpha_{q|n|} \epsilon^q, \qquad |n| = 0, 1, \ldots, \infty,
$$

$$
a_{|n|-1} - b_{|n|-1} = \sum_q \beta_{q|n|} \epsilon^q, \qquad |n| = 1, 2, \ldots, \infty, \qquad (64)
$$

$$
a_{n+p} + p b_{n+p} = \sum_q \alpha_{qn}^{(p)} \epsilon^q, \qquad n = 0, \pm 1, \ldots, \pm\infty,
$$

where $q - |n| = 0, 2, 4, \ldots, \infty$ and the coefficients of ϵ^q are pure numbers. If we rewrite (62) by means of the last of equations (64) and add the result to the new version of (61) multiplied by mm', we finally obtain for the term inside the square brackets in (22),

$$
\frac{mm'}{\Delta} - (XX' + YY' + ZZ') = mm' \sum_{i=0}^{\infty} \sum_{j,n,n'}^{\infty} \sum_{k,q,q'} \Big[\Pi_{nn'}^{qq'} \sigma'^{-1} b_j^{(i)}
$$

---

[*] The coefficients $a_i + b_i$ and $a_i - b_i$ are also listed in Cayley's tables.

$$+ \frac{1}{8} a_{qn}^{(j+k_1)} \cdot a_{q'n'}^{(-j+k_2)} (\nu\nu'\sigma\sigma')^{-1/2} \Bigg] c_k^{(i)} s^{|k_3|} s'^{|k_4|} {}_\epsilon q_\epsilon, q'$$

$$\times \cos [n\ell + n'\ell' + (j+k_1)\lambda + (-j+k_2)\lambda' + k_3\Omega + k_4\Omega'] \quad , \tag{65}$$

where, for each i, k is summed over a finite number of values of k_1, k_2, k_3, and k_4, and $q - |n|$ and $q' - |n'|$ take on the values 0, 2, 4, ..., ∞, and where $b_j^{(i)}(a) = b_{-j}^{(i)}(a)$, $c_k^{(i)}(s, s') = c_{-k}^{(i)}(s, s')$, and $\Pi_{nn'}^{qq'}(D, j, k_1, k_2) = \Pi_{-n, -n'}^{qq'}(D, -j, -k_1, -k_2)$ are defined by (53), (54), and (60) with

$$a = \sigma/\sigma' \quad , \qquad D = a\partial/\partial a \quad . \tag{66}$$

Furthermore, the a-coefficients are given by

$$a_{qn}^{(p)} = 0 \text{ if } |p| \neq 1 \text{ or } i > 1 \quad ,$$

$$= p a_{q|n|} \text{ if } pn \geq 0 \text{ and } |p| = 1 \text{ and } i \leq 1 \quad ,$$

$$= -p\beta_{q|n|} \text{ if } pn \leq -1 \text{ and } |p| = 1 \text{ and } i \leq 1 \quad , \tag{67}$$

where the only nonzero values of $a_{q|n|}$ and $\beta_{q|n|}$, and with q and $|n|$ below five, are easily found to be, by means of (31), (44), and (64):

$$a_{00} = 2 \quad , \qquad a_{20} = -2 \quad , \qquad a_{40} = 55/32 \quad ,$$

$$a_{11} = 2 \quad , \qquad a_{31} = -13/4 \quad ,$$

$$a_{22} = 9/4 \quad , \qquad a_{42} = -81/16 \quad , \qquad \beta_{22} = 1/4 \quad , \qquad \beta_{42} = -11/48$$

$$a_{33} = 8/3 \quad , \qquad \beta_{33} = 1/6 \quad ,$$

$$a_{44} = 625/192 \quad , \qquad \beta_{44} = 9/64 \quad . \tag{68}$$

In view of (40), we need include only the terms with $i \leq 2$ and $q + q' + |k_3| + |k_4| \leq 4$ in (65), and de Sitter's variation orbits depend on only the secular and critical terms which do not involve the inclinations, i.e., $i = k_1 = k_2 = k_3 = k_4 = 0$ and $c_k^{(i)} = c_0^{(0)} = 1$. The secular and critical terms in addition satisfy the respective conditions $n = n' = j = 0$ and $n\ell + n'\ell' + j\lambda - j\lambda' = 0$.

Since $\sigma/r = 1 + e \cos f = 1 - e^2 + \sum_{i=1}^{\infty} eb_i \cos i\ell$, we obtain readily the following expansion for the term under the first summation sign in (22),

$$\mu\beta^2 \frac{m^*}{r} = \frac{\mu m}{\sigma(1 - \mu)} \Bigg[1 - \epsilon^2 + \frac{3}{4}\epsilon^4 + (\epsilon - \frac{3}{2}\epsilon^3) \cos \ell$$

$$+ (\epsilon^2 - \frac{3}{4}\epsilon^4) \cos 2\ell + \frac{9}{8}\epsilon^3 \cos 3\ell + \frac{4}{3}\epsilon^4 \cos 4\ell + 0(\epsilon^5) \Bigg] \quad , \tag{69}$$

where μ, like m, is of the second order, according to de Sitter.

We do not give here an expansion for the disturbing potential (34) due to Jupiter's oblateness since it may turn out to be simpler to obtain the resulting perturbations from Brouwer's (1959) theory for an artificial satellite.

ACKNOWLEDGMENTS

It is a great pleasure to thank Dr. Kozai and the Japan Society for Promotion of Science whose kind invitation to the Tokyo Astronomical Observatory and financial support made this study possible. My sincere thanks also go to Dr. Hori and Dr. Yuasa of Tokyo University. I have benefited a great deal from discussions with them. Partial support by National Aeronautics and Space Administration grant NGR 09-015-213 is also gratefully acknowledged.

NOTE ADDED IN PROOF

After this work had been completed, I became aware that Poincare's canonical relative coordinates had, in fact, been considered for use on the planetary problem by Izsak et al. (1965), who showed how to develop the associated disturbing function. Although their results do not apply here, since the development was made in terms of the mutual inclination, they pointed out that the indirect term in (22) can be written

$$-(XX' + YY' + ZZ') = -mm'(\nu\nu'aa')^{-1/2} \frac{\partial^2}{\partial\ell\,\partial\ell'} \left(\frac{rr'}{aa'}\cos\psi\right) , \qquad (70)$$

and hence can be expanded by means of the Newcomb operators, affording a somewhat simpler derivation than that presented here.

REFERENCES

Aksnes, K., and Franklin, F.A. (1975). Nature 258, 503.
Aksnes, K., and Franklin, F.A. (1976). Astron. J. 81, 464.
Brouwer, D. (1959). Astron. J. 64, 378.
Cayley, A. (1861). Mem. R. Astron. Soc. 29, 191.
Charlier, C.L. (1902). Die Mechanik Des Himmels, Vol. I (Von Veit & Co., Leipzig).
de Sitter, W. (1918). Leiden Annals 12, Part I.
de Sitter, W. (1931). Mon. Not. R. Astron. Soc. 91, 705.
Hori, G. (1966). Publ. Astron. Soc. Japan 18, 287.
Izsak, I.G., and Benima, B. (1963). Smiths. Astrophys. Obs. Spec. Rep. No. 129.
Izsak, I.G., Benima, B., and Mills, S.B. (1965). Smiths. Astrophys. Obs. Spec. Rep. No. 164.
Izsak, I.G., Gerard, J.M., Efimba, R., and Barnett, M.P. (1964). Smiths. Astrophys. Obs. Spec. Rep. No. 140.
Lieske, J. (1975). Cel. Mech. 12, 5.
Marsden, B.G. (1964). Doctoral Thesis, Yale Univ., New Haven, Conn.
Poincaré, H. (1897). Bull. Astron. 14, 53.
Sampson, R.A. (1910). Tables of the Four Great Satellites of Jupiter (Wesley, London).
Yuasa, M., and Hori, G. (1975). In Proceedings of a Symposium on Celestial Mechanics, held in Tokyo, February 12-13, 1975 (Ed. by Hori and Yuasa). To be published in Publ. Astron. Soc. Japan.

THEORY OF MOTION OF JUPITER'S GALILEAN SATELLITES

J.H. Lieske
Jet Propulsion Laboratory

ABSTRACT

The final results for the theory enabling one to calculate the
positions of the Galilean satellites and their partial derivatives
are presented, following the techniques outlined in earlier papers.
Extensive use of algebraic manipulation software on a digital computer
is employed to generate the final expressions. The new theory is, in
effect, a revitalization of Sampson's theory in which we (a) remove
algebraic and mathematical errors existing in Sampson's work, (b)
introduce some neglected effects due to solar interactions and the 3-7
commensurability, (c) allow for non-zero amplitude and phase of the
free libration, (d) express the final results as analytic functions
of variations in 49 arbitrary constants of integration and physical
parameters, (e) construct the theory in a manner which readily allows
for future revision, and (f) provide analytic expressions for the
partial derivatives with respect to the 49 parameters.

A SECOND-ORDER THEORY OF THE GALILEAN SATELLITES OF JUPITER

S. FERRAZ-MELLO
University of São Paulo, Brazil

ABSTRACT. The theory of the motion of the Galilean satellites of Jupiter is developed up to the second-order terms. The disturbing forces are those due to mutual attractions, to the non-symmetrical internal mass distribution of Jupiter and to the attraction from the Sun. The mean equator of Jupiter is taken as the reference plane and its motion is considered. The integration of the equations is performed. The geometric equations are solved for the case in which the amplitude of libration is zero. The perturbation method is shortly commented on the grounds of some recent advances in non-linear mechanics.
    In a previous paper (Ferraz-Mello, 1974) one perturbation theory has been constructed with special regard to the problem of the motion of the Galilean satellites of Jupiter. In this problem, the motions are nearly circular and coplanar; on the other hand the quasi-resonances lead to strong perturbations. The main characteristic of the theory is that it allows the main frequencies to be kept fixed from the earlier stages, and so, to have a purely trigonometric solution.

1. THE EQUATIONS

    The equations were derived for a second-order theory; the small parameters are the satellite masses, the eccentricities and the inclinations. For each satellite the variables are the radius vector, the longitude, and two pairs of variables P,Q, and K,H, built respectively from Laplace's and the area's first integrals. These variables are close to Poincaré's variables in exponential form: $e.exp-i(1-\varpi)$ and $I.exp-i(1-\Omega)$ and their complex conjugates.
    The equations for the variables P_j and Q_j are separated; they are integro-differential linear equations;

$$DP_j + \kappa_j P_j = \frac{\lambda_j - \kappa_j^2}{\kappa_j} [1 + \frac{1}{4} \kappa_j D^{-1}(P_j + Q_j)' - \frac{3}{4} \kappa_j^2 D^{-2}(P_j - Q_j) - $$

$$- \kappa_j D^{-1}(P_j - Q_j)] + \frac{1}{\kappa_j} \mathcal{R}_j + \frac{1}{16} D^{-1} X_j \quad 7P_j - 50Q_j + 7\kappa_j D^{-1}(P_j - Q_j) +$$

209

V. Szebehely (ed.), Dynamics of Planets and Satellites and Theories of Their Motion, 209-236.

$$+ \frac{1}{4\kappa_j} X_j \left[1 + \frac{1}{2} \kappa_j D^{-1} (P_j - Q_j) + \frac{1}{4} \kappa_j D^{-1} (P_j + Q_j) - \frac{3}{2} \kappa_j^2 D^{-2} (P_j - Q_j) \right] ;$$

$$DQ_j - \kappa_j Q_j = \frac{\lambda_j - \kappa_j^2}{\kappa_j} \left[-1 + \frac{1}{4} \kappa_j D^{-1} (P_j + Q_j) - \frac{3}{4} \kappa_j^2 D^{-2} (P_j - Q_j) + \right.$$

$$\left. + \kappa_j D^{-1} (P_j - Q_j) \right] - \frac{1}{\kappa_j} \tau_j - \frac{1}{16} D^{-1} X_j \left[7Q_j - 5P_j + 7\kappa_j D^{-1} (P_j - Q_j) \right.$$

$$- \frac{1}{4\kappa_j} X_j \left[1 + \frac{1}{2} \kappa_j D^{-1} (P_j - Q_j) - \frac{1}{4} \kappa_j D^{-1} (P_j + Q_j) + \frac{3}{2} \kappa_j^2 D^{-2} (P_j - Q_j) \right].$$

·In these equations we have

$$\kappa_j = \frac{n_j}{\nu_3}$$

and

$$\lambda_j = \frac{k^2 m_0 (1 + m_j)}{\nu_3^2 a_j^3}$$

where the mean motion ν_3 is that of the third satellite with respect to a rotating frame in which the mean motions of the three inner satellites are exactly commensurable; the mean motion n_j is that of the jth satellite with respect to a Galilean (inertial axes) frame and a_j is the mean distance of the jth satellite to the planet. D is the same differential operator as in Hill's moon theory

$$D = \zeta \frac{d}{d\zeta} \qquad (\zeta = \exp i\nu_3 t).$$

The equations for K_J and H_j are also separated; they are linear:

$$(DK_j + \kappa_j K_j) = \frac{1}{\kappa_j} \mathcal{V}_J + \frac{1}{2\kappa_j} (K_j + H_j) \mathcal{R}_j$$

$$(DH_j - \kappa_j H_j) = - \frac{1}{\kappa_j} \mathcal{V}_j - \frac{1}{2\kappa_j} (K_j + H_j) \mathcal{T}_j .$$

Still we have the geometric equations

$$D \varepsilon_j = \frac{1}{2} \kappa_j (P_j - Q_j) - \frac{1}{4} \kappa_j (P_j + Q_j) \Gamma_j + \frac{1}{4} \kappa_j (K_j + H_j)(K_j - H_j)$$

$$D\Gamma_j = - 3\kappa_j \varepsilon_j + 2\kappa_j \varepsilon_j^2 + \frac{1}{2} \kappa_j (P_j + Q_j) - \frac{3}{2} \kappa_j (P_j + Q_j) \varepsilon_j - \qquad (3.19)$$

$$- \frac{1}{4} \kappa_j (P_j - Q_j) \cdot \Gamma_j + \frac{1}{4} \kappa_j \left[(K_j - H_j)^2 - \frac{1}{2} (K_j + H_j)^2 \right]$$

The variables Γ_j and ε_j are closely related to the perturbations in longitude and vector radius as defined in Ferraz-Mello (1974). The cylindrical coordinates of each satellite in one Galilean frame of reference are easily obtained from the solutions of the above set of equations through

$$\rho_j = a_j (1 + \varepsilon_j)$$

$$\phi_j = \phi_{0j} + n_j t + \frac{1}{2} i \Gamma_j$$

$$z_j = - \frac{1}{2} a_j (K_j + H_j) - \frac{1}{4} a_j \Gamma_j (K_j - H_j) - \frac{1}{4} a_j \varepsilon_j (K_j + H_j) +$$

$$+ \frac{1}{8} a_j (K_j + H_j)(P_j = Q_j) .$$

All these equations except those for K_j, H_j and z_j are exact. The equations for the space variables are approximated up to the second-order terms.

The disturbing forces enter the equations for P_j, Q_j, K_j, H_j through $\mathcal{R}, \tau, \vartheta$ and χ. The main disturbing forces acting on the satellites arise from their mutual attractions, form the non-symmetrical distribution of the masses inside the planet and from the Sun.

We could also consider the forces arising from other planets as well as those which give account of the corrections to Newton's equations due to the general relativity. The effects of the disturbing forces arising for other planets are not sensible in reason of their differential action; other planets may be only considered through their perturbations on the orbit of Jupiter, which leads to modulations of the solar action. The only noticeable relativistic effects are the advance of the perijoves (Ferraz-Mello, 1966); these effects do not need to be considered now and will be introduced in a later paper.

2. THE MUTUAL ATTRACTIONS

The disturbing actions due to the mutual attractions are given by

$$\mathcal{R}_j = \sum_{i \neq j} \mathcal{R}_{ji} = \sum_{i \neq j} \frac{k^2 m_i}{v_3^2} \left\{ \frac{(1+U_j) - \alpha_{ij}(1+U_i)\xi_{ij}}{r_{ij}^3} + \frac{\alpha_{ij}(1+U_i)\xi_{ij}}{r_i^3} \right\}$$

$$\mathcal{T}_j = \sum_{i \neq j} \mathcal{T}_{ji} = \sum_{i \neq j} \frac{k^2 m_i}{v_3^2} \left\{ \frac{(1+S_j) - \alpha_{ij}(1+S_i)\xi_{ij}^{-1}}{r_{ij}^3} + \frac{\alpha_{ij}(1+S_i)\xi_{ij}^{-1}}{r_i^3} \right\}$$

$$\vartheta_j = \sum_{i \neq j} \vartheta_{ji} = \sum_{i \neq j} \frac{k^2 m_j}{v_3^2} \left\{ \frac{Z_j - \alpha_{ij} Z_i}{r_{ij}^3} + \frac{\zeta_{ij} Z_i}{r_i^3} \right\} ,$$

where r_{ij} is the mutual distance between the satellites P_i and P_j, α_{ij} is the ratio a_i/a_j and

$$U_j = \frac{3}{4} \kappa_j D^{-1} P_j - \frac{1}{4} \kappa_j D^{-1} Q_j - \frac{3}{4} \kappa_j^2 D^{-2} (P_j - Q_j)$$

$$S_j = \frac{1}{4} \kappa_j D^{-1} P_j - \frac{3}{4} \kappa_j D^{-1} Q_j + \frac{3}{4} \kappa_j^2 D^{-2} (P_j - Q_j)$$

$$Z_j = -\frac{1}{2} (K_j + H_j).$$

These three relations are exact up to the first order only; they are always multiplied by first-order parameters and then the error will be of third order.

The ξ_{ij} are

$$\xi_{ij} = \sigma_i \sigma_j^* \xi^{g_i - g_j},$$

where $g_j = \nu_j/\nu_3$ (ratio of Eulerian mean motions), σ_j i exp iθ_{0j} (position of the jth satellite at the epoch) and σ_j^* its complex conjugate (Ferraz-Mello, 1974).

The distances r_i may be written

$$r_i = a_i (1 + U_i + S_i + U_i S_i + z_i^2)^{1/2}$$

and then, to the first order,

$$r_i^{-3} = a_i^{-3} [1 - \frac{3}{2} (U_i + S_i)].$$

The mutual distances r_{ij}^{-3} may be easily calculated to the same order; we have

$$R_{ij}^{-3} = a_i^{-3} [1 - \alpha_{ji} (\xi_{ij} + \xi_{ij}^{-1}) + \alpha_{ji}^2]^{-3/2}$$

$$- \frac{3}{2} a_i^{-3} [1 - \alpha_{ji} (\xi_{ij} + \xi_{ij}^{-1}) + \alpha_{ji}^2]^{-5/2} \cdot [(U_i + S_i) -$$

$$- \alpha_{ji} \xi_{ij} (U_i + S_j) - \alpha_{ji} \xi_{ij}^{-1} (S_i + U_j) + \alpha_{ji}^2 (U_j + S_j)].$$

For the developments we follow Sagnier (1973) and we introduce the coefficients $\gamma_s^{(k)}$ through

$$[1 - \alpha_{ji} (\xi_{ij} + \xi_{ij}^{-1}) + \alpha_{ji}^2]^{-s/2} = \sum_{-\infty}^{+\infty} \gamma_s^{(k)} (\alpha_{ji})^{|k|} \cdot \xi_{ij}^k.$$

These coefficients are related to Laplace's $b_{s/2}^{(k)}$ (Brouwer and Clemence, 1961, p. 495) through the relations

$$\gamma_s^{(k)}(\alpha) = \frac{1}{2} b_{s/2}^{(k)}(\alpha) \qquad \text{if } \alpha < 1$$

$$\gamma_s^{(k)}(\alpha) = \frac{1}{2} \alpha^{-s} b_{s/2}^{(k)}(\alpha) \qquad \text{if } \alpha > 1.$$

For the brackets which appear in the actual equations we write

$$[1 - \alpha_{ji}(\xi_{ij} + \xi_{ij}^{-1}) + \alpha_{ji}^2]^{-3/2} = \sum_{-\infty}^{+\infty} \mathscr{G}_{ji}^{(k)} \varepsilon_{ij}^k$$

$$[1 - \alpha_{ji}(\xi_{ij} + \xi_{ij}^{-1}) + \alpha_{ji}]^{-5/2} = \sum_{-\infty}^{+\infty} \mathscr{H}_{ji}^{(k)} \xi_{ij}^k.$$

The disturbing forces arising from the mutual attractions are

$$\mathscr{R}_{ji} = \sum_{-\infty}^{+\infty} \{R_{1(ji)}^{(k)} + R_{2(ji)}^{(k)} D^{-1} P_j + R_{3(ji)}^{(k)} D^{-1} Q_j + R_{4(ji)}^{(k)} D^{-2}(P_j - Q_j) +$$

$$+ R_{5(ji)}^{(k)} D^{-1} P_i + R_{6(ji)}^{(k)} D^{-1} Q_i + R_{7(ji)}^{(k)} D^{-2}(P_i - Q_i)\} \cdot \xi_{ij}^k$$

$$\mathscr{T}_{ji} = \sum_{-\infty}^{+\infty} \{T_{1(ji)}^{(k)} + T_{2(ji)}^{(k)} D^{-1} P_j + T_{3(ji)}^{(k)} D^{-1} Q_j + T_{4(ji)}^{(k)} D^{-2}(P_j - Q_j) +$$

$$+ T_{5(ji)}^{(k)} D^{-1} P_i + T_{6(ji)}^{(k)} D^{-1} Q_u + T_{7(ji)}^{(k)} D^{-2}(P_i - Q_i)\} \cdot \xi_{ij}^k$$

$$\mathscr{V}_{ji} = \sum_{-\infty}^{+\infty} \{V_{1(ji)}^{(k)} (K_j + H_j) + V_{2(ji)}^{(k)} (K_i + H_i)\} \xi_{ij}^k,$$

where the coefficients $R_{m(ji)}^{(k)}$, $T_{m(ji)}^{(k)}$, $V_{m(ji)}^{(k)}$ are numerical factors that depend only on the mean distances a_j. The function χ_j which appears as a factor in the integro-differential equations may be calculated after its definition

$$\chi_j = (1 - \frac{1}{2} c_j^2)^{-1} [(1 + S_j) \mathscr{R}_j - (1 + U_j) \mathscr{T}_j],$$

where

$$c_j^2 = (1 + U_j)(1 + S_j).$$

It follows

$$\chi_j = \sum_{i \neq j} \chi_{ji}$$

and

$$X_{ji} = \sum_{-\infty}^{+\infty} \{X_{1(ji)}^{(k)} + X_{2(ji)}^{(k)} D^{-1}P_j + X_{3(ji)}^{(k)} D^{-1}Q_j + X_{4(ji)}^{(k)} D^{-2}(P_j - Q_j) +$$

$$+ X_{5(ji)}^{(k)} D^{-1}P_i + X_{6(ji)}^{(k)} D^{-1}Q_i + X_{7(ji)}^{(k)} D^{-2}(P_i - Q_i)\} \cdot \xi_{ij}^k.$$

When calculating $D^{-1}X_j$ it is enough to restrict it to the first-order. Then, since the $X_{m(ji)}^{(k)}$ are of first-order in the disturbing mass, it follows

$$D^{-1}X_{ji} = \sum_{-\infty}^{+\infty} \frac{X_{1(ji)}^{(k)} \cdot \xi_{ij}}{k(g_i - g_j)}.$$

This primitive function in singular when $k = 0$ unless $X_{1(ji)}^{(0)}$ is also equal to zero; this fact indeed happens as a consequence of $g_{ji}^{(-1)} = g_{ji}^{(+1)}$.

3. JUPITER's SHAPE ACTIONS

In rectangular coordinates the equations of the motion of the j-th satellite under the action of the planet writes

$$\ddot{x}_j - 2N\dot{y}_j - N^2 x_j = \frac{\partial}{\partial x_j} k^2 (1 + m_j) (\frac{1}{r_j} + \Omega)$$

$$\ddot{y}_j + 2N\dot{x}_j - N^2 y_j = \frac{\partial}{\partial y_j} k^2 (1 + m_j) (\frac{1}{r_j} + \partial)$$

$$\ddot{z}_j = \frac{\partial}{\partial z_j} k^2 (1 + m_j) (\frac{1}{r_j} + \Omega),$$

where

$$\Omega = -\frac{1}{r_j} \sum_{k=2}^{\infty} \frac{4 J_k R^k}{r_j^k} P_k (\sin \beta_j).$$

and N is the angular velocity of the rotating Eulerian frame. The zonal harmonics of subscripts greater than 4 are neglected, as well as the tesseral harmonics and the harmonics related to the shape of the satellites (Ferraz-Mello, 1966). J_2 is considered as a first-order parameter while J_3 and J_4 are considered as second-order parameters. It follows

$$\ddot{x}_j - 2N\dot{y}_j - N^2 x_j = -k^2 (1+m_j) \{\frac{x_j}{r_j^3} + \frac{3}{2} \frac{J_2 R^2 x_j}{r_j^5} - \frac{15}{8} \frac{J_4 R^4 x_j}{r_j^7}\}$$

$$\ddot{y}_j + 2N\dot{x}_j \; N^2 y_j = -k^2(1+m_j)\;\{\frac{y_j}{r_j^3} + \frac{3}{2}\;\frac{J_2 R^2 y_j}{r_j^5} - \frac{15}{8}\;\frac{J_4 R^4 y_j}{r_j^7}\}$$

$$\ddot{z}_j = -k^2(1+m_j)\;\{\frac{z_j a_j}{r_j^3} + \frac{3}{2}\;\frac{J_2 R^2 z_j a_j}{r_j^5} + \frac{3J_2 R^2 z_p \sin\beta_j}{r_j^4} -$$

$$- \frac{3J_3 R^3 z_p}{2r_j^5}\}.$$

In these equations β_j is the latitude of the satellite over the planet's figure equator and z_p is the third coordinate of the planet's figure pole. If the special coordinates U_j and S_j are introduced instead of x_j, y_j we may derive the values of the force components $\tilde{\mathscr{R}}_j$, $\tilde{\mathscr{T}}_j$ and $\tilde{\mathscr{V}}_j$ due to the non-sphericity of Jupiter. We have

$$\tilde{\mathscr{R}}_j = \lambda_j a_j^3 \; [\frac{3J_2 R^2}{2r_j^5} - \frac{15J_4 R^4}{8r_j^7}] \; (1 + U_j)$$

$$\tilde{\mathscr{T}}_j = \lambda_j a_j^3 \; [\frac{3J_2 R^2}{2r_j^5} - \frac{15J_4 R^4}{8r_j^7}] \; (1 + S_j)$$

$$\tilde{\mathscr{V}}_j = \lambda_j a_j^3 \; \frac{3J_2 R^2}{2r_j^5} \; z_j + \lambda_j a_j^2 \; [\frac{3J_2 R^2 \sin\beta_j}{r_j^4} - \frac{3J_3 R^3}{2r_j^5}] \; z_p.$$

These disturbing forces may be expanded as in the previous section. It follows then:

$$\tilde{\mathscr{R}}_j = \tilde{R}_{1(j)} + \tilde{R}_{2(j)} D^{-1} P_j + \tilde{R}_{3(j)} D^{-1} Q_j + \tilde{R}_{4(j)} D^{-2}(P_j - Q_j)$$

$$\tilde{\mathscr{T}}_j = \tilde{T}_{1(j)} + \tilde{T}_{2(j)} D^{-1} P_j + \tilde{T}_{3(j)} D^{-1} Q_j + \tilde{T}_{4(j)} D^{-2}(P_j - Q_j)$$

$$\tilde{\mathscr{V}}_j = \tilde{V}_{0p(j)} + \tilde{V}_{1(j)} (K_j + H_j) + \tilde{V}_{3(j)} K_p + \tilde{V}_{4(j)} H_p.$$

The coefficients $\tilde{R}_{m(j)}$, $\tilde{T}_{m(j)}$ and $\tilde{V}_{m(j)}$ are numerical factors depending only in the mean distances. X_j in the order of approximation of this theory is zero. In the calculation of $\mathscr{V}_j$ the coordinates of the pole of inertia of Jupiter intervenes also through

$$\sin\beta_j = \frac{x_j x_p + y_j y_p + z_j z_p}{r_j}.$$

If I_p and Ω_p are the inclination and the longitude of the ascending node of the figure's equator, then

$$x_p = \sin I_p \sin \Omega_p \qquad Y_p = -\sin I_p \cos \Omega_p \qquad Z_p = Z_p = 1$$

except for terms of the third-order in the equations for $x_p Y_p$ and of the second-order in the equation for z_p. We also introduce two new parameters

$$K_p = \sin I_p \exp \quad i(\Omega_p + \frac{3\pi}{2})$$

$$H_p = \sin I_p \exp - i(\Omega_p + \frac{3\pi}{2}) \; .$$

4. THE MEAN EQUATOR OF JUPITER AS REFERENCE

The inclination of the equator of Jupiter over the planet's orbital plane is $3\overset{0}{.}07$. If the orbital plane is conserved as a reference plane, we have

$$|K_j| \sim 0.06 \qquad |K_j|^2 \sim 0.004 \qquad |K_j|^3 \sim 0.0002$$

since the satellites move very close to the plane of the equator of the planet. We may compare these figures to those for the eccentricities:

$$|P_j| \sim 10^{-2} \qquad |P_j|^2 \sim 10^{-4}.$$

These figures allow us to see that if the orbital plane of Jupiter is kept as the reference plane, we need the third powers of the K_j to get the same precision as that of the second powers of the P_j. To reestablish homogeneity it is convenient to adopt the mean equator of Jupiter as the reference plane; in this case we will have $|K_j|^2 \sim 5\times10^{-5}$.

Let x be the matrix of the coordinates with respect to the mean equator of Jupiter and **X** the matrix of the coordinates with respect to an inertial plane close to the plane of the orbital motion of the planet. We have

$$x = (1 + R) . \mathbf{X},$$

where

$$R = \begin{Bmatrix} \sin^2\theta_p (\cos I - 1) & \sin\theta_p \cos\theta_p (1 - \cos I) & -\sin\theta_p \sin I \\ \sin\theta_p \cos\theta_p (1 - \cos I) & \cos^2\theta_p (\cos I - 1) & \cos\theta_p \sin I \\ \sin\theta_p \sin I & - \cos\theta_p \sin I & \cos I - 1 \end{Bmatrix}$$

The equations of the motion of a satellite in reactangular coordinates writes (Section 3)

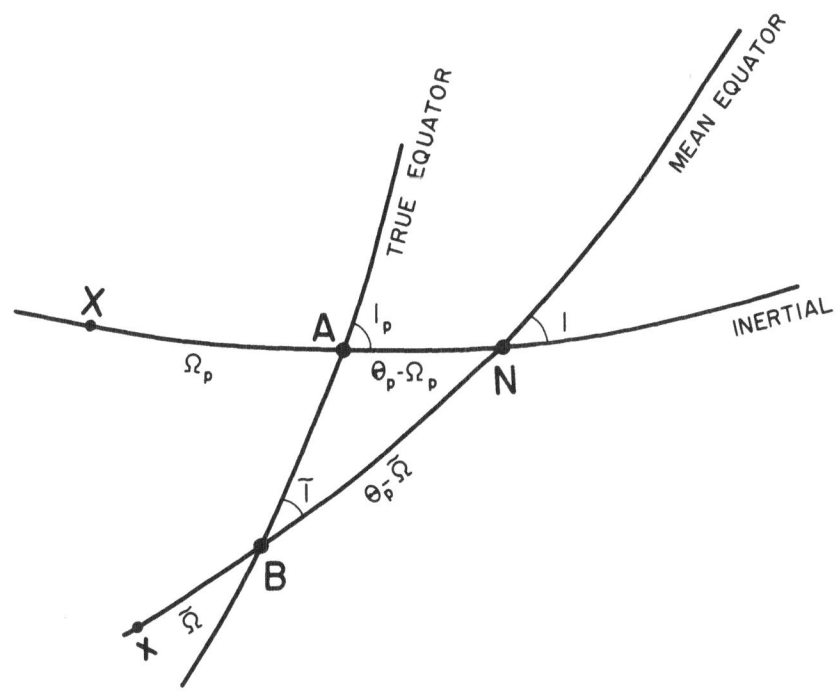

$$A\ddot{X} + B\dot{X} + Cx = \mathscr{L}(x)$$

where A is the unit-matrix,

$$B = \begin{Bmatrix} 0 & -2N & 0 \\ 2N & 0 & 0 \\ 0 & 0 & 0 \end{Bmatrix}, \qquad C = \begin{Bmatrix} -N^2 & 0 & 0 \\ 0 & -N^2 & 0 \\ 0 & 0 & 0 \end{Bmatrix}$$

and $\mathscr{L}(X)$ is a vector function of X. It is important to observe that $\mathscr{L}(X)$ depends on X in two different ways: directly through linear relations and indirectly through intrinsic parameters like distances, mutual distances and latitudes. Then

$$(1 + R) . \mathscr{L}(X) = \mathscr{L}(X).$$

If the rotation matrix (1-R) is applied to both sides of the equations of the motion it follows that

$$A\ddot{x} + B\dot{x} + Cx = \mathscr{L}(x) + B\dot{R}(1 + R)^{-1}x + 2A\dot{R}(1 + R)^{-1}\dot{x} -$$

$$- 2A[\dot{R}(1 + R)^{-1}2]x + A\ddot{R}(1 + R)^{-1}x.$$

The matrix $\dot{R}$ is

$$\dot{R} = \dot{\theta}_p \begin{Bmatrix} -\sin 2\theta_p(1 - \cos I) & \cos 2\theta_p(1 - \cos I) & -\cos\theta_p \sin I \\ \cos 2\theta_p(1 - \cos I) & \sin 2\theta_p(1 - \cos I) & -\sin\theta_p \sin I \\ \cos \theta_p \sin I & \sin \theta_p \sin I & 0 \end{Bmatrix}$$

Since θ_p is very small, 2.185×10^{-6} degrees per day (Tisserand, 1896), we may neglect the term in which $\dot{R}^2$ appears; also, since $\ddot{\theta}_p = 0$ we may also neglect the term in $\ddot{R}$. The new equations are

$$A\ddot{x} + B\dot{x} + Cx = (x) + B\dot{R}(1 + R)^{-1}x + 2A\dot{R}(1 + R)^{-1}\dot{x}.$$

The effect of the adoption of the mean equator of Jupiter as the reference plane is the introduction of the disturbing force $B\dot{R}(1 + R)^{-1}x +$ $+ 2A\dot{R}(1 + R)^{-1}\dot{x}$. The magnitude of this force is at most equal to 3×10^{-6} AU/y^2 per unit mass. It may be compared to the magnitude of the mutual interactions between two satellites (10^{-2} to 10^{-1} in the same units). They will be neglected, and we will adopt in the new frame the same equations as in the inertially oriental frame. As a consequence, this theory will not be able to show the eventual existence of inequal-ities of very long periods (4.5×10^{r} yr) arising from the precession of the equator of the planet.

In the preceding section, in the expansion giving $\mathscr{V}_j$ the parameters K_p and H_p were related to the inclination and longitude of the ascending node of the true equator over the reference frame, i.e., the orbital plane. In the new reference frame we consider.

$$\tilde{K} = \tilde{I}.\exp i(\tilde{\Omega} + \frac{3}{2}\pi)$$

$$\tilde{H} = \tilde{I}.\exp -i(\tilde{\Omega} + \frac{3}{2}\pi)$$

and then

$$\mathscr{V}_j = \tilde{V}_{0p(j)} + \tilde{V}_{1(j)}(\tilde{K}_j + \tilde{H}_j) + \tilde{V}_{3(j)}\tilde{K} + \tilde{V}_{(4j)}\tilde{H}.$$

The fundamental relations of the spherical triangle NAB allow us to write

$$\sin I.\exp i(\theta_p - \tilde{\Omega}) = \sin I_p \exp i(\theta_p - \tilde{\Omega}_p) - \cos I_p.\sin I +$$

$$+ \sin I_p\cos(\theta_p - \tilde{\Omega}_p) (\cos I-1)$$

or, neglecting all terms of second or higher order with respect to $\tilde{I}$,

$$\tilde{I}.\exp(-i\tilde{\Omega}) = \sin I_p.\exp(-i\Omega_p) - \sin I.\exp - i\theta_p,$$

and then

$$\tilde{H} = H_p - \sin I.\exp - i(\theta_p + \frac{3\pi}{2})$$

$$\tilde{K} = K_p - \sin I.\exp i(\theta_p + \frac{3\pi}{2}).$$

The new expression for $\tilde{\mathscr{V}}_j$ is then

$$\tilde{\mathscr{V}}_j = \tilde{V}_{0(j)} + \tilde{V}_{1(j)}(K_j + H_j) + \tilde{V}_{3(j)}K_p + \tilde{V}_{4(j)}H_p,$$

where

$$\tilde{V}_{0(j)} = V_{0p(j)} - \frac{3}{2}\lambda_j J_2 \frac{R^2}{a_j^2} \sin I.(\xi_{jp} - \xi_{jp}^{-1}).$$

In analogy to former parameters we introduce

$$\xi_{jp} = \sigma_j \sigma_p^* \zeta^{g_j - g_p}$$

where

$$\sigma_p = i \exp i\theta_{op} \qquad g_p \nu_3 = \nu_p \qquad \text{and} \qquad \nu_p = \dot{\theta}_p - N.$$

5. MOTION OF THE POLE OF JUPITER

The researches of Souillart and Laplace (Tisserand, 1896) show that it is not possible to consider the motion of the orbital plane of the satellites and the motion of the equator of the planet separately. These motions appear in the approximation considered here, as a linear system with 5 degrees of freedom.

If the planet is supposed to have axial symmetry, we have (Ferraz-Mello, 1972)

$$\sin I_p \dot{\Omega}_p = -\sum_j \frac{1}{rC} (\frac{\partial W}{\partial z_j}) \bar{y}_j$$

$$\dot{I}_p = -\sum_j \frac{1}{rC} (\frac{\partial W}{\partial z_j}) \bar{x}_j$$

where C is the polar momentum of inertia of the planet, r the angular velocity of the planetary rotation, and x_j, y_j are the rectangular coordinates of the disturbing bodies in a frame in which the true equator of Jupiter is the fundamental plane of reference, and, in which the

x-axis is directed to point A (see Figure 1 in the preceding section)
ascending node of the true equator over the inertial frame. We have

$$\bar{x}_j = x_j.\cos \theta_p + y_j.\sin \theta_p$$

$$\bar{y}_j = -x_j.\sin \theta_p + y_j.\cos \theta_p$$

the other components of the rotation, the amplitudes of which are I, $-I_p$
and $\theta_p - \Omega_p$, give rise to corrections small enough to be neglected. We
still have

$$\frac{\partial W}{\partial z_j} + \frac{4k^2 m_j J_2 R^2}{3r_j^4} \sin \beta_j$$

which is proportional to the torque of the attraction of the satellite
about the polar axis of the planet. These equations and the quantities
K_p and H_p already defined allow us to write the equations of the motion
of the equator of Jupiter in a suitable form. From the definition of K_p
we have

$$\dot{K}_p = \cos I_p.\dot{I}_p \exp i(\Omega_p + \frac{3\pi}{2}) + i.\sin I_p.\dot{\Omega}_p.\exp i(\Omega_p + \frac{3\pi}{2})$$

and then

$$K_p = \sum_j \frac{4k^2 m_j J_2 R^2}{3rCr_j^4} a_j i\sigma_j \zeta^{g_j} \sin\beta_j.$$

If we observe that

$$\sin \beta_j = \frac{1}{2} \sigma_j \zeta^{g_j} [H_p - \sin I \exp-i(\theta + \frac{3\pi}{2})] +$$

$$+ \frac{1}{2} \sigma_j^* \zeta^{-g_j} [K_p - \sin I \exp i(\theta + \frac{3\pi}{2})] - \frac{1}{2} (K_j + H_j)$$

we obtain

$$DK_p + \kappa_p K_p = \tilde{V}_{0(p)} + \tilde{B}_{1(p)} H_p + \sum_j \tilde{V}_{3(p)} (K_j + H_j)$$

and in a similar way

$$DH_p - \kappa_p H_p = -\tilde{V}_{0(p)} - V_{1(p)}^* K_p - \sum_j \tilde{V}_{3(p)}^* (K_j + H_j)$$

where

$$\tilde{V}_{0(p)} = \sum_j - \frac{2k^2 m_j J_2 R^2}{3Crv_3^2 a_j^3} \sin I.\exp-i(\theta + \frac{3\pi}{2})\sigma_j^2 \zeta^{2g_j} + \kappa_p \sin I$$

$$\exp i(\theta + \frac{3\pi}{2})$$

$$V_{1(p)} = \sum_j \frac{2k^2 m_j J_2 R^2}{3Cr v_3^2 a_j^3} \sigma_j^2 \zeta^{2g_j}$$

$$\tilde{V}_{3(p)} = - \sum_j \frac{2k^2 m_j J_2 R^2}{3Cr v_3^2 a_j^3} \sigma_j \zeta^{g_j}$$

and

$$\kappa_p = - \sum_j \frac{2k^2 m_j J_2 R^2}{3Cr v_3 a_j^3}$$

6. SOLAR PERTURBATIONS

The disturbing actions due to the Sun are given by the same relations as for the satellites when i=o (see Section 2). The difference is
that in this case the α_{j0} are very small (5×10^{-4} to 3×10^{-3}). The quantity
r_{0j}^{-3} may be expanded in the powers of α_{j0} without the need of introducing
new Laplace coefficients. Up to the second-order we have

$$r_{0j}^{-3} = r_0^{-3} \{ 1 + \frac{3}{2} \alpha_{j0} (\frac{a_0}{r_0})^2 (u_0 s_j + s_0 u_j) - \frac{3}{2} \alpha_{j0}^2 (\frac{a_0}{r_0})^2 u_j s_j +$$

$$+ 3\alpha_{j0} (\frac{a_0}{r_0})^2 z_0 z_j + \frac{15}{8} \alpha_{j0}^2 (\frac{a_0}{r_0})^4 (u_0 s_j + s_0 u_j)^2 \},$$

where

$$u_j = \sigma_j \zeta^{g_j} (1 + U_j); \qquad s_j = \sigma_j^* \zeta^{-g_j} (1 + S_j) \qquad j=0,1,2,3,4.$$

After the substitutions

$$\mathcal{R}_{j0} = \sum_{k \neq 0,2} R_{1(j0)}^{(k)} + R_{2(j0)}^{(k)} D^{-1} P_j + R_{3(j0)}^{(k)} D^{-1} Q_j + R_{4(j0)}^{(k)} D^{-2}$$

$$(P_j - Q_j) \} \xi_{0j}^k$$

$$\mathcal{T}_{j0} = \sum_{k \neq 0,-2} \{ T_{1(j0)}^{(k)} + T_{2(j0)}^{(k)} D^{-1} P_j + T_{3(j0)}^{(k)} D^{-1} Q_j +$$

$$+ T_{4(j0)}^{(k)} D^{-2} (P_j - Q_j) \} \xi_{0j}^k$$

$$\mathcal{V}_{j0} = - \frac{3}{2} \frac{k^2 m_0 z_0}{v_3^2 r_0^3} (\frac{a_0}{r_0})^2 [(1 + U_0) \xi_{0j} + (1 + S_0) \xi_{0j}^{-1}] - \frac{1}{2} \frac{k^2 m_0}{v_3^2 r_0^3} (K_j + H_j)$$

and also

$$X_{j0} = \frac{3}{2}(\mathscr{R}_{j0} - \mathscr{T}_{j0}) + \frac{1}{2} \kappa_j D^{-1} (P_j - Q_j)(\mathscr{R}_{j0}^I - \mathscr{T}_{j0}^I) +$$

$$+ \frac{3}{8} [\kappa_j D^{-1} P_j - 3\kappa_j D^{-1} Q_j + 3\kappa_j^2 D^{-2}(P_j - Q_j)]\mathscr{R}_{j0}^I -$$

$$- \frac{3}{8} [3\kappa_j D^{-1} P_j - \kappa_j D^{-1} Q_j - 3\kappa_j^2 D^{-2}(P_j - Q_j)]\mathscr{T}_{j0}^I$$

where $\mathscr{R}_{j0}^I$ and $\mathscr{T}_{j0}^I$ represent the first-order terms of $\mathscr{R}_{j0}$ and $\mathscr{T}_{j0}$. The $R_{m(j0)}^{(k)}$ and $T_{m(j0)}^{(k)}$ are numerical coefficients.

The introduction of the solar perturbations does not give rise to any difficulties except for the complexity of the formulae giving the relative motion of Jupiter around the Sun. In this theory U_0, S_0 and Z_0 are considered as known functions of the time through the Theory of Hill (Hill, 1890) and they are written in a form compatible with the kind of algebra adopted here. All terms which give rise to long-period inequalities of amplitude larger than 0.1 arcsecond, and especially those that give rise to secular inequalities must be kept. In this paper, for the sake of simplicity, we restrain ourselves to

$$\mathscr{R}_{j0} = R_{1(j0)}^{(0)} + R_{1(j0)}^{(2)} \xi_{0j}^2$$

$$\mathscr{T}_{j0} = R_{1(j0)}^{(0)} + R_{1(j0)}^{(2)} \xi_{0j}^{-2}$$

These restrictions are justified because $k^2 m_0/\nu_3^2 r_0^3$ is a second-order quantity; the terms which are not considered are quantities of third-order. These restrictions do not affect the presentation of the theory itself but only the actual calculations.

7. THE INTEGRO-DIFFERENTIAL EQUATIONS. FINAL FORM

In order to obtain the final form of the integro-differential equations for P_j and Q_j we must collect the partial contributions of $\mathscr{R}_j$, $\mathscr{T}_j$ and χ_j and to introduce them in the equations of Section 1. So we obtain

$$\begin{pmatrix} D\bar{P} \\ D\bar{Q} \end{pmatrix} + \begin{pmatrix} X_0 & 0 \\ 0 & -X_0 \end{pmatrix} \begin{pmatrix} \bar{P} \\ \bar{Q} \end{pmatrix} + \frac{1}{8} \begin{pmatrix} 7A^I & -5A^I \\ 5A^I & -7A^I \end{pmatrix} \begin{pmatrix} \bar{P} \\ \bar{Q} \end{pmatrix} + \begin{pmatrix} V_{11} \\ V_{12} \end{pmatrix} + \begin{pmatrix} V_{21} \\ V_{22} \end{pmatrix} +$$

$$+ \begin{pmatrix} M_{11} & M_{12} \\ M_{21} & M_{22} \end{pmatrix} \begin{pmatrix} D^{-1}\bar{P} \\ D^{-1}\bar{Q} \end{pmatrix} + \begin{pmatrix} N_{11} & N_{12} \\ N_{21} & N_{22} \end{pmatrix} \begin{pmatrix} D^{-2}\bar{P} \\ D^{-2}\bar{Q} \end{pmatrix} = 0$$

where $\bar{P}$ and $\bar{Q}$ are 4-vectors whose components are respectively $P_1, P_2, P_3,$ P_4 and Q_1, Q_2, Q_3, Q_4; $X_0 = \text{diag}(\kappa_j)$;

$$A^I = \text{diag} \left\{ \sum_{i \neq j} \sum_{-\infty}^{+\infty} X_{0(ji)}^{(k)} \; \xi_{ij}^k \right\};$$

V_{11} and V_{12} are the 4-vectors whose components are

$$V_{11(j)} = -\frac{\lambda_j - \kappa_j^2}{\kappa_j} - \frac{1}{\kappa_j} R_{1(j)} - \sum_{i \neq j} \sum_{-\infty}^{+\infty} \left(\frac{1}{\kappa_j} R_{1(ji)}^{(k)} + \frac{1}{4\kappa_j} X_{1(ji)}^{(k)} \right) \xi_{ij}^k$$

$$V_{21(j)} = -\frac{1}{\kappa_j} \mathcal{R}_{j0} - \frac{1}{4\kappa_j} \mathcal{T}_{j0}$$

and M_{11}, M_{12} and M_{11} are matrices which components are

$$M_{11(jj)} = \frac{3}{4}(\lambda_j - \kappa_j^2) - \frac{1}{\kappa_j} \tilde{R}_{2(j)} - \sum_{i \neq j} \sum_{-\infty}^{+\infty} \left([2;3] - \frac{7}{8}\kappa_j X_{0(ji)}^{(k)} \right) \xi_{ij}^k$$

$$M_{11(ji)} = -\sum_{-\infty}^{+\infty} \left(\frac{1}{\kappa_j} R_{5(ji)}^{(k)} + \frac{1}{4\kappa_j} X_{5(ji)}^{(k)} \right) \xi_{ij}^k \qquad (j \neq i)$$

$$M_{12(jj)} = -\frac{5}{4}(\lambda_j - \kappa_j^2) - \frac{1}{\kappa_j} \tilde{R}_{3(j)} - \sum_{i \neq j} \sum_{-\infty}^{+\infty} \left([3;-1] + \frac{7}{8}\kappa_j X_{0(ji)}^k \right) \xi_{ij}^k$$

$$M_{12(ji)} = -\sum_{-\infty}^{+\infty} \left(\frac{2}{\kappa_j} R_{6(ji)}^{(6)} + \frac{1}{4\kappa_j} X_{6(ji)}^{(k)} \right) \xi_{ij}^k \qquad (j \neq i)$$

$$N_{11(jj)} = \frac{3}{4}\kappa_j(\lambda_j - \kappa_j^2) - \frac{1}{\kappa_j} \tilde{R}_{4(j)} - \sum_{i \neq j} \sum_{-\infty}^{+\infty} [4;-3\kappa_j] \xi_{ij}^k$$

$$N_{11(ji)} = -\sum_{-\infty}^{+\infty} \left(\frac{1}{\kappa_j} R_7^{(k)}{}_{(ji)} + \frac{1}{4\kappa_j} X_{7(ji)}^{(k)} \right) \xi_{ij}^k \qquad (j \neq i)$$

where the brackets mean

$$[a;b] = \frac{1}{\kappa_j} R_{a(ji)}^{(k)} + \frac{1}{4\kappa_j} X_{a(ji)}^{(k)} + \frac{1}{16} b X_{1(ji)}^{(k)}.$$

We still have $V_{12} = V_{11}^*$, $V_{22} = V_{21}^*$, $N_{21} = M_{11}^*$, $M_{22} = M_{12}^*$, $N_{12} = -N_{11}$, $N_{21} = N_{11}^*$ and $N_{22} = -N_{11}^*$ where the asterisks indicate complex conjugation.

This equation still may be written using vectors and matrices of rang 8:

$$XP + \hat{X}_0 P + L_1 P + V_1 + V_2 + M_1 D^{-1} P + N_1 D^{-2} P = 0$$

where the symbols introduced have obvious meanings; still, in every case the subscripts give the order of magnitude of the elements they represent.

8. THE AVERAGING METHOD OF KRASINSKY

In order to reduce the integro-differential linear equation we introduce the linear transformation (the subscripts indicate the order of the term)

$$P = \hat{P} + B_1 + B_2 + C_1 \hat{P} + E_1 D^{-1} \hat{P} + T_1 D^{-2} \hat{P}.$$

It follows the new integro-differential equation:

$$D\hat{P} + DB_1 + DB_2 + DC_1 . \hat{P} + C_1 . D\hat{P} + DE_1 . D^{-1} \hat{P} + E_1 . \hat{P} + DT_1 . D^{-2} \hat{P} +$$

$$+ T_1 D^{-1} \hat{P} + \hat{X}_0 \hat{P} + \hat{X}_0 B_1 + \hat{X}_0 B_2 + \hat{X}_0 C_1 \hat{P} + \hat{X}_0 E_1 D^{-1} \hat{P} + \hat{X}_0 T_1 D^{-2} \hat{P} +$$

$$+ L_1 \hat{P} + L_1 B_1 + V_1 + V_2 + M_1 D^{-1} \hat{P} + M_1 D^{-1} B_1 + N_1 D^{-2} \hat{P} +$$

$$+ N_1 D^{-2} B_1 = 0.$$

The optimum transformation would be obtained if we could find constant vectors b_1 and b_2, and constant matrices e_1, f_1 and g_1 such that

$$(I+C_1)^{-1} (DB_1 + DB_2 + \hat{X}_0 B_1 + \hat{X}_0 B_2 + L_1 B_1 + V_1 + V_2 + M_1 D^{-1} B_1 + N_1 D^{-2} B_1) = b_1 + b_2 = b_1$$

$$(I+C_1)^{-1} (DC_1 + E_1 + \hat{X}_0 + \hat{X}_0 C_1 + L_1) = \hat{X}_0 + e_1$$

$$(I+C_1)^{-1} (DE_1 + T_1 + \hat{X}_0 E_1 + M_1) = f_1 = 0$$

$$(I+C_1)^{-1} (DT_1 + \hat{X}_0 T_1 + N_1) = g_1 = 0$$

if we restrict these equations to the second order and if we equate separately the terms of different orders, it follows that

$$DB_1 + \hat{X}_0 B_1 + V_1 = b_1$$

$$DT_1 + \hat{X}_0 T_2 + N_1 = g_1 = 0$$

$$DE_1 + T_1 + \hat{X}_0 E_1 + M_1 = f_1 = 0$$

$$DC_1 + \hat{X}_0 C_1 - C_1 \hat{X}_0 + E_1 + L_1 = e_1$$

$$DB_2 + \hat{X}_0 B_2 + L_1 B_1 + M_1 D^{-1} B_1 + N_1 D^{-2} B_1 + V_2 - C_1 b_1 = b_2 = 0$$

These equations have been written following their increasing difficulty. If they are solved adequately the integro-differential equation becomes

$$D\hat{P} \mid \hat{X}_0 \hat{P} \mid e_1 \hat{P} \mid b_1 = 0,$$

and it may be solved by elementary methods.

The fact that g_1, f_1 and b_2 are made equal to zero will become clear in the following section.

8. DISCUSSION OF THE EQUATIONS

To discuss the first equation let it be decomposed into its two parts of rang 4:

$$DB_{11} + X_0 B_{11} + V_{11} = b_{11}$$

$$DB_{12} - X_0 B_{12} + V_{12} = b_{12};$$

the complex conjugate of the first part is

$$DB_{11}^* - X_0 B_{11}^* - V_{11}^* = -b_{11}$$

$(b_{11} = b_{11}^*$ since this vector is constant). The comparison lead to

$$B_{12} = B_{11}^* \qquad \text{and} \qquad b_{12} = -b_{11}.$$

The problem is reduced to the solution of only one of the two rang-4 equations. Equating the elements for the first of them we have

$$DB_{11(j)} + \kappa_j B_{11(j)} + V_{11(j)} = b_{11(j)}$$

and we adopt the solution

$$B_{11(j)} = \sum_{i \neq j} \sum_{\substack{-\infty \\ k \neq 0}}^{+\infty} \frac{1}{\kappa_j + k(g_i - g_j)} \left(\frac{1}{\kappa_j} R_{1(ji)}^{(k)} + \frac{1}{4\kappa_j} X_{1(ji)}^{(k)} \right) \xi_{ij}^k$$

$$b_{11(j)} = -\frac{1}{\kappa_j} (\lambda_j - \kappa_j^2 + \tilde{R}_{1(j)}) - \sum_{i \neq j} \frac{1}{4\kappa_j} (4R_{1(ji)}^{(0)} + X_{1(ji)}^{(0)})$$

i.e., we put the secular terms in b_1 and the periodic terms in B_1.
   Let the second equation be decomposed:

$$DT_{11} + X_0 T_{11} + N_{11} = g_{11}$$

$$DT_{21} - X_0 T_{21} + N_{11}^* = g_{21}$$

$$DT_{12} + X_0 T_{12} - N_{11} = g_{12}$$

$$DT_{22} - X_0 T_{22} - N_{11}^* = g_{22}.$$

As for the first equation we obtain

$$T_{21} = -T_{11}^* \qquad T_{22} = -T_{12}^* \qquad T_{12} = -T_{11}$$

$$g_{21} = g_{11} \qquad g_{22} = g_{12} \qquad \dot{g}_{12} = -g_{11}$$

The question is then reduced to the solution of only one equation

$$DT_{11} + X_0 T_{11} + N_{11} = g_{11}$$

and we adopt the solution

$$T_{11(jj)} = -\frac{3}{4}(\lambda_j - \kappa_j^2) + \frac{1}{\kappa_j^2} R_{4(j)} - \sum_{i \neq j} \sum_{-\infty}^{+\infty} \frac{[4;-3\kappa_j]\xi_{ij}^k}{k(g_i - g_j) + \kappa_j}$$

$$T_{11(ji)} = \sum_{-\infty}^{+\infty} \frac{4R_{7(ji)}^{(k)} + X_{7(ji)}^{(k)}}{4\kappa_j[k(g_i - g_j) + \kappa_j]} \xi_{ij}^k \qquad (j \neq i)$$

$$g_{11(jj)} = g_{11(ji)} = 0.$$

The condition of validity of the above solution is

$$k(g_i - g_j) + \kappa_j \neq 0 \qquad\qquad\qquad (i \neq j)$$

or, since $\kappa_j = g_j + m \ (m=-0.01448)$:

$$k(g_i - g_j) + g_j + m \neq 0. \qquad\qquad\qquad (i \neq j)$$

For the three inner satellites g_j are integers (4,2, and 1); the condition of validity reduces then to $m \notin Z$. For the fourth satellite $g_4=0.437$.
It is easy to see that we do not have a small divisor. In the problem of
the motion of an asteroid disturbed by Jupiter this leads to exclude the
commensurabilities of the kind

$$a\kappa_i + b\kappa_j = 0; \qquad |a + b| = 1. \qquad (a,b \in Z).$$

The third equation may be discussed exactly in the same manner as the former two. The fourth equation shows special features and will be explicitly discussed. After decomposition we have for this equation two independent relations:

$$DC_{11} + X_0 C_{11} - C_{11} X_0 + E_{11} + L_{11} = e_{11}$$

$$DC_{12} + X_0 C_{12} + C_{12} X_0 + E_{12} + L_{12} = e_{12}.$$

The other two relations do not need to be considered since we can prove that $C_{22} = C_{11}^*$, $C_{21} = C_{12}^*$, $e_{22} = -e_{11}^*$ and $e_{21} = -e_{12}^*$.

The basic difference for the above equation is that the method of Krasinsky does not allow us to eliminate all periodic terms. For instance, for each (j,i) from the first equation we have

$$DC_{11(ji)} + (\kappa_j - \kappa_i) C_{11(ji)} + E_{11(ji)} + L_{11(ji)} = e_{11(ji)}$$

therefore, if terms factored by ξ_{ij} exist in the non-homogeneous part of the equation, since their motion is exactly $(\kappa_j - \kappa_i)$, they lead to null divisors. These terms must then be excluded from the equation. We adopt

$$e_{11(jj)} = -\frac{1}{\kappa^3} \tilde{R}_{4(j)} + \frac{1}{\kappa^2} \tilde{R}_{2(j)} + \sum_{i \neq j} \{\frac{1}{\kappa_j^2} [4; -\frac{3}{\kappa_j}] + \frac{1}{\kappa_j^3}[2; \frac{3}{\kappa_j}] -$$
$$- \frac{7}{8} X_{0(ji)}^{(0)}\} = \tilde{e}_{jj}$$

$$e_{11(ji)} = -\frac{4R_{7(ji)}^{(1)} + X_{7(ji)}^{(1)}}{4\kappa_j \kappa_i^2} \xi_{ij} - \frac{4R_{5(ji)}^{(1)} + X_{5(ji)}^{(1)}}{4\kappa_j \kappa_i} \xi_{ij} = \tilde{e}_{ji} \xi_{ij}$$

$$e_{12(jj)} = e_{12(ji)} = 0.$$

Now the system giving the components of the matrices C_{11} and C_{12} may be solved. The condition for the validity of the solutions are

$$k(g_i - g_j) + \kappa_i + \kappa_j \neq 0$$

$$k(g_i - g_j) + 2\kappa_j \neq 0$$

$$g_i - g_j \neq 0.$$

These relations exclude two well-known families of commensurabilities:
(a) the trojan commensurabilities

$$a\kappa_i + b\kappa_j ; \qquad a + b = 0 \qquad (a,b \in Z)$$

(b) the commensurabilities

$$a\kappa_i + b\kappa_j ; \qquad |a + b| = 2 \qquad (a,b \in Z).$$

9. THE GALILEAN CRITICAL TERMS

·The fifth equation arising from Krasinsky's method may be decomposed intwo two 4-vector equations which are complex conjugates. Let one
of them be considered.

$$DB_{21} + X_0 B_{21} + \frac{7}{8} A^I B_{11} - \frac{5}{8} A^I B_{12} + M_{11} D^{-1} B_{11} + M_{12} D^{-1} B_{12} +$$

$$+ N_{11} D^{-2} B_{11} - N_{11} D^{-2} B_{12} + V_{21} - C_{11} b_{11} - C_{12} b_{12} = b_{21}$$

The solution which is adopted is

$$b_{21} = 0$$

$$B_{21(j)} = \sum_s \delta_s (B_{21(j)})$$

(sum of parts arising from different terms of the non-homogeneous side
of the equation). The parts which arise respectively from $(\frac{7}{8} A^i B_{11})_{(j)}$
and $(M_{11} D^{-1} B_{11})_{(j)}$ have the form

$$\delta_1 (B_{21(j)}) = \sum_{i \neq j} \sum_{l \neq j} \sum_{-\infty}^{+\infty} \sum_{-\infty}^{+\infty} \frac{A^{kk'}_{jil} \xi^k_{ij} \xi^{k'}_{lj}}{\kappa_j + k(g_i - g_j) + k'(g_1 - g_j)}$$
$$(\neq 0)$$

and

$$\delta_2 (B_{21(j)}) = \sum_{i \neq j} \sum_{l \neq j} \sum_{-\infty}^{+\infty} \sum_{+\infty}^{+\infty} \frac{B^{kk'}_{jil} \xi^k_{ij} \xi^{k'}_{li}}{\kappa_j + k(g_i - g_j) + k'(g_1 - g_i)}$$

They introduce as new conditions of validity

$$\kappa_j + k(g_i - g_j) + k'(g_1 - g_j) \neq 0$$

$$\kappa_j + k(g_i - g_j) + k'(g_1 - g_i) \neq 0$$

$$\delta_9(B_{21(j)}) = (D + \kappa_j)^{-1} \{\frac{1}{\kappa_j}\mathcal{R}_{j0} + \frac{1}{4\kappa_j}\chi_{j0}\}.$$

10. SOLUTION OF THE EQUATIONS FOR P_j and Q_j

The 'averaged' equation is

$$D\ddot{P}_j + \kappa_j\dot{\hat{P}}_j + \sum_{i=1}^{4} e_{11(ij)}\hat{P}_i + b_{1j} = 0,$$

this equation is not actually averaged since

$$e_{11(ij)} = \tilde{e}_{ij}\xi_{ji}.$$

If the variables $\hat{p}_j = \hat{P}_j\sigma_j\zeta^{\kappa_j}$ are introduced, we have the linear differential equation with constant coefficients

$$D\hat{p}_j + \sum_i e_{ij}\hat{p}_i + \sigma_j b_{ij}\xi^{\kappa_j} = 0$$

whose general solution is

$$\hat{p}_j = \sum_{i=1}^{4} A_{ji}\zeta^{-\beta_i};$$

the constants A_{ji} may be known if four among them are supposed as known integration constants. The motions β_i are the roots of the characteristic equation

$$\det(\tilde{e}_{ij} - \beta\delta_{ij}) = 0.$$

These characteristic roots are the motions of the proper perijoves of the satellites in unities of the Eulerian mean motion ν_3.
        The particular solution of the complete equations is easily obtained; it gives rise to terms having the same frequencies as those of the satellites. The mean feature of the solution obtained in this way is the Laplace's result after which it is not possible to separate the proper oscillations. The system oscillates as a whole: every satellite shows in its motion the four proper oscillations.

11. THE EQUATIONS FOR K_j AND H_j

In order to obtain the final form of the differential equations for K_j and H_j we must collect the disturbing terms and substitute them in the space equations given in Section 1. It follows which are satisfied in the problem of the motion of the Galilean satellites because $m \notin Z$.

The Galilean critical terms are those for which

$$k(g_i - g_j) + k'(g_1 - g_j) = 0$$

$$k(g_i - g_j) + k'(1 - g_i) = 0.$$

They may arise in two different ways± (a) the trojan resonance, which happens in the case $i = 1$, when $g_i = g_j$, i.e. in the case

$$ak_i + bk_j = 0; \qquad a + b = 0 \qquad (a,b \in Z),$$

(b) the Galilean resonance, which happens in the case $i \neq 1$, when

$$ak_i + bk_j + ck_1 = 0; \qquad a + b + c = 0 \qquad (a,b,c \in Z).$$

In this case it is possible to find a system of rotating axes for which k_i, k_j, k_1 reduce to three integers g_i, g_j, g_1 (as it has been made in this theory). Indeed, if $k_i = g_i + m$, since $a+b+c = 0$, we have

$$ag_i + bg_j + ch_1 = 0$$

or

$$a(g_i - g_1) + b(g_j - g_1) = 0;$$

the last equation has always integer solutions. We may still choose the rotating axes so that

$$ag_i - bg_1 = ag_1 - bg_j = 0$$

i.e.

$$g_1^2 = g_i \cdot g_j$$

which gives rise to the following situation

$$g_i = 1, \qquad g_1 = a, \qquad g_j = a^2 \qquad (a \in Z);$$

the Galilean satellites are such that $m=0.01448$ and $a = 2$.

The Galilean critical terms do not disturb the integration of the integro-differential equation; they will actually become critical when the geometric equations are considered.

The other parts of the vector equation will introduce terms exactly like the two discussed above. The only part which introduces terms which are different, is that which contains the solar perturbations:

$$DK_j + \kappa_j K_j = \frac{1}{\kappa_j} \sum_{i \neq j} \sum_{-\infty}^{+\infty} \{ (V_{1(ji)}^{(k)} + \frac{1}{2} R_{1(ji)}^{(k)}) \xi_{ij}^k + \frac{2}{3\kappa_j} \tilde{V}_{1(j)} \} (K_j + H_j) +$$

$$+ \frac{1}{\kappa_j} \sum_{i \neq j} \sum_{-\infty}^{+\infty} V_{2(ji)}^{(k)} \xi_{ij}^k (K_i + H_i) +$$

$$+ \frac{1}{\kappa_j} \{ \tilde{V}_{3(j)} K_p + \tilde{V}_{4(j)} H_p + \tilde{V}_{0(j)} + \mathscr{V}_{jo} \}$$

and its complex conjugate for $-DH_j + \kappa_j H_j$. To these equations we must add the differential equations giving the motion of the pole of Jupiter:

$$DK_p + \kappa_p K_p = \tilde{V}_{0(p)} + \tilde{V}_{1(p)} H_p + \sum_i \tilde{V}_{3(p)} (K_i + H_i)$$

$$DH_p - \kappa_p H_p = -\tilde{V}_{0(p)} - \tilde{V}_{1p}^* K_p - \sum_i \tilde{V}_{3(p)}^* (K_i + H_i).$$

All these equations may be written using vectors and matrices of rang 10

$$DX + \hat{X}_0' K + L_1' K + V_1' + V_2' = 0,$$

where the new symbols have evident meanings. This vector equation may be averaged by means of Krasinsky's method, by using the transformation

$$K = \hat{K} + B_1' + B_2' + C_1' \hat{K}.$$

The averaged equation will have the form

$$D\hat{K} + \hat{X}_0' \hat{K} + e_1 \hat{K}` = 0$$

and may be solved by elementary methods like those used in the preceding section. The main feature of the solution thus obtained is Laplace's result after which it is not possible to separate the proper oscillations in latitude as well as the proper oscillation of the planetary equator. The systems oscillates as a whole and the plane of the motion of each satellite and the plane of Jupiter's equator show the five proper oscillations.

The details of the actual calculations are nothing but similar to those for P_j and Q_j.

12. THE GEOMETRIC EQUATIONS

At least we need to consider the geometric equations

$$D\varepsilon_j = \frac{1}{2} \kappa_j (P_j - Q_j) - \frac{1}{4} \kappa_j (P_j + Q_j) \Gamma_j + \frac{1}{4} \kappa_j (K_j + H_j)(K_j - H_j)$$

$$D\Gamma_j = -3\kappa_j\epsilon_j + \frac{1}{2}\kappa_j(P_j+Q_j) - \frac{3}{2}\kappa_j(P_j+Q_j)\epsilon_j -$$

$$-\frac{1}{4}\kappa_j(P_j-Q_j)\cdot\Gamma_j + \frac{1}{4}\kappa_j[(K_j-H_j)^2 - \frac{1}{2}(H_j+H_j)^2] + 2\kappa_j\epsilon_j^2.$$

Now P_j, Q_j, K_j, H_j are known functions since these variables have been supposed to be integrated in the preceding sections. These functions contain the four circulatory frequencies g_j, the four oscillatory frequencies $\tilde{w}_j = \beta_j v_3$ and the five oscillatory frequencies $\Omega_j = \beta'_j v_3$. The geometric equations may be considered separately for each satellite. For this reason in what follows subscripts will be omitted. We have, after substitutions

$$D\epsilon = S_1 + S_2\Gamma$$

$$D\Gamma = S_4 + (S_5 - 3\kappa)\epsilon + S_6\Gamma + 2\kappa\epsilon^2$$

where

$$S_i = \sum_k S_{ik}\zeta^{(k)};$$

$\zeta^{(k)}$ means ζ to the power arising in the k-th term of the series. Two particular cases are of importance: $\zeta^{(k)} = \zeta^0$ and $\zeta^{(k)} = \zeta^G$ where G is the critical Galilean frequency

$$G = g_1 - 3g_2 + 2g_3.$$

All series are first-order and the series S_1 and S_4 contain also second-order parts. The galilean critical frequency may appear only in these second-order parts. The averaging method of Krasinsky is used notwithstanding the fact that the geometric equations are not linear. Such use is possible since the coefficient of the non-linear term is a constant. We consider the transformation

$$\epsilon = Y_1 + (1 + Y_2)\hat{\epsilon} + Y_3\hat{\Gamma}$$

$$\Gamma = Y_4 + (1 + Y_5)\hat{\Gamma} + Y_6\hat{\epsilon},$$

where the Y_j are all first-order and Y_1 and Y_4 have also second-order parts. After substitution we have

$$D\hat{\epsilon} = S_1 - DY_1 + S_2\hat{\Gamma} + S_2Y_4 - \hat{\epsilon}DY_2 - \hat{\Gamma}DY_3 - Y_2(S_1-DY_1) - Y_3(S_4-DY_4)$$

$$+ 3\kappa Y_3(Y_1+\hat{\epsilon})$$

$$D\hat{\Gamma}=S_4-DY_4+S_5\hat{\epsilon}=S_6\hat{\Gamma}+S_5Y_1+S_6Y_4+2\kappa\hat{\epsilon}^2+2\kappa Y_1^2 +$$

$$+4\kappa\hat{\epsilon}Y_1-\hat{\Gamma}DY_5-\hat{\epsilon}DY_6-3\kappa\hat{\epsilon}-Y_5(S_4-DY_4)-Y_6(S_1-DY_1) +$$

$$+3\kappa Y_5(Y_1+\hat{\epsilon})-3\kappa Y_1+3\kappa Y_2\hat{\epsilon}-3\kappa Y_3\hat{\Gamma},$$

to reduce these equations to

$$D\hat{\epsilon}=W_1+W_2\hat{\Gamma}+W_3\hat{\epsilon}$$

$$D\hat{\Gamma}=W_4-(3\kappa+W_5)\hat{\epsilon}+W_6\hat{\Gamma}+2\kappa\hat{\epsilon}^2,$$

where the W_j are free from non-libratory periodic oscillations, it is enough to solve

$$S_{11}-DY_{11}=W_{11}$$

$$S_{12}-DY_{12}+S_2Y_{41}-Y_2W_{11}-Y_3W_{41}=W_{12}$$

$$S_2-DY_3=W_2$$

$$3\kappa Y_3-DY_2=W_3$$

$$S_{41}-DY_{41}-3\kappa Y_{11}=W_{41}$$

$$S_{42}-DY_{42}+S_5Y_{11}+2\kappa Y_{11}^2-Y_5W_{41}-Y_6W_{11}-3\kappa Y_{12}+S_6Y_{41}=W_{42}$$

$$S_5-DY_6+4\kappa Y_{11}+3\kappa Y_5+3\kappa Y_2=-W_5$$

$$S_6-3\kappa Y_3-DY_5=W_6$$

in which the series where the indices were 1 and 4 have been split in first-order (indices 11 and 41) and second-order (indices 12 and 42) parts. The integration is easy. The first equation, for example, leads to

$$W_{11}=<S_{11}>$$

$$Y_{11}=D^{-1}(S_{11}-W_{11}),$$

where $<S_{11}>$ means the constant term in S_{11}. In the equations for the second-order parts we need to include in the average not only the constant but also the critical terms.

The difficulties in the integration of the averaged equations, in reason of the critical terms, is the same which appear in Laplace's equations for the inequalities of the mean longitudes. As we know, the critical terms lead to libration in the mean longitudes whose period is close to 6 years and whose amplitude is an integration constant to be determined from the observations. This determination is a very difficult problem since the amplitude is very small. In 1907 de Sitter found $0°\!.158 \pm 0°\!.033$. In 1928, after a new discussion (de Sitter, 1931) he found $0°\!.0247 \pm 0°\!.0075$. In a discussion using only the first satellite he found an amplitude 4 times greater, and using only the second satellite the amplitude remained unchanged but the phase shifted by $100°\!.$ These determinations are very uncertain and they did not affect the mean residuals of the observations. At the same epoch Brouwer (1928), from a different set of observations found $0°\!.0309 \pm 0°\!.0058$ for the amplitude of the libration. Also, when using separately the first or the second satellite he obtained contradictory results. He concluded that the libration is much too small to be determined from the observations.

This fact justifies omitting the libration and integrating the averaged equations using G=0 identically. The solution in this case may be easily obtained.

(a) $W_{11} = 0$; indeed $W_{11} = <S_{11}> = \frac{1}{2}\kappa_j<P_j-Q_j> = 0$, since P_j and Q_j are complex conjugates and so P_j-Q_j is a sinus series.

(b) $W_3 = 0$; the solution Y_3 of the equation $S_2-DY_3=W_2$ is an odd function because the parity of the series is changed by the operator D^{-1}. The result then follows as in (a).

(c) $W_{12} = 0$; the proof of this statement is longer. By Krasinsky's method we have $W_{12} = <S_{12} + S_2Y_{41} - Y_2W_{11} - Y_3W_{41}>$; the analysis of the parities must be made with details but it is algebraically elementary. W_{12} will have the form $A(\zeta^G-\zeta^{-G})$ and its average is zero under the hypotheses previously adopted.

(d) $W_6 = 0$; this result follows as in (b).

Thus, the remaining system is

$$D\hat{\varepsilon}= W_2\hat{\Gamma}$$

$$D\hat{\Gamma}= W_4-(3\kappa-W_5)\hat{\varepsilon}+2\kappa\hat{\varepsilon}^2$$

and we have

$$D^2\hat{\varepsilon} = -\nu_3^2 \frac{d^2\hat{\varepsilon}}{dt^2} = W_2W_4 - (3\kappa-W_5)W_2\hat{\varepsilon} + 2\kappa W_2\hat{\varepsilon}^2.$$

This equation is the equation of a non-linear oscillator. Its stationary solution is

$$\hat{\varepsilon} = \hat{\varepsilon}_E = \frac{W_4}{3\kappa} [1 + \frac{W_5}{3\kappa} + \frac{2W_4}{9\kappa}].$$

If this stationary solution is substituted in the equation for $\hat{\Gamma}$ it follows that $D\hat{\Gamma}$ is equal to a non-zero constant: $-2W_4^2/9$. Then $\hat{\Gamma}(t)$ and $\phi(t)$, will have one term which is linear with respect to time. The existence of such linear term contradicts the working hypothesis that the frequencies n_i are exactly the observed mean motions. Thus, this linear term may not exist. We impose $W_4=0$, in order to have the solutions $\hat{\Gamma}=0$, $\hat{\epsilon}=0$. The equations $W_4=0$ determine the normalization parameters a_j, mean distances from the satellites to the planet.

This technique may be compared to the technique recently suggested by Eminhizer et al. (1976). Indeed the variational equations built with respect to a given set of circular orbits (intermediate orbits) allow us to introduce a fixed set of frequencies; these equations are a 'relocated' version of the equations of motion. Note that the 'relocation' in this case is restricted only to the mean motions. The Eminhizer et al.'s 'forward scheme' would consist in the improvement of the introduced frequencies while the mean distances a_j were kept fixed (frequencies normalization). The technique adopted in this theory is similar to Eminizer et al.'s 'backward scheme'; the frequencies are kept fixed and the value of the amplitudes a_j is improved (amplitudes renormalization). In both cases the non-zero solutions of the averaged geometric equations correspond to orbits in the neighbourhood of one central orbit. This theory allows us to obtain solutions without secular terms. It is hoped that the rate of formal convergence of the solution is better than in solutions founded on the osculating frequencies

CONCLUSION

The theory outlined in a preceding paper has been developed up to the second order. The integration has been performed and the possibility of doing it completely has been demonstrated. The techniques used are able to show the main known features of the motion. They may also show new features especially when the perturbations arising from the Sun are completely considered. New features may also arise when the coupling of the oscillations in longitude and in latitude are considered - indeed, on several occasions very weak coupling terms arose and have been neglected since they are of higher orders.

ACKNOWLEDGEMENTS

This research has been supported partly by the Brazilian Council for Scientific and Technological Development, procs. CPNq 13364/72 and 3728/73 and by the Research Foundation of São Paulo, proc. FAPESP 76/0166.

REFERENCES

Brouwer, D.: 1928, *Ann. Sterrew. Leiden* 16(1).
Brouwer, D. and Clemence, G.M.: 1961, *Methods of Celestial Mechanics*, Academic Press, New York.

de Sitter, W.: 1931, *Monthly Notices Roy. Astron. Soc.* 91, 706.

Emihhizer, C.R., Helleman, R.H.G., and Montroll, E.W.: 1976, *J. Math. Phys.* 17, 121.

Ferraz-Mello, S.: 1966, *Bull. Astron. (3e série)*, 1, 287.

Ferraz-Mello, S.: 1972, *Dinâmica dos Sistema Galileano*, Inst. Tecn. Aeron., Sao José dos Campos.

Ferraz-Mello, S.: 1974, in Y. Kozai (ed.), *The Stability of the Solar System and Small Stellar Systems*, D. Reidel, Dordrecht, p. 167.

Hill, G.W.: 1890, *Astron. Papers Amer. Ephemeris* 4.

Sganier, J.L.: 1973, *Astron. Astrophys.* 25, 113.

Tisserand, F.: 1896, *Traité de mécanique céleste*, Vol. 4, Gauthier-Villars, Paris.

SOLAR PERTURBATIONS IN SATURNIAN SATELLITE MOTIONS AND IAPETUS-TITAN
INTERACTIONS

Yoshihide Kozai
Tokyo Astronomical Observatory

ABSTRACT

Solar perturbations in Saturnian satellite motions are computed
with accuracy of 10^{-5} to try to analyze observed data of orbital elements for Titan. Perturbations due to Iapetus in Titan's orbit are
also developed by taking into account the motion of the orbital plane
of Iapetus. Then the mass of Iapetus is determined by the motion of
the orbital plane of Titan. Also oblateness parameters of Saturn and
the mass of Rhea are determined by seven secular motions for six
satellites. It is also found in the analysis that G. Struve's values
for the semi-major axes adopted in almanacs differ from the computed
values by using the new data.

The full text of this paper was submitted to the Publications of the
Astronomical Society of Japan under the title "Masses of Satellites
and Oblateness Parameters of Saturn."

IMPROVEMENT OF ORBITS OF SATELLITES OF SATURN USING PHOTOGRAPHIC
OBSERVATIONS

A.T. Sinclair
Royal Greenwich Observatory

ABSTRACT

A programme of photographic observations of the satellites of
Saturn, and the method of reduction and analysis are described. The
observations have been used to update existing theories of Tethys,
Dione, Rhea and Titan. A new theory of the motion of Iapetus has been
fitted to the observations.

NEW ORBITS FOR ENCELADUS AND DIONE BASED ON THE PHOTOGRAPHIC
OBSERVATIONS

W.H. Jefferys, J.D. Mulholland and L.M. Ries
The University of Texas at Austin

ABSTRACT

A program is underway at the McDonald Observatory to extend the
series of photographic observations of the satellites of the outer
planets (Abbot, Mulholland and Shelus, A.J. 80, 1975), and concurrent
theoretical studies have led to a new orbital theory for the resonant
pair of satellites, Enceladus and Dione (Jefferys and Ries, A.J. 80,
1975). The construction of the new theory, using the computer software system TRIGMAN, has provided Fortran subroutines for the computation of the planetocentric coordinates of the two satellites, as
well as partial derivatives for the orbit elements and certain other
physical parameters of the orbit problem, including some of the harmonics of the gravitational field of Saturn. The available photographic observations for these two objects are currently being discussed with the new theory, and improved values of the orbital parameters are expected in the near future.

LONG PERIODIC VARIATION OF ORBITAL ELEMENTS OF A SATELLITE PERTURBED BY DISCRETE GRAVITY ANOMALIES

M.P. Ananda
Jet Propulsion Laboratory
Pasadena, California

ABSTRACT

A method for generating long periodic variations in satellite orbital elements when perturbed by discrete gravity anomalies is presented. The method consists of developing a disturbing potential as a function of orbital and gravity anomaly parameters, and generatii partial derivatives of the potential with respect to the orbital elements. The partials are averaged over the period of the satellite to eliminate the short periodic variations. The averaged partials are substituted into the variation of parameter equations to give the meaɪ orbital rates. Classically orbital elements are used in generating gravity field and thus the method is dynamic in nature. The problem is extremely cumbersome and complex when multi-state parameters have to be estimated from a considerably large data set. However, when mean orbital rates are used, the problem reduces to a simple linear static case, where only the gravity parameters have to be estimated, and it is a simple matrix inversion problem. Thus the method developɛ here was utilized in reducing Appolo 15 and 16 subsatellite radio tracking data to produce a lunar gravity field represented by point masses.

THIRD-ORDER SOLUTION OF AN ARTIFICIAL-SATELLITE THEORY

HIROSHI KINOSHITA
Center for Astrophysics, Harvard College Observatory and
Smithsonian Astrophysical Observatory, Cambridge, Massachusetts
02138

ABSTRACT. A third-order solution is developed for the motions of artificial satellites moving in the gravitational field of the Earth, whose
potential includes the second-, third-, and fourth-order zonal harmonics. Third-order periodic perturbations with fourth-order secular
perturbations are derived by Hori's perturbations method. All quantities
are expanded into power series of the eccentricity, but the solution is
obtained so as to be closed with respect to the inclination. A comparison with the results of numerical integration of the equations of
motion indicates that the solution can predict the position of a close-
earth satellite with a small eccentricity with an accuracy of better
than 1 cm over 1 month.

1. INTRODUCTION

     Second-order theories of artificial satellites have been established
by many authors during the past 15 years; an excellent review is given
by Hori and Kozai (1975). A third-order solution was derived by Deprit
and Rom (1970), but their solution does not include J_3 and J_4. According to Aksnes's numerical experiments comparing his second-order solution (Aksnes, 1970) with numerical integration, residuals of a few
decimeters remain in position, most of which come from the third-order
interaction among J_2, J_3, and J_4. On the other hand, the accuracy of a
recently launched geodetic satellite equipped with retroreflectors for
laser ranging will reach 3 to 5 cm. Therefore, we can expect to obtain
more accurate information on satellite motions by taking into account
the third-order periodic perturbations. Here we consider only the
second-, third-, and fourth-order zonal-harmonics perturbations. We
chose a Keplerian motion as an intermediate orbit and adopted Hori's
(1966) perturbation method. We assumed that the eccentricity of a
geodetic satellite requiring highly accurate solutions is usually low;
therefore, all quantities in the present solution are expanded into
power series of the eccentricity, but the solution is obtained in
closed form with respect to the inclination. Delaunay variables were
selected as the canonical elements used to construct the new Hamiltonian

V. Szebehely (ed.), Dynamics of Planets and Satellites and Theories of Their Motion, 241-257.
All Rights Reserved. Copyright © 1978 by D. Reidel Publishing Company, Dordrecht, Holland.

and the determining functions eliminating periodic terms, which do not
depend on the chosen canonical variables as long as Keplerian motion
is adopted as an intermediate orbit. However, the final expressions of
the periodic perturbations are given in $\ell + g$, h, e cos g, e sin g, and
L, which are not singular at zero eccentricity. All literal calculations
were carried out by means of the computer algebra program Smithsonian
Package for Algebra and Symbolic Mathematics (SPASM) (Hall and Cherniack,
1969). Final results were checked analytically in various ways and were
compared with the results of numerical integration.

2. OUTLINE OF THE METHOD OF SOLUTION

Let us consider an artificial satellite orbiting in an axially
symmetric gravitational field of the earth, whose force function is of
the form

$$U = \frac{\mu}{r} \left[1 - \sum_{n=2}^{4} \left(\frac{a_e}{r} \right)^n J_n P_n (\sin \beta) \right],$$ (1)

where a_e is the equatorial radius of the Earth, r the radius vector of
the satellite, and β the declination. In this paper, the coefficient J_2
is assumed to be a small quantity of the first order and J_3 and J_4 are
of the second order.

To solve for the motion of the satellite under the force function
(1), we adopted Hori's (1966) perturbation method, which utilizes Lie
transformation; all formulas are canonically invariant, and the perturbations of any quantity are given by simple formulas and in explicit
form. Because of the generality of Hori's method, we can choose any set
of canonical variables. In the present paper, we use Delaunay variables
as a canonical set for their simplicity, where

$$L = \sqrt{\mu a}, \qquad \ell = \text{mean anomaly,}$$

$$G = L\sqrt{1 - e^2}, \qquad g = \text{argument of perigee,}$$

$$H = G \cos i, \qquad h = \text{longitude of ascending node.}$$

The equations of motions are

$$\frac{d}{dt} (L,G,H) = \frac{\partial F}{\partial (\ell,g,h)} , \quad \frac{d}{dt} (\ell,g,h) = - \frac{\partial F}{\partial (L,G,H)},$$ (2)

where

$$F = F_0 + F_1 + F_2,$$

$$F_0 = \frac{\mu^2}{2L},$$

$$F_1 = -\frac{\mu a_e^2}{r^3} J_2 P_2 (\sin \beta),$$ (3)

$$F_2 = -\frac{\mu a_e^3}{r^4} J_3 P_3 (\sin \beta) - \frac{\mu a_e^4}{r^5} J_4 P_4 (\sin \beta).$$

Under the assumption that the eccentricity is small, we expanded the disturbing functions F_1 and F_2 into a power series of the eccentricity:

$$F_1 = \frac{\mu^4 J_2}{L^6} \sum_{p=0}^{1} \sum_{k=-\infty}^{\infty} X_k^{2,2p} (e) B_{2,2p}(i) \cos (k\ell + 2pg),$$

$$F_2 = \frac{\mu^5 J_3}{L^8} \sum_{p=0}^{1} \sum_{k=-\infty}^{\infty} X_k^{3,2p+1} (e) B_{3,2p+1}(i) \sin [k\ell + (2p +1) g]$$

$$+ \frac{\mu^6 J_4}{L^{10}} \sum_{p=0}^{2} \sum_{k=-\infty}^{\infty} X_k^{4,2p} (e) B_{4,2p}(i) \cos (k\ell + 2pg),$$ (4)

where $X_k^{n,m}(e)$ is a Hansen coefficient,

$$B_{20} = -\frac{1}{4} (-2 +3 \sin^2 i), \qquad B_{22} = \frac{3}{4} \sin^2 i,$$

$$B_{31} = -\frac{3}{8} \sin i (-4 + 5 \sin^2 i), \qquad B_{33} = \frac{5}{8} \sin^3 i,$$

$$B_{40} = -\frac{3}{8} (8 - 40 \sin^2 i + 35 \sin^4 i),$$

$$B_{42} = \frac{5}{16} \sin^2 i (-6 + 7 \sin^2 i), \qquad B_{44} = -\frac{35}{64} \sin^4 i.$$

When the lowest powers of e and sin i of the coefficient of $\frac{\sin}{\cos}(k\ell + qg)$ in the trigonometric series (4) are α and δ, we have the following relations:

$$\alpha = |k - q| \quad (\text{mod. } 2),$$

$$\delta = |q| \quad (\text{mod. } 2),$$

which are called the d'Alembert characteristics.

The algorithm for deriving the new Hamiltonian F* (Hori, 1966) and the determining function S eliminating short-period terms is

zeroth order:

$$F_0^* = F_0,$$ (5a)

first order:

$$F_1^* = F_{1s},$$

$$S_1 = \frac{L'^3}{\mu^3} \int F_{1p} \, d\ell',$$

second order:

$$F_2^* = F_{2s} + \frac{1}{2} \{F_1 + F_1^*, S_1\}_s,$$

$$S_2 = \frac{L'^3}{\mu^3} \int (F_{2p} + \frac{1}{2} \{F_1 + F_1^*, S_1\}_p) d\ell', \qquad (5b)$$

third order:

$$F_3^* = \frac{1}{12} \{\{F_{1p}, S_1\}, S_1\}_s + \frac{1}{2} \{F_2 + F_2^*, S_1\}_s + \frac{1}{2} \{F_1 + F_1^*, S_2\}_s,$$

$$S_3 = \frac{L'^3}{\mu^2} \int (\frac{1}{12} \{\{F_{1p}, S_1\}, S_1\}_p + \frac{1}{2} \{F_2 + F_2^*, S_1\}_p \qquad (5d)$$

$$+ \frac{1}{2} \{F_1 + F_1^*, S_2\}_p) d\ell',$$

fourth order:

$$F_4^* = \frac{1}{2} \{F_1 + F_1^*, S_3\}_s + \frac{1}{2} \{F_2 + F_2^*, S_2\}_s + \frac{1}{2} \{F_3 + F_3^*, S_1\}_s$$

$$+ \frac{1}{12} \{\{F_{1p}, S_1\}, S_2\}_s + \frac{1}{12} \{\{F_{1p}, S_2\}, S_1\}_s \qquad (5e)$$

$$+ \frac{1}{12} \{\{F_{2p}, S_1\}, S_1\}_s,$$

where the braces represent Poisson brackets and the subscripts s and p indicate the constant and periodic parts in ℓ'. It should be noted that we can add any function of L', G', H', and g' to S.

The algorithm for calculating the new Hamiltonian F** and the determining function S* eliminating long-period terms is given by the following equations:

first order:

$$F_1^{**} = F_1^*, \qquad (6a)$$

second order:

$$F_2^{**} = F_{2s}^*, \qquad (6b)$$

$$S_1^* = - \left(\frac{\partial F_1}{\partial G''}\right)^{-1} \int F_{2p}^* dg'',$$

third order:

$$F_3^{**} = \frac{1}{2} \{F_2^* + F_2^{**}, S_1^*\}_s, \qquad (6c)$$

$$S_2^* = - \frac{1}{2} \left(\frac{\partial F_1^*}{G''}\right)^{-1} \int \{F_2^* + F_2^{**}, S_1^*\}_p dg'',$$

fourth order:

$$F_4^{**} = \frac{1}{12} \{\{F_{2p}^*, S_1^*\}, S_1^*\}_s + \frac{1}{2} \{F_3^* + F_3^{**}, S_1^*\}_s$$

$$+ \frac{1}{2} \{F_2^* + F_2^{**}, S_2^*\}_s \,,$$

$$S_3^* = - (\frac{\partial F_1^*}{\partial G \,''})^{-1} \int (\frac{1}{12} \{\{F_{2p}^*, S_1^*\}, S_1^*\}_p + \frac{1}{2} \{F_3^* + F_3^{**}, S_1^*\}_p$$

$$+ \frac{1}{2} \{F_2^* + F_2^{**}, S_2^*\}_p) \, dg'', \tag{6d}$$

where the subscripts s and p indicate the constant and periodic parts
in g",
  These algorithms are actually very simple, but calculating them
by hand is laborious. Therefore, all computations were carried out by
the computer program SPASM (Hall and Cherniack, 1969). SPASM handles
the operations in (5) and (6) easily, keeping rational fractions for
coefficients.

  A key operation in (5) and (6) is evaluating the Poisson bracket
$\{A,B\}$:

$$\{A,B\} = \frac{\partial A}{\partial L} \frac{\partial B}{\partial \ell} - \frac{\partial A}{\partial \ell} \frac{\partial B}{\partial L} + \frac{\partial A}{\partial G} \frac{\partial B}{\partial g} - \frac{\partial A}{\partial g} \frac{\partial B}{\partial G} + \frac{\partial A}{\partial H} \frac{\partial B}{\partial h} + \frac{\partial A}{\partial h} \frac{\partial B}{\partial H} . \tag{7}$$

In the present theory, as we check to obtain a solution that is closed
with respect to the inclination, the atomic variables in the computer
algebra are L, e, s = sin i, c = cos i, and $\gamma = 1 - 5 \cos^2 i$ (γ appears
in the denominator of S^*). The derivatives with respect to L, G, and H
are

$$\frac{\partial}{\partial L} = (\frac{\partial}{\partial L}) + \frac{1 - e^2}{eL} \frac{\partial}{\partial e} \,,$$

$$\frac{\partial}{\partial G} = \frac{1}{L\sqrt{1-e^2}} (- \frac{1 - e^2}{e} \frac{\partial}{\partial e} + \frac{1 - \sin^2 i}{\sin i} \frac{\partial}{\partial s} - \cos i \frac{\partial}{\partial c}$$

$$+ 10(1 - \sin^2 i) \frac{\partial}{\partial \gamma}) \,,$$

$$\frac{\partial}{\partial H} = - \frac{1}{L\sqrt{1 - e^2}} (\frac{\cos i}{\sin i} \frac{\partial}{\partial s} + \frac{\partial}{\partial c} + 10 \cos i \frac{\partial}{\partial \gamma}) \,,$$

where the factor $(1 - e^2)^{-1/2}$ will be replaced by a Taylor expansion in
powers of e. If both A and B satisfy the d'Alembert characteristics, then
even if these derivatives include the terms 1/e and 1/s in i, the
Poisson bracket $\{A,B\}$ keeps the d'Alembert characteristics and does not
have 1/e and 1/sin i in the expression. This serves as a good check on
literal manipulation by a computer. In Deprit and Rom's (1970)theory,
cos i = $H/L(1 - e^2)^{1/2}$ is developed in power series of e. Therefore,
their determining function W apparently loses the d'Alembert characteristics with respect to the inclination. If A and B are expanded into
power series of e and truncated at e^n, the derivatives with respect to

L and G are correct up to e^{n-2}; however $\{A,B\}$ is correct up to e^{n-1}. In other words, with one operation of the Poisson bracket, only one degree in e (not two) is lost. Our program, which takes this fact into consideration, saves a lot of computer time.

Complete to J_2^3, the analytical solution must take into account all the following terms:

first order:

$$J_2, \quad J_3/J_2, \quad J_4/J_2,$$

second order:

$$J_3, \quad J_3 J_4/J_2^2, \quad J_2^2, \quad J_4,$$

$$(J_3/J_2)^2, \quad (J_4/J_2)^2,$$

third order:

$$J_2 J_3, \quad J_3 J_4^2/J_2^2, \quad (J_3/J_2)^3, \quad J_3 J_4/J_2,$$

$$J_2^3, \quad J_2 J_4, \quad (J_4/J_2)^3, \quad J_3^2 J_4/J_2^3,$$

$$J_3^2/J_2, \quad J_4^2/J_2.$$

Most of these terms arise from the interaction among J_2, J_3, and J_4. Tables I and II list the numbers of terms involved in S_2 and S_3, respectively. The total number of terms with the factor J_2^3 in S^* is 106. On the other hand, Deprit and Rom's W_3 contains 192 terms up to e4, partly because they chose $\eta = H/L$ as one of the arguments. Figures 1 and 2 show parts of S_3 and S_3^*.

Table I
Numbers of terms in the determining function S_3

	J_2^3	$J_2 J_3$	$J_2 J_4$	
e^0	6	6	6	18
e^1	16	8	16	40
e^2	19	18	18	55
e^3	32	19	32	38
e^4	33	30	31	94
Total	106	81	103	290

Table II
Numbers of terms in the determining function S_3^*

	$J_3 J_4^2 J_2^3$	$(J_3/J_2)^3$	$J_3 J_4/J_2$	$J_2 J_3$	$(J_4/J_2)^3$	$(J_3^2 J_4/J_2^3)$	J_4^2/J_2	J_3^2/J_2	$J_2 J_4$	J_2^3	Total
e	5	4	8	9	–	–	–	–	–	–	26
e^2	–	–	–	–	6	7	15	7	15	15	65
e^3	19	14	26	26	–|	–|	–|	–|	–|	–|	86
Total	24	18	34	35	6	7	15	7	15	15	176

Fig. 1. Determining function S_3 with factor J_2^3.

COEFFICIENT OF COS(SG)

$$
\begin{aligned}
& 1/A2^3\ *A3*A4^2\ *E/L^9\ *SI*(-50/ING\ +3325/8/ING^2\ *SI^2\ -616575/512/ING^2\ *SI^4\ +723275/512/ING^2\ *SI^6\ -3543925/6144/ING^2\ *SI^8\ +1125/32*E^2\ /ING^3 \\
& *SI^2\ -120375/512*E^2\ /ING^3\ *SI^4\ +16625/32*E^2\ /ING^3\ *SI^6\ -2925125/6144*E^2\ /ING^3\ *SI^8\ +961625/6144**E^2\ /ING^3\ *SI^{10}\ -2925/8*E^2\ /ING^2 \\
& /128*E^2\ /ING^2\ *SI^2\ -71025/8*E^2\ /ING^2\ *SI^4\ +7975975/768*E^2\ /ING^2\ *SI^6\ -12986225/3072*E^2\ /ING^2\ *SI^8\ ) \\[4pt]
& +\ 1/A2^3\ *A3^3\ *E/L^5\ *SI*(35/32/ING*SI^4\ +1/6-35/48*SI^2\ -25/64*E^2\ /ING^2\ *SI^2\ +25/32*E^2\ /ING^2\ *SI^4\ -35/32*E^2 \\
& /ING+1055/128*E^2\ /ING*SI^2\ -915/128*E^2\ /ING*SI^4\ +47/48*E^2\ -575/192*E^2\ *SI^2\ ) \\[4pt]
& +\ 1/A2*A3*A4*E/L^9\ *SI*(150/ING\ -15825/16/ING^2\ *SI^2\ +574175/256/ING^2\ *SI^4\ -1640825/768/ING^2\ *SI^6\ +750575/1024/ING^2\ *SI^8\ +285/4/ING \\
& -21705/64/ING*SI^2\ +125555/256/ING*SI^4\ -115605/512/ING*SI^6\ +75/32*E^2\ /ING^3\ *SI^2\ -13225/256*E^2\ /ING^3\ *SI^4\ +240175/1536**E^2\ /ING^3\ *SI^6 \\
& -258925/1536*E^2\ /ING^3\ *SI^8\ +7875/128*E^2\ /ING^3\ *SI^{10}\ +13825/16*E^2\ /ING\ -707445/128*E^2\ /ING\ *SI^2\ +12423585/1024*E^2\ /ING\ *SI^4\ -68666125 \\
& /6144*E^2\ /ING\ *SI^6\ +15171975/4096**E^2\ /ING\ *SI^8\ +25725/64*E^2\ /ING-934995/512*E^2\ /ING*SI^2\ +2583775/1024*E^2\ /ING*SI^4\ -569695/512*E^2\ /ING \\
& *SI^6\ ) \\[4pt]
& +\ A2*A3*E/L^9\ *SI*(-108/ING\ +8619/16/ING^2\ *SI^2\ -1519933/1536/ING^2\ *SI^4\ +407615/512/ING^2\ *SI^6\ -30375/128/ING^2\ *SI^8\ -99/4/ING+4329/64/ING \\
& *SI^2\ -7535/128/ING*SI^4\ +7875/512/ING*SI^6\ -195/8*E^2\ /ING^3\ *SI^2\ +201475/1536*E^2\ /ING^3\ *SI^4\ -386335/1536*E^2\ /ING^3\ *SI^6\ +105975/512*E^2\ /ING^3 \\
& *SI^8\ -31875/512*E^2\ /ING^3\ *SI^{10}\ -7323/16*E^2\ /ING\ +51769/24*E^2\ /ING\ *SI^2\ -1443601/3072*E^2\ /ING\ *SI^4\ +5738015/2048*E^2\ /ING\ *SI^6\ -3178525 \\
& /4096*E^2\ /ING\ *SI^8\ -6593/64*E^2\ /ING+129189/512*E^2\ /ING*SI^2\ -366959/2048*E^2\ /ING*SI^4\ +108245/4096*E^2\ /ING*SI^6 \\
& *SI^6\ )
\end{aligned}
$$

Fig. 2. Determining function S_3^*.

When both short- and long-period terms have been eliminated, the new Hamiltonian F** contains no angular variables. The action variables L'', G'', and H'' are constant, and the angular variables ℓ'', g'', and h'' are expressed as follows:

$$\ell'' = -\frac{\partial}{\partial L''} F^{**} + \ell_0^{\,\prime},$$

$$g'' = -\frac{\partial}{\partial G''} F^{**} + g_0^{\,\prime}, \qquad\qquad (9)$$

$$h'' = -\frac{\partial}{\partial H''} F^{**} + h_0^{\,\prime\prime}.$$

The generating functions S and S* determine a completely canonical transformation from the osculating elements (L, G, H, ℓ, g, and h) to the mean elements (L'', G'', H'', ℓ'', g'', and h''). Because ℓ, g, and e do not satisfy the d'Alembert characteristics with respect to the eccentricity, we selected $\ell + g$, h, e cos i, e sin i, L, and H as the set of elements for calculating the ephemeris of a satellite, as Deprit and Rom (1970) did. If ε stands for one of these elements, the osculating element ε is obtained from the following formulas through the third order of J_2^3:

$$\varepsilon = \varepsilon' + \{\varepsilon', S\} + \frac{1}{2}\{\{\dot\varepsilon', S\}, S\} - \frac{1}{6}\{\{\{\varepsilon', S\}, S\}, S\}$$
$$+ O(J_2^4),$$

$$\varepsilon' = \varepsilon'' + \{\varepsilon'', S^*\} + \frac{1}{2}\{\{\varepsilon'', S^*\}, S^*\} + \frac{1}{6}\{\{\{\varepsilon'', S^*\}, S^*\}, S^*$$
$$+ O(J_2^4). \qquad\qquad (10)$$

These expressions do not contain negative powers of the eccentricity. Even though neither $\ell + g$ nor h satisfies the d'Alembert characteristics with respect to the inclination, the sum of $\ell + g + h$ does, thus providing another good check on the lengthy calculations. Figures 3 and 4 show parts of the third-order short- and long-period perturbations of $\ell + g$. The method of calculating position and velocity from these elements is given in Deprit and Rom (1970).

3. ANALYTICAL AND NUMERICAL CHECK

Hori (1970a) and Yuasa (1971) showed that both Hori's and von Zeipel's perturbation theories give the same canonical transformation, together with the same new Hamiltonian, up to the third order. This allows us to compare the present solution, based on Hori's theory, with Kozai's (1962) solution, based on von Zeipel's theory. Our F_3^{**} completely agress with Kozai's F_3^{**}. Although Hori (1966) gave a relation between the determining functions of his and von Zeipel's theories, we have not compared these functions, because of the complexity of Kozai's ex-

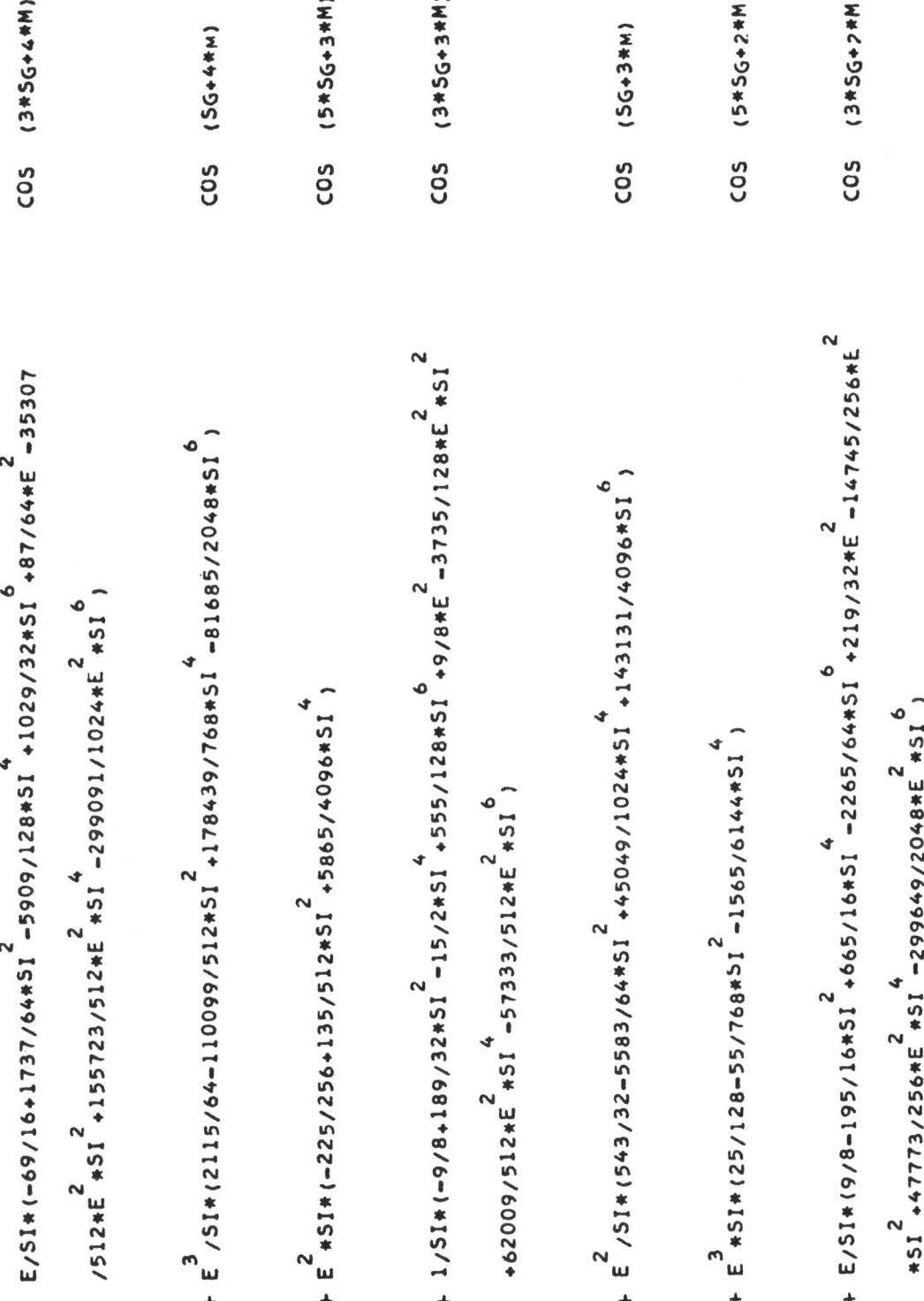

Fig. 3. Third-order short-period perturbation of $\ell + g$ with the factor $J_2 J_3$.

COEFFICIENT OF COS(SG)

$1/A2^3 *A3*A4^2 *E/L^{10} *(-1000/ING *SI+37625/4/ING *SI-527375/16/ING^3 *SI^3 +1712375/32/ING^3 *SI^5 -1306375/32/ING^3 *SI^7 +189875/16/ING^3 *SI^9$
$*SI^{11} +50/ING^2 /SI-7125/4/ING^2 /SI-2897125/256/ING^2 *SI-14174475/512/ING^2 *SI^3 +29614375/1024/ING^2 *SI^5 -22217825/2048/ING^2 *SI^7 *SI^9)$
$+ 1/A2^3 *A3^6 *E/L^{10} *(25/2/ING^2 *SI^3 -25/ING^2 *SI^5 +25/2/ING^2 *SI^{,5} -55/16/ING*SI+505/32/ING*SI^3 -395/32/ING*SI^5 -1/8/SI+53/16*SI-413/64$
$*SI^3)$

$+ 1/A2*A3*A4*E/L^{10} *(3000/ING^3 *SI-9175/4/ING^3 *SI^3 +519325/8/ING^3 *SI^5 -5623675/64/ING^3 *SI^7 +7356475/128/ING^3 *SI^9 -1875125/128/ING^3 *SI^{11}$
$*SI^{11} -150/ING^2 /SI+42085/8/ING^2 *SI-3546125/128/ING^2 *SI^3 +14268245/256/ING^2 *SI^5 -25027215/512/ING^2 *SI^7 +16119775/1024/ING^2 *SI^9 *SI$
$-285/4/ING/SI+56355/32/ING*SI-847595/128/ING*SI^3 +546425/64/ING*SI^5 -1861195/512/ING*SI^7)$

$+ A2*A3*E/L^{10} *(-2160/ING^3 *SI+25575/2/ING^3 *SI^3 -478465/16/ING^3 *SI^5 +2214225/64/ING^3 *SI^7 -1269275/64/ING^3 *SI^9 +144375/32/ING^3 *SI^{11}$
$+108/ING^2 /SI-23883/8/ING^2 *SI+3200581/256/ING^2 *SI^3 -10716335/512/ING^2 *SI^5 +8069435/512/ING^2 *SI^7 -2581125/512/ING^2 *SI^9 +99/4/ING$
$/SI-14905/32/ING*SI+151543/128/ING*SI^3 -538593/512/ING*SI^5 +311625/1024/ING*SI^7)$

Fig. 4. Third-order long-period perturbation of ℓ + g.

pression of S_2 and the tediousness of the calculations. Instead, we compared our $\check{S}_2$ with that derived later by Hori (1970b). In determining S, we have an ambiguity and may add an arbitrary function of L, G, H, and g to S, giving us $\bar{S}$. In the present theory, the disturbing function is expanded into a Fourier series with arguments of ℓ and g; therefore, it is natural to determine S in such a manner that its mean vlue with respect to the mean anomaly is zero. On the other hand, the mean value of Hori's S is not zero. The relation between S and $\bar{S}$, both of which are determining functions eliminating short-period terms in the Hamiltonian, is

$$\bar{S}_1 = S_1 + f_1,$$

$$\bar{S}_2 = S_2 + \frac{1}{2} \{S_1, f_1\} + f_2,$$

$$(11)$$

where f_1 and f_2 are arbitrary functions of L, G, H, and g. Then, the relation between S* and $\bar{S}$*, which are determining functions eliminating long-period terms, is

$$\bar{S}_1^* = S_1^* - f_1,$$

$$\bar{S}_2^* = S_2^* + \frac{1}{2} \{S_1^*, f_1\} - f_2.$$

$$(12)$$

Even if the functional forms of these determining functions are different, the composite canonical transformation of (S, S*) is identical to that of $(\bar{S}, \bar{S}*)$. The second-order determining function S_2 fo the present theory and that derived by Hori are found to satisfy relation (11).

It is of interest to compare the present solution with that due to Deprit and Rom (1970), in which J_3 and J_4 are zeros. Their solution was obtained by their own perturbation method, which, like Hori's, is based on Lie transformation. Campbell and Jefferys (1970) showed that the perturbation theories of Hori and Deprit are equivalent and derived explicit relations between the determining functions for the two:

$$W_1 = -S_1,$$

$$W_2 = -2S_2,$$

$$W_3 = -6S_3 - \{S_1, S_2\},$$

$$(13)$$

where W_n are the determining functions in Deprit's theory. Using these relations, we compared our solution with Deprit and Rom's, which also serves as an independent check. We found only a few disagreements with their terms, which all seemed to be typographical errors. Kutuzov (1975) also discovered some discrepancies between his solution, obtained by computer algebra, and theirs.

As an internal consistency check, we wanted to make sure that a small divisor $(1-5 \cos^2 i)$ disappeared for the even-order harmonics when the geopotential was equal to that in Vinti (1959) $[J_{2n} = (-1)^{n+1} J_2^n,$ $n \geq 2]$. In checking the third order, we had to add long-period perturbations arising from J_6 and J_8.

Finally, the present solution was compared with the results obtained from numerical integration. A Taylor-type integrator was adopted,
in which the positions and velocities are expanded into a power series
of time and the coefficients of the series are determined by recurrent
formulas (Rabe, 1961; Deprit and Zahar, 1966). The order of the power
series and the step size of time were chosen so as to maintain about 12
significant figures in the integral of energy; the integration step is
roughly one-fifth of the convergence radius τ of the two-body problem,
$\tau = (1/n) \{\ln [(1/e) + \sqrt{(1/e^2) - 1}] - \sqrt{1 - e^2}\}$, when the degree of the
Taylor series is 16. The numerical calculations were carried out by a
CDC 6400 computer in double precision in order to avoid roundoff errors.
It takes about 0.7 s to evaluate the series of the present theory and
about 0.3 s to integrate the equation of motion for one step.

The initial conditions were computed from the present theory from
a set of mean elements a" , e" , i"', ℓ" , g" , and h" . It should be
noted that the mean motion of the mean anomaly of the integrated orbit
is expected to be different from the computed mean motion by an order
of J_2^4 because the accuracy of the periodic terms is of the order of J_2^3.
This discrepancy can be avoided by adjusting the semimajor axis so as
to remove the secular term in the residuals of the mean anomaly.

Such comparisons were done for the artificial satellites Geos 3,
Starlette, and Lageos. The results for Starlette are plotted in Figures
5 and 6. Figures 5a, 5b, and 5c show in-track, along-track, and across-
track errors over two revolutions. The deviations are less than 2×10^{-4}
m and are totally negligible. Figures 6a, 6b, and 6c show the errors
over about 600 revolutions. The deviations along and across the track
are totally insignificant, but the in-track error has a secular trend
that seems to be proportional to the square of time. The error first
increases, then decreases, and finally vanishes at about t = 22 days, at
which time the semimajor axis was adjusted. This secular trend can be
explained by the accumulation of truncation errors, which are caused
by replacing infinitesimal operations with finite operations. The
round-off errors do not accumulate significantly, because our calculations were carried out in double precision, which amounts to an accuracy
of about 30 significant figures in decimal notation. The accumulation
error due to discretization (Kinoshita, 1968) is

$$\Delta r_{\text{in-track}} \propto a\ell^2 h^p, \qquad (14)$$

where a is the semimajor axis, ℓ the mean anomaly, h the step size, and
p the degree of Taylor expansion. About 1 cm of the in-track error at
t = 40 days (1 = 3600 radians) is obtained from using Equation (14) with
$a = 7.335 \times 10^6$ m and h = 600 s = 1/10 radians, the order of the error
agreeing with that of the numerical experiment. We can avoid discretization errors by employing a much more accurate integrator, but it seems
to be an unnecessary use of computer time. The comparisons for Geos 3
and Lageos gave roughly the same results as for Starlette.

We are now confident that the present solution can predict the
position of a geodetic satellite with a small eccentricity with an
accuracy of better than 1 cm over 1 month.

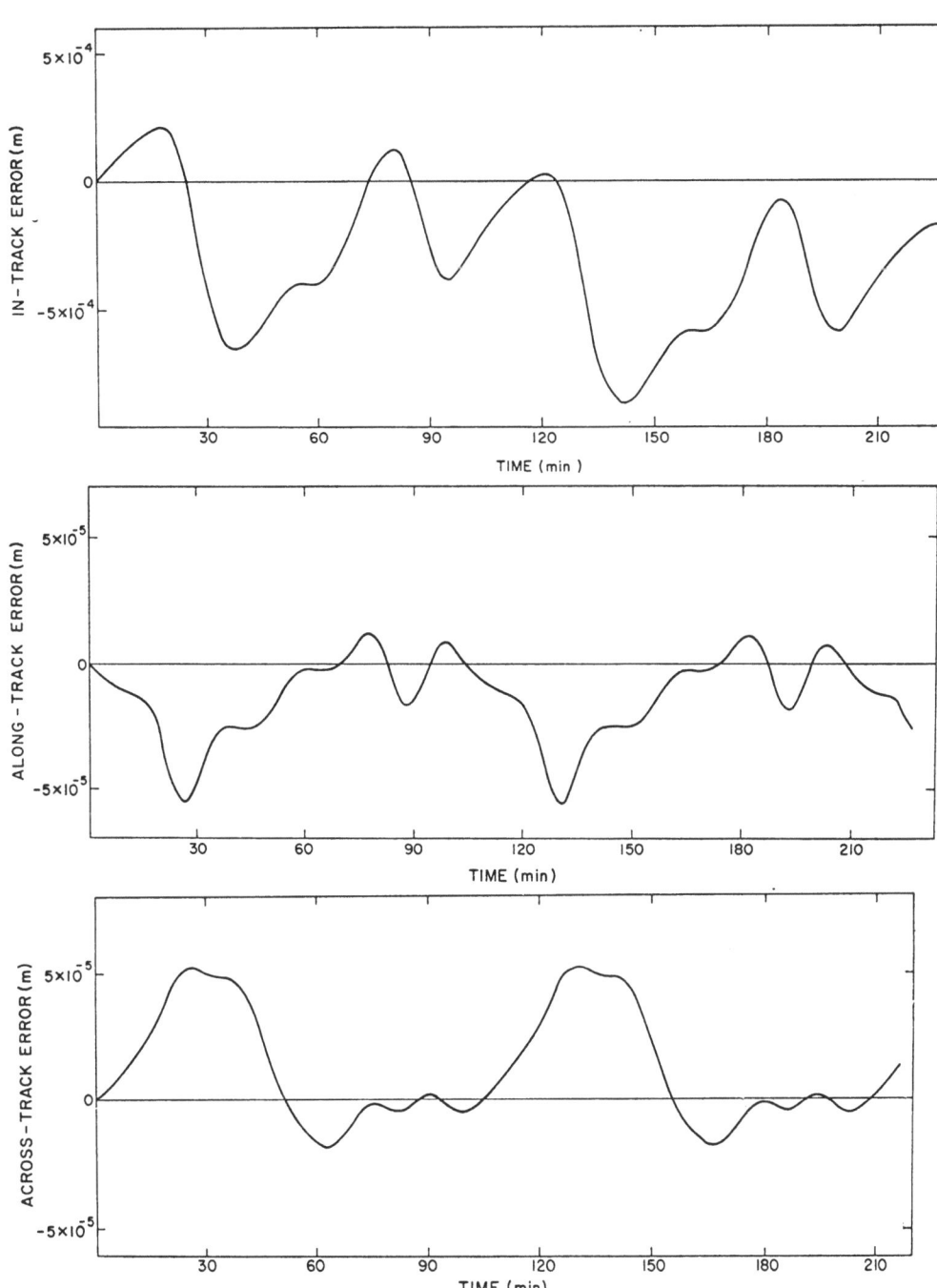

Fig. 5. Prediction error of the third-order solution for Starlette
plotted over two revolutions: $J_2 - 1.082 \times 10^{-3}$, $J_3 = -2.54 \times 10^{-6}$,
$J_4 = -1.619 \times 10^{-6}$, $a'' = 7.335 \times 10^6$ m, $e'' = 0.020636$, $i'' =$
49.8223, $l_0' = 350.23968$, $g_0'' = 82.7702$, $h_0'' = 125.0266$, $P =$
104.2 min.
(a) in-track error.
(b) along-track error.

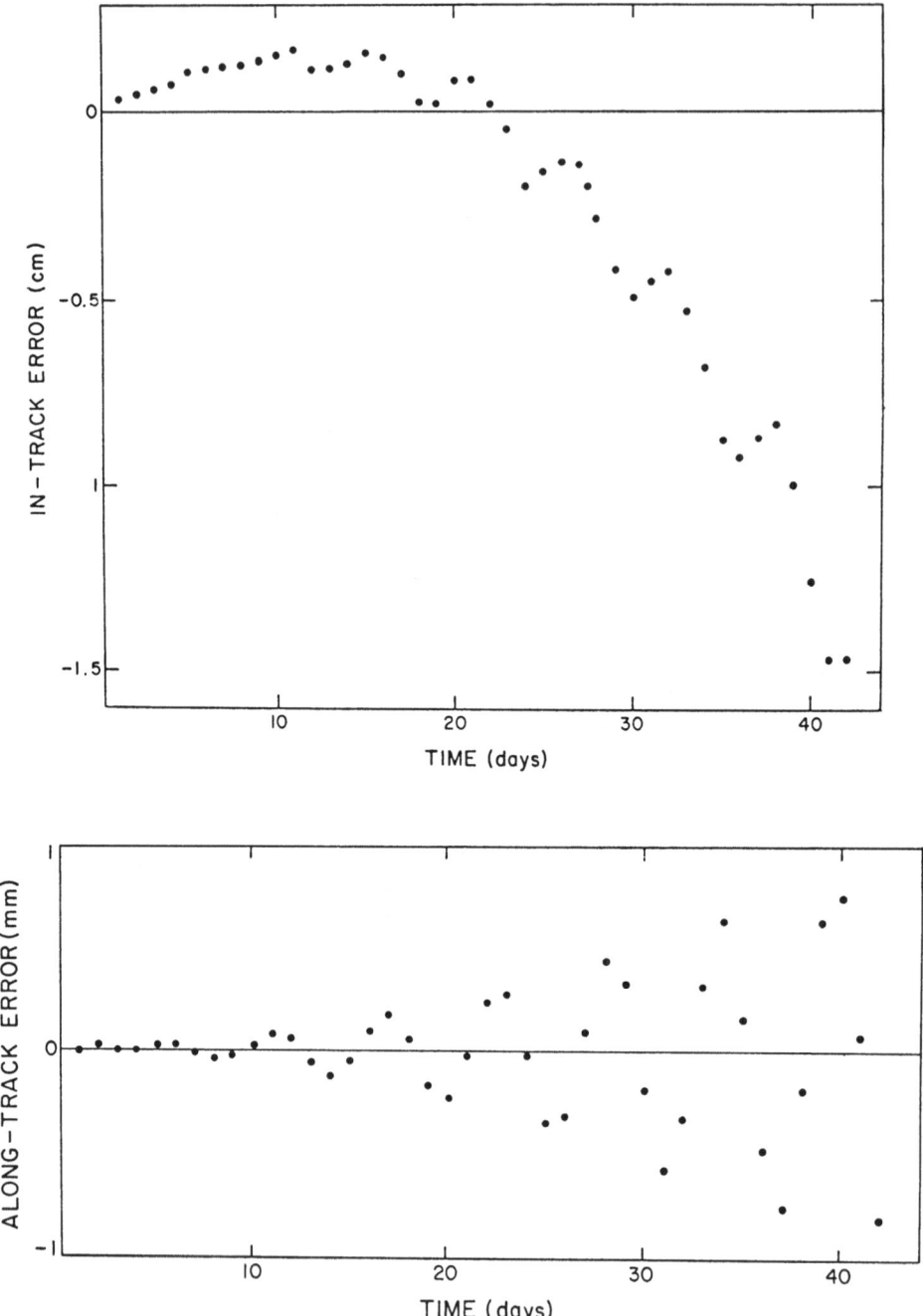

Fig. 6. The same as Figure 5 plotted over about 600 revolutions.

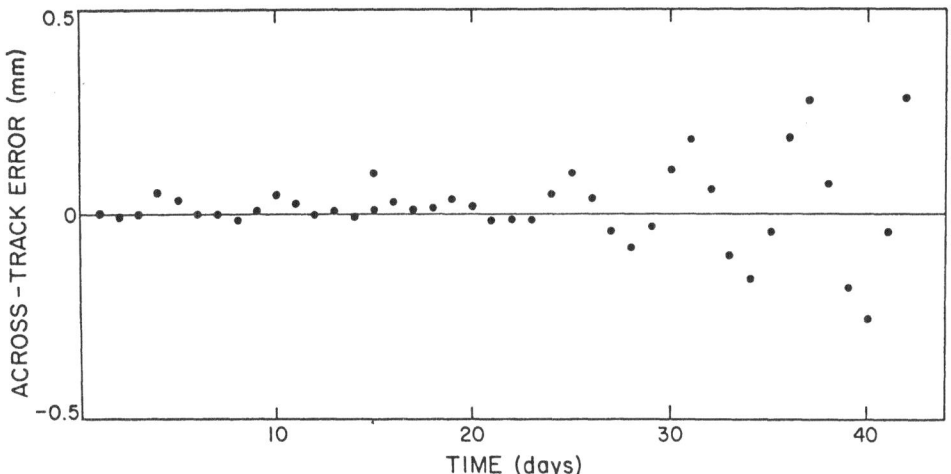

(c) across-track error.

ACKNOWLEDGMENTS

The author would like to acknowledge the continuing encouragement of Dr E..M. Gaposchkin and expresses his appreciation to Dr K. Aksnes for valuable discussions. The author is also grateful to Mr J. R. Cherniack and Mr R. T. Poland for their occasional patience in explaining the mysteries of SPASM. This work was supported in part by Grant NGR 09-015-022 from the National Aeronatuics and Space Administration.

REFERENCES

Aksnes, K.: 1970, *Astron. J.* 75, 1066-1076.
Campbell, A and Jefferys, W. H.: 1970, *Celes. Mech.* 2, 467-473.
Deprit, A. and Rom, A.: 1970, *Celes. Mech.* 2, 166-206.
Deprit, A. and Zahar, R.: 1966, *Z. angew. Math. Phys.* 17, 426-430.
Hall, N. M. and Cherniack, J. R.: 1969, *Smithsonian Astrophys. Obs. Special Report No. 291*, 49 pp.
Hori, G.: 1966, *Publ. Astron. Soc. Japan* 18, 287-296.
Hori, G.: 1970a, *Publ. Astron. Soc. Japan* 22, 191-198.
Hori, G.: 1970b, unpublished (paper read by B. Garfinkel at IAU General Assembly, Commission No. 7, Brighton, England, August 1970).
Hori, G. and Kozai, Y.: 1975, *Satellite Dynamics* (ed. by G.E.O. Giacaglia), Springer-Verlag, Berlin, pp. 1-15.
Kinoshita, H.: 1968, *Publ. Astron. Soc. Japan* 20, 1-23.
Kozai, Y.: 1962, *Astron. J.* 67, 446-461.
Kutuzov, A. L.: 1975, *Soviet Astron. Lett.* 1, 42-44.
Rabe, E.: 1961, *Astron. J.* 66, 500-513.
Vinti, J. P.: 1959, *J. Res. Nat. Bur. Standards*, 63B, 105-116.
Yuasa, M.: 1971, *Publ. Astron. Soc. Japan* 23, 399-403.

SOME CONSIDERATIONS ON THE THEORETICAL DETERMINATION OF THE POTENTIAL
BY THE MOTION OF ARTIFICIAL SATELLITES IN THE PLANE CASE

J.J. Martinez-Benjamin
The University of Texas at Austin and University of Barcelona

ABSTRACT

We consider the determination of the potential from the orbit of
a satellite. The potential obtained is not unique because the satellite might describe the same orbit with different forces. The classical problem of determination of orbits has a unique solution, for
given initial conditions. The problem is reduced to the planar case
with a conservative force. The mathematical results depend on the
arbitrary functions of the system of coordinates used.

PART V

GRAVITATIONAL PROBLEM OF THREE OR MORE BODIES

FAMILIES OF PERIODIC PLANETARY-TYPE ORBITS IN THE N-BODY PROBLEM AND THEIR APPLICATION TO THE SOLAR SYSTEM

J. D. Hadjidemetriou and M. Michalodimitrakis
University of Thessaloniki, Thessaloniki, GREECE

ABSTRACT. A new approach to the study of the Solar System and planetary systems in general is proposed, through the use of periodic planetary-type orbits of the general N-body problem. In such an orbit, one body (called Sun) has a large mass and the rest N-1 bodies (called planets) have small but not negligible masses and it can be proved that monoparametric families of periodic orbits of the N-body problem exist in a rotating frame of reference , all being of the planetary type

Two cases are studied in detail, N=3 and N=4. In N=3, apart from a general discussion, we present a detailed analysis of the Sun-Jupiter-Saturn system and a study is made on which configurations with the masses of these two planets, or a multiple of them, are stable or unstable. Also, part of a family is shown to represent the Jupiter family of comets. It was found that commensurabilities are not in general associated with instabilities. For N=4 we present three families of periodic orbits. The motion corresponding to a branch of one of the above families has many similarities with the actual motion of the three inner satellites of Jupiter.

It is shown that there exist many commensurable cases in the obtained periodic orbits and that the resonant orbits increase as the number of bodies increases. Based on these results, an attempt is made to explain the existence of commensurabilities in the Solar System.

Finally, it is mentioned that a periodic motion of the planetary type can be used as a reference orbit for accurate computations for the actual motions of the planets or satellites of the Solar System. In this way the small divisor difficulties existing in the classical approach will not appear.

1. INTRODUCTION

The purpose of this paper is to present a new approach to the study of the solar system and of planetary systems in general, based on the study

V. Szebehely (ed.), Dynamics of Planets and Satellites and Theories of Their Motion, 263-281.
All Rights Reserved. Copyright © 1978 by D. Reidel Publishing Company, Dordrecht, Holland.

of families of periodic orbits of the general N-body problem (N>3). It can be proved (Hadjidemetriou 1975a, 1976b,c) that families of periodic orbits exist in the general planar N-body problem, in a rotating frame or reference, for fixed values of the masses of all the bodies. In particular, we shall consider here the case where only one body, say P_2, has a large mass and the rest N-1 bodies P_1, P_3,..., P_N have small but not negligible masses. Thus, this system represents a planetary system with the body P_2 being the Sun and the bodies P_1, P_3,..., P_N the planets (or comets) or the body P_2 being a planet and the bodies P_1, P_3,..., P_N being its satellites.

The rotating frame mentioned above is defined as follows: The origin O coincides with the center of mass of the bodies P_1 and P_2 and the x axis contains always these bodies, the positive direction being from P_2 to P_1. This system is rotating with a non constant angular velocity and it can be proved (Hadjidemetriou 1975a, 1976b,c) that the motion of the N-body system in this rotating frame can be studied independently of the motion of the rotating system with respect to an inertial frame. This separation is possible because of the existence of the angular momentum integral which introduces the angular position of the rotating frame as an ignorable coordinate. Thus, a qualitative study of the motion can be made in the rotating frame only and this simplifies the analysis. The equations of motion in the rotating frame are given in the above mentioned papers. In what follows we shall restrict ourselves, to the study of the motion in the rotating frame only.

In this approach to the study of a planetary system no approximation is made and the gravitational effect of each planet on the other is completely taken into account. Also, this method applies equally well to planetary orbits with small and with large eccentricities and thus one can study planetary and cometary orbits by using the same method. And indeed, as we shall see in the following, there is a continuous transition from circular (planetary) orbits to elliptic (cometary) orbits. Also, any existing comensurabilities do not present any problem at all and the same method is used for all cases. Another advantage is that the smallness of the masses of the planets (or comets, or satellites) is not required and the same method can be applied to planetary systems with large masses of the planets. Thus, one can study the evolution of a planetary system, particularly its stability, by increasing the masses of the planets.

The purpose of this paper is to present an overall qualitative view of planetary systems, as obtained through the study of periodic orbits of the general N-body problem, based on the work made so far at the University of Thessaloniki. For this reason we shall not present detailed numerical computations. In some particular cases however which have a special interest, as the motion of the Sun-Jupiter-Saturn system or the three inner satellites of Jupiter, we present the exact numerical data. The applications are made for three and four bodies, but the same method can be applied for any number of bodies.

2. PLANETARY-TYPE ORBITS IN THE GENERAL THREE-BODY PROBLEM

(a). A family for zero masses of the two planets

In order to obtain approximate initial conditions for a periodic orbit we assume that the mass m_2 of the body P_2 (which we shall call "Sun") is equal to 1 and the masses of P_1 and P_3 (which we shall call planets) are equal to zero and that they describe circular orbits around P_2 in the same direction, in the plane. We can assume, without loss of generality, that the distance R_1 between P_2 and P_1 is equal to unity. Then, for any value of the distance $R_3 > 1$ between P_2 and P_3 we have a periodic motion in the rotating frame x0y, defined in the previous section. The ratio of the periods of the orbits of P_1 and P_3 around P_2 is equal to

$$T_1/T_3 = R_3^{-3/2} \tag{1}$$

and the period of the periodic motion in the rotating frame x0y is equal to

$$T = 2\pi/(1-T_1/T_3). \tag{2}$$

We note that T varies between $T=2\pi$ (for $R=\infty$) to $T=\infty$ (for $R_3=1$). Thus, we have a degenerate monoparametric family of symmetric periodic orbits of three bodies, with respect to the x axis, where the two bodies P_1 and P_3 have zero masses. One can use the relative period T, given by (2), as a parameter along the family. We shall always take the two planets and the Sun to lie on the same straight line at $t=0$.

The initial conditions corresponding to the above family can be used as approximate initial conditions to obtain a symmetric periodic orbit in the rotating frame when the masses of P_1 and P_3 are increased. This continuation is possible, as shown by Hadjidemetriou (1976a), for all members of the degenerate family except those corresponding to a resonance of the form

$$T_1/T_3 = n/(n+1), \tag{3}$$

because in that case the period T is a multiple of 2π. This has as a consequence the generation of an infinite number of families of periodic orbits, all corresponding to the same (nonzero) masses of the three bodies. In this continuation the ratio m_1/m_3 may have any prespecified value.

This method has been applied by Hadjidemetriou (1976a) to obtain families of periodic planetary-type orbits for the case $m_1=m_3=0.001$ and $m_2=0.998$. This work was extended by Delibaltas (1976) who obtained families of periodic orbits for the case where the three bodies have masses equal to the mass of Sun, Jupiter and Saturn, respectively. Also, the evolution of these families is studied, when the masses of the planets are increased.

To summarize the results obtained so far, we note at this point that a symmetric periodic orbit can be specified by its initial conditions x_{10}, x_{30}, $\dot{y}_{30}$ in the rotating frame x0y, since $y_{30}=\dot{x}_{10}=\dot{x}_{30}=0$ provided a certain normalization scheme is used. We have used in our calculations the normalization

$$G = 1, \quad m_1+m_2+m_3 = 1, \quad \dot{\vartheta}_o = 1, \tag{4}$$

where $\dot{\vartheta}_o$ is the initial value of the angular velocity of x0y, (Hadjidemetriou and Christides, 1975). Thus, we see that a family of periodic orbits, for fixed masses of all the bodies, can be represented by a conti- nuous curve in the space $x_{10}x_{30}\dot{y}_{30}$. In this paper, in order to present the results in the simplest possible way, we shall use the projection of the above curve in the $x_{10}x_{30}$ plane only. Evidently, the above mentioned degenerate family   of periodic orbits is represented in this plane by the straight line $x_{10}=1$.

We also note that we can take, without loss of generality, $R_3>1$, which implies that $|x_{30}|>1$, i.e. the orbit of P_3 is outside the orbit of P_1. If we had taken $0<R_3<1$ we would obtain the same family but with the roles of P_1 and P_3 interchanged, i.e. P_1 is the outer planet and P_3 the inner planet.

(b). Families for nonzero masses of the planets

In Fig. 1 the straight line $x_{10}=1$ represents the degenerate family for zero masses of the two planets and the points A_1, A_2, A_3,..., are the resonant orbits 1/2, 2/3, 3/4,..., respectively. These points have an

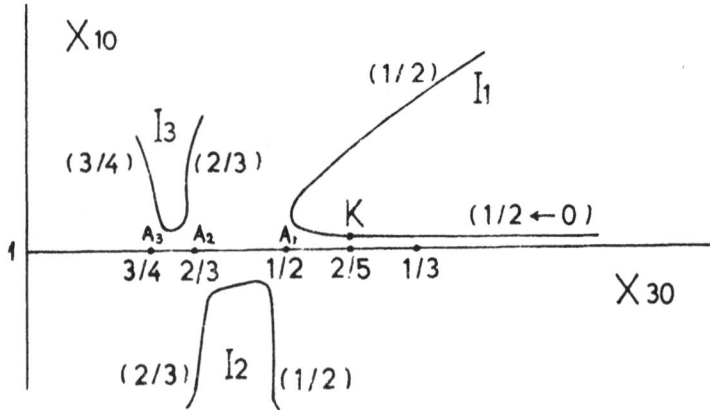

Fig. 1: The degenerate family $x_{10}=1$ and the families I_1, I_2, I_3,... generated from it by increasing the masses of the planets (schematically). The points A_1, A_2, A_3 represent resonant orbits of the form 1/2, 2/3, 3/4, respectively, of the dege- nerate family. The point K, at the resonance 2/5, represents the Sun-Jupiter-Saturn system. All these families correspond to a fixed value of m_2 and m_1/m_3. The picture is qualitative- ly the same for all values of m_1/m_3.

accumulation point at $x_{30}=1$. When this degenerate family is extended
by increasing the masses of the two planets, it breaks down to an infinity of families, all corresponding to the same fixed masses for the
three bodies. Each family lies approximately between two consequtive
resonant orbits of the form $n/(n+1)$ and $(n+1)/(n+2)$, respectively
$(n=1,2,3,\ldots)$.

The continuation of the degenerate family, as given in Fig.1, is
qualitatively the same for all ratios of the masses of the two planets.
The part of the families I_1, I_2, I_3,... which is nearly parallel to the
line $x_{10}=1$ corresponds to almost circular orbits of the two planets
around the Sun and the rest part corresponds to an elliptic orbit of at
least one planet. An interesting result is that all along the branch
of a family which is not parallel to the line $x_{10}=1$, there is an almost
constant resonance T_1/T_3 of the osculating periods of the two planets.
These constant values of the resonance are shown in Fig.1, in the parentheses near each branch. The association of this resonance with the
resonance of the degenerate orbits A_1, A_2... is evident.

The actual Sun-Jupiter-Saturn system corresponds to the resonant
orbit K of Figure 1 for the resonance 2/5. The initial conditions
of a periodic orbit closely representing the Sun-Jupiter-Saturn system
are given (Hadjidemetriou 1976a) by

$$x_{10}=0.99915744, \quad x_{30}=-1.84094099, \quad \dot{y}_{30}=1.10378309, \qquad (5)$$

for the masses

$$m_1=0.0009508, \qquad m_3=0.9987640, \qquad m_3=0.0002852, \qquad (6)$$

according to the normalization (4). The ratio T_1/T_3 at $t=0$ is equal
to 0.406 and the osculating elements of the orbits of the two planets
vary during one period as follows:

$$1.000550 \leq a_1 < 1.000950, \qquad 1.8367 \leq a_3 < 1.8492$$

$$0.000611 \leq e_1 < 0.000677, \qquad 0.00075 \leq e_3 < 0.00368 \ .$$

We note that the eccentricities are smaller than in the actual case.
Perhaps the addition of more bodies in the system has as a consequence
an increase in the eccentricities. This seems to be confirmed by numerical results in the 4-body problem (the resonant branches of families
B,C in Fig.5, for the resonance 2/5 are found to have eccentricities
of the order of 10^{-2}).

In the families of the type I_1 the upper branch (not parallel to
the line $x_{10}=1$) corresponds to an almost circular orbit of the outer
planet, P_3, and an elliptic orbit of the inner planet, P_1, (more
appropriately called now a comet). The eccentricity increases as we
proceed to larger values of x_{30} along this branch of the family. This
picture is qualitatively the same for equal masses of the two planets

and also for the actual masses (6) of Jupiter and Saturn and it seems
that it is the same for any value of m_1/m_3. However, we have important
differences as far as stability is concerned, as we shall describe below.

As far as the families I_2 and I_3 are concerned, we note that these
are mostly almost resonant families, corresponding to the resonances
1/2 and 2/3 for I_2, and 2/3 and 3/4 for I_3. These resonances appear
in the parts of the families not parallel to the line $x_{10}=1$.   The
transition from one resonance to the other in each family is along the
lower part of the family in Fig.1, which is nearly parallel to the
line $x_{10}=1$. This latter segment which does not correspond to resonant
motion is small, and becomes smaller and smaller as we proceed to the
families I_4, I_5... (not shown in Fig.1), corresponding to higher
resonances.

A general remark is that all the resonant orbits of the form
$n/(n+1)$, $n=1,2,3,...$ correspond to elliptic motion of at least one
planet, though the eccentricities are in most cases small and only
towards the end of the branches the eccentricities have large values.
The values of the elements of the orbits of families I_1, I_2, I_3 are
presented in Hadjidemetriou (1976a) and Delibaltas (1976).

(b). Stability of planetary-type orbits

We shall discuss families of the type I_1, I_2, I_3 (Figure 1). From the
available numerical results (Hadjidemetriou 1976a, Delibaltas 1976) we
can draw the following conclusions:
-      The stability character of an orbit is not necessarily associated
with commensurabilities in the periods of the two planets. Indeed,
almost all resonant periodic orbits of family I_1 with nearly circular
orbits of the two planets are stable, for all values m_1/m_3. Also the
resonant orbits 1/2 in I_1, 2/3 in I_2 and 3/4 in I_3, for $m_1 \geq m_3$ are
stable while the resonant orbits 1/2 and 2/3 in I_3 are unstable.
-      The resonant periodic orbit corresponding to the commensurability
1/3 is the only unstable resonant motion with nearly circular orbits
of the two planets. This seems to be true for all values of m_1/m_3,
even for vanishingly small values of the masses of the two planets.

For finite masses of the two planets there is a small instability
region corresponding to the resonance 1/3. In particular, the resonant
orbit 2/5, corresponding to the actual Sun-Jupiter-Saturn system is
found to be outside this unstable region.
-      The upper branch of the family I_1, corresponding to an almost
circular orbit of the outer planet P_3 and an elliptic orbit of the
inner planet P_1 is stable only when the outer planet is the more massive
one (we remind that in the degenerate family in 2(a) $R_3>1$).   When the
mass of the inner planet (comet) becomes larger than the mass of the
outer planet, the system becomes unstable. The exact point of transition
from stability to instability along the upper branch of I_1 depends on
the ratio m_1/m_3.

As mentioned above, the upper branch of I_3 corresponds to a circular orbit of the outer planet P_3, and an elliptic orbit of the inner planet P_1. Also, the period of P_1 is just larger than half the period of P_3. All these facts suggest clearly that this upper branch of I_1 can be considered as the Jupiter family of comets. Consequently, the Jupiter family of comets could not exist if the masses of the comets were larger than the mass of Jupiter.

Another interesting aspect concerning stability is the study of the evolution of the families shown in Fig.1 by increasing the masses of the two planets. For example, we can take a certain set of families as in Fig. 1, for a fixed ratio of the masses m_1/m_3 and continue all these families by increasing the masses of the two planets, keeping the ratio m_1/m_3 fixed. The study of the stability of the obtained families will give interesting information for the generation and evolution of a planetary system as a whole. Of course, this procedure must be

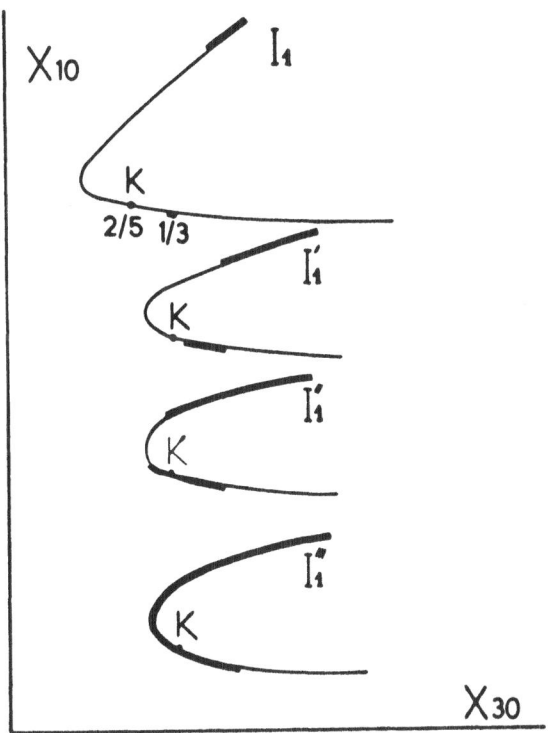

Fig. 2: The evolution of the family I_1 by increasing the mass of both planets, for a fixed ratio of the masses (schematically). The orbit K is the resonant orbit 2/5 and the region in bold line represents unstable motion. It is assumed that $m_3 < m_1$. If $m_3 \geq m_1$ then the unstable region in the upper branch of I_1 does not appear when m_1, $m_3 << 1$.

repeated for several values of m_1/m_3 in order to obtain a clear picture of the whole problem.

We have done this in detail for the case $m_1=m_3$ (Hadjidemetriou 1976a) and also, to a lesser extent, for the case where the values of m_1 and m_3 are those of Jupiter and Saturn, respectively. The results are shown, qualitatively, in Fig.2. Only the family I_1 has been used for this continuation.

We note that when the masses of the two planets are small there is a very small unstable region corresponding to the resonance 1/3. An additional unstable region in the upper branch of I_1 exists when $m_3<m_1$. For the actual masses of Jupiter and Saturn and also for the masses of these two planets equal to 0.001 (in normalized units), the resonant orbit 2/5 is outside the unstable region and consequently it is stable. As the masses of the two planets increase, their ratio being fixed, we obtain the families I'_1, I''_1, I'''_1.. etc. We note that the unstable region due to the resonance 1/3 extends and also the unstable region in the upper branch of I_1 extends (this latter unstable region would not be present for very small values of m_1, m_2 if $m_1<m_3$ but would appear as m_1, m_3 increase). Eventually, the unstable region due to the resonance 1/3 extends and covers the resonant orbit 2/5 which corresponds to the Sun-Jupiter-Saturn system. For still larger values of the masses, the above mentioned two unstable regions merge and we are left with a family whose stable region corresponds to nearly circular orbits of the two planets not very near to each other.

It was also found that this continuation can be carried out until the mass of the Sun becomes equal to zero, i.e. we end up to the circular restricted three-body problem.

The transition from stability to instability for the actual Sun-Jupiter-Saturn system, by incrasing the masses of both planets (keeping their ratio fixed) is shown in Tables I and II. In Table I we present a part of a family (family A) of periodic orbits corresponding to the masses

$$m_1=0.0342888, \qquad m_2=0.9554260, \qquad m_3=0.0102852 \qquad (7)$$

and in Table II a part of family B, corresponding to the masses

$$m_1=0.0409564, \qquad m_2=0.9467594, \qquad m_3=0.0122952. \qquad (8)$$

The masses of the planets in family A are about 36 times the masses of Jupiter and Saturn and those of family B are about 43 times the masses of these planets. As a parameter along the family we have used the ratio $T_{JUPITER}/T_{SATURN}$ of the osculating elements of the two planets at t=0 and we present the stability index b_1 (of Hadjidemetriou 1975b), which is the first which becomes unstable, as a function of T_{JUP}/T_{SAT}. The value of b_1 when $T_{JUP}/T_{SAT}=2/5$ is obtained by linear interpolation between the adjacent values.

TABLE I
A part of the family A, for the masses (7).

T_{JUP}/T_{SAT}	b_1	
.414	1.865	Stable
.402	1.973	"
.400	1.985	"
.398	1.996	"
.394	2.015	Unstable
.390	2.029	"
.379	2.043	"

TABLE II
A part of the family B, for the masses (8).

T_{JUP}/T_{SAT}	b_1	
.426	1.878	Stable
.419	1.947	"
.415	1.974	"
.412	1.998	"
.410	2.008	Unstable
.405	2.032	"
.400	2.049	"
.399	2.052	"
.391	2.061	"

From the numerical results presented we can find that the transition from stability to instability for the Sun-Jupiter-Saturn system (defined as that system corresponding to the resonance 2/5) takes place when the masses of the two planets are increased by about 38 times the actual masses. We must note however that the numerical value of this factor is quite sensitive to the definition of the "Sun-Jupiter-Saturn" system. For example, if we allow for a variation in the value of T_{JUP}/T_{SAT}, in the definition of the "Sun-Jupiter-Saturn" system, this factor will change appreciably.

The above stability analysis is based on a linear theory. In order to study the nonlinear effects we have computed the intersections of a perturbed orbit to the Sun-Jupiter-Saturn system, given by (7), with the plane $y_3=0$, by a method described in Hadjidemetriou (1975b). At each point of intersection, in the same direction, we have computed the osculating semimajor axes and eccentricities of the two planets. The results are shown in Fig.3 for the semimajor axes and in Fig.4 for the eccentricities. We note that these points lie on smooth curves in such a way that every third point of intersection lies on the same curve. There does not seem to exist any secular change in the semimajor axes. This result does not contradict Poisson's theorem on the invariability

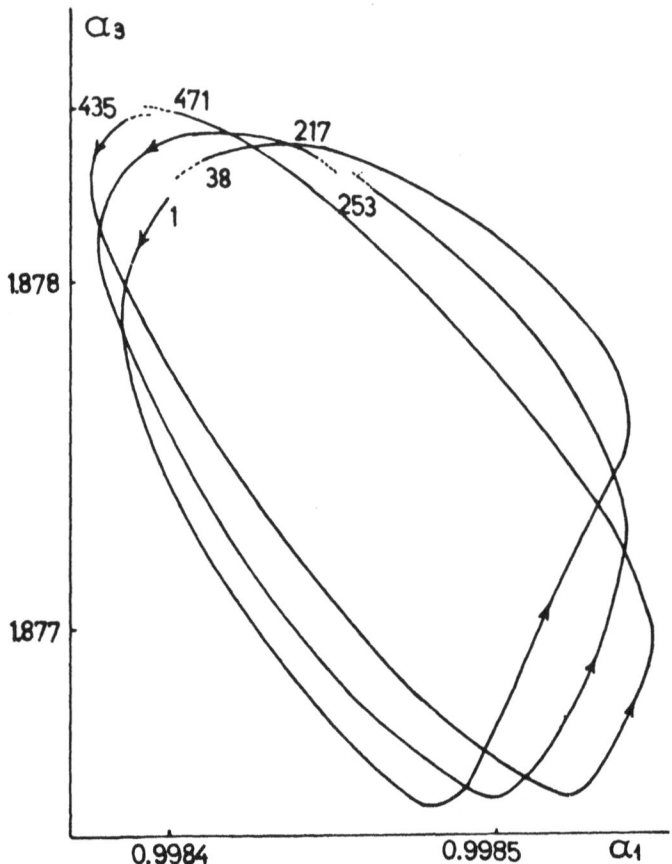

Fig. 3: Curves defined by the values of a_1 and a_3 at the consecutive
points of intersection with $y_3=0$ (in the same direction) for
a perturbed orbit to the periodic orbit (8) for the isoenerge-
tic perturbation $\Delta\dot{x}_{10}=0.04$, $\Delta\dot{x}_{30}=0$, $\Delta\dot{x}_{10}=0.025$, $\Delta\dot{x}_{30}=0.04$.
Every third point of intersection is shown only. The computa-
tions correspond to about 500 periods. Only the curves defined by the points 1-38, 217-253, 435-471 are shown.

of the semimajor axes (Hagihara, 1961, p.101). As far as the eccentricities are concerned, we noted appreciable changes. However, although
we did not carry out the computations very far, we believe that these
changes are also quasiperiodic, with very long periods. The obtained
points are an indication (but not a proof) that the orbit is stable to
all orders. These results can be also concidered as an indication that
additional integrals exist, at least locally (see also Hagihara, 1961,
p.107).

The stability studied above is with respect to perturbations in the
plane of motion. As far as vertical stability is concerned (i.e. with
respect to perturbations normal to the plane of motion) it was found by

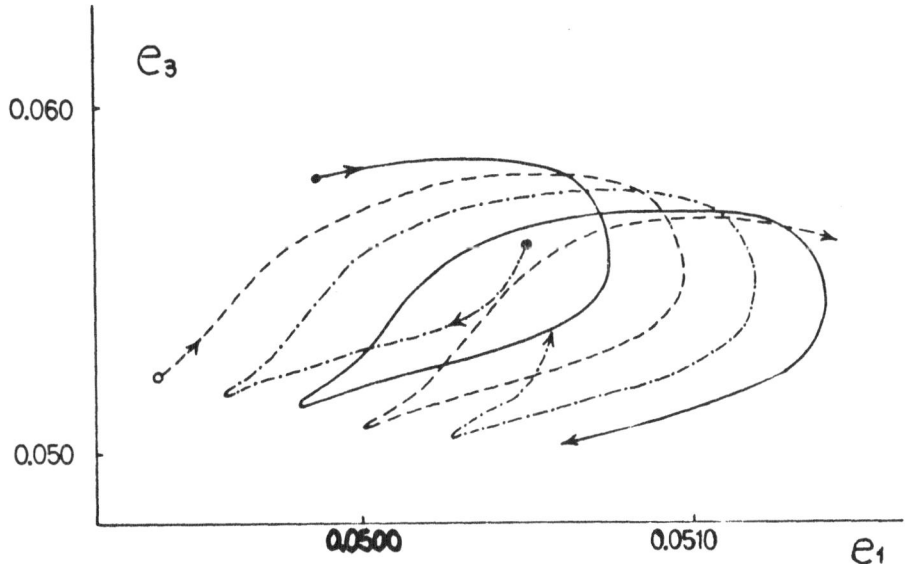

Fig.4: The same as in Fig.3 for the eccentricities e_1 and e_3. The three smooth curves mentioned in the text are shown. Only 60 points have been used in this Figure. The value of e_1 for the 500 periods increased from 0.050 to 0.058 and the value of e_3 decreased from 0.058 to 0.031.

Delibaltas (1976) that the family I_1, for the masses (6) of Jupiter and Saturn, is vertically stable. Thus, there are no critical orbits which would generate three-dimensional periodic orbits. This means that the extension of planetary type orbits of the kind I_1 to three dimensions is not possible. Hence, the actual three-dimensional motion of the Sun-Jupiter-Saturn system must be considered as a perturbed motion to a <u>planar</u> periodic orbit.

3. PLANETARY-TYPE ORBITS IN THE GENERAL 4-BODY PROBLEM

(a). Families for zero masses of the three planets

We consider the mass of the body P_2 to be equal to unity and the masses of the bodies P_1, P_3, P_4 equal to zero, and assume that the bodies P_1, P_3, P_4 (called <u>planets</u>) describe circular orbits in the same plane around the body P_2 (called <u>Sun</u>), in the same direction. Let

$$\omega_1 = 2\pi/T_1, \qquad \omega_3 = 2\pi/T_3, \qquad \omega_4 = 2\pi/T_4, \tag{9}$$

be the angular velocities of rotation of the three planets whose periods are T_1, T_3, T_4, respectively. We define a rotating frame x0y such that the origin is at P_2 and the x axis contains always the body P_1 and normalize the unit of length in such a way that the radius R_1 of P_1 is

equal to unity. We note now that if we take an arbitrary radius R_3 for
the orbit of P_3 and select the radius R_4 of P_4 in such a way that

$$(\omega_3 - \omega_1)/(\omega_4 - \omega_1) = p/q, \tag{10}$$

where p,q are integers, the system is periodic with respect to the
rotating frame x0y with a period equal to

$$T = \frac{q}{1 - T_1/T_4} T_1. \tag{11}$$

Using (9) and (10) we can find that the ratio T_1/T_4 is expressed in
terms of T_1/T_3 by the relation

$$\frac{T_1}{T_4} = 1 + \frac{q}{p} \left(\frac{T_1}{T_3} - 1 \right). \tag{12}$$

If now we normalize the units in such a way that $(P_2 P_1) = R_1 = 1$ and
keep p/q fixed (p<q), we have for each value of $R_3 > 1$, or equivalent-
ly of T_1/T_3, a periodic motion of the four bodies. In this normalization we have $T_1 = 2\pi$. Thus, we obtain a degenerate family of periodic
orbits of the four bodies, with the ratio T_1/T_3 (or the distance R_3)
as a parameter. This family is characterized by a particular value of
the ratio p/q and consequently we have several families, one for each
value of p/q. Thus, we can characterize a degenerate family of 4 bodies
by its ratio p/q. We note that if $R_3 > 1$ and p/q<1, then $T_1/T_4 < T_1/T_3$
which implies that the radii of P_1, P_3, P_4 increase in this order, i.e.
P_4 is the outer planet, P_3 the intermediate planet and P_1 the inner
planet. In this case the ratio T_1/T_3 varies between the values 0 (for
$R_3 = \infty$) and 1 (for $R_3 = 1$). It can be verified that there is no loss of
generality in selecting $R_1 = 1$, $R_3 > 1$ and p/q<1. For the other possible
values of R_1, R_3 and p/q we would obtain the same family, but with
the roles of P_1, P_3, P_4 interchanged in their hierarchigal order.

We must also note that, apart from all the above parameters, a
particular orbit depends also on the relative positions of the 3 planets
with respect to the Sun, at t=0. In the present study we consider symmetric periodic orbits with respect the x0y only. For this reason we
have taken all three planets to lie on the x axis at t=0. Moreover,
we have restricted ourselves to positive values for x_1, x_3, x_4 respectively, at t=0. If, other things being the same, we had taken $x_1 > 0$, $x_3 < 0$
and $x_4 > 0$ ($x_1 < |x_3| < x_4$) then we would obtain a different family.

It can be proved (Hadjidemetriou 1976b,c) that the above mentioned
degenerate families of periodic orbits can be continued as monoparametric
families of symmetric periodic orbits of the general planar 4-body
problem, in a rotating frame of reference, by increasing the masses of
the planets. In this way we can obtain a family for fixed masses. In
this continuation we may have any value for the ratio $m_1 : m_3 : m_4$ of the
masses of the three planets. The continuation is unique for all the
orbits of the degenerate family except for those orbits whose period is

a multiple of 2π. In this latter case the continuation theorem is not applicable and this results to a situation similar to that in the three-body problem (Figure 1). And we may note from (11) and (12) that there is an infinity of such resonant orbits when T_1/T_3 varies between 1 and zero. Thus, a continuous degenerate family corresponding to a ratio p/q is extended to an infinity of families of periodic orbits for fixed non-zero masses of the three planets.

As an example we give below, in Table III, the values of T_1/T_4 and T as a function of T_1/T_3, for a degenerate family corresponding to $p/q=2/3$.

TABLE III

Some characteristic orbits of the degenerate family for $p/q=2/3$

N_o	T_1/T_3	T_1/T_4	T	corresponding family
1	1/3	0	$3\times2\pi$	
2	7/15	1/5	$(15/4)\times2\pi$	A
3	1/2	1/4	$4\times2\pi$	
4	5/9	1/3	$(9/2)\times2\pi$	B
5	3/5	2/5	$5\times2\pi$	
6	2/3	1/2	$6\times2\pi$	C

We note that the 1st, 3rd, 5th and 6th cases in Table III correspond to resonant periodic orbits of the corresponding degenerate family ($m_1=m_3=m_4=0$), where the period is a multiple of 2π. Evidently, there is an infinite number of such resonant orbits as T_1/T_3 increases to unity. As a consequence of the existence of these orbits in the degenerate family mentioned above, we obtain an infinite number of families of periodic orbits (for a fixed value p/q) when the masses of the three planets are increased. Each one of these latter families can be thought of as lying "between" two consecutive resonant orbits of the degenerate case with periods $n\times2\pi$ and $(n+1)\times2\pi$, respectively ($n=3,4...$).

The degenerate family $p/q=2/3$ can be associated with the three inner Galilean satellites of Jupiter, because the family for nonzero masses of P_1, P_3 and P_4 generated from the points of the degenerate family between the 3rd and 5th point in Table III has a branch which approximates the actual motion of Jupiter's satellites. (i.e. the periods are in the ratio 1:2:4).

(b). Planetary-type families of periodic orbits of four bodies with nonzero masses

Detailed calculations of planetary-type orbits in the general 4-body problem will be given elsewhere (Hadjidemetriou and Michalodimitrakis 1976). We shall present here the main qualitative features of the calculations obtained so far, for the family corresponding to $p/q=2/3$.

We extended the degenerate family $p/q=2/3$ for the masses

$$m_1=0.0000379946, \qquad m_2=0.9998570204,$$

$$m_3=0.0000249964, \qquad m_3=0.0000799886 .$$

(13)

These masses, normalized so that $m_1+m_2+m_3+m_4=1$, correspond to the mass of Jupiter (P_2) and its three inner satellites (P_1, P_3, P_4) (Reek, 1958). As mentioned above, this degenerate family will be ex- tended, for the masses (13), to an infinity of families. We have comput- ed three of these families, the family A lying between the degenerate resonant orbits

$$(T_1/T_3=1/3, \quad T_1/T_4=0) \quad \text{and} \quad (T_1/T_3=1/2, \quad T_1/T_4=1/4),$$

the family B lying between

$$(T_1/T_3=1/2, \quad T_1/T_4=1/4) \quad \text{and} \quad (T_1/T_3=3/5, \quad T_1/T_4=2/5)$$

and the family C lying between

$$(T_1/T_3=3/5, \quad T_1/T_4=2/5) \quad \text{and} \quad (T_1/T_3=2/3, \quad T_1/T_4=1/2),$$

as can be seen from table III.

To present the results, we note (Hadjidemetriou 1976b,c) that a symmetric periodic orbit for $N=4$ can be specified, in the rotating frame of reference whose origin coincides with the center of mass of P_1, P_2 and its x axis contains always these bodies, by the initial conditions

$$x_{10}, \quad x_{30}, \quad x_{40}, \quad \dot{y}_{30}, \quad \dot{y}_{40},$$

(14)

provided a certain normalization scheme is used. Thus, a family for fixed masses is represented by a continuous curve in the space x_{10} x_{30} x_{40} $\dot{y}_{30}$ $\dot{y}_{40}$. To simplify things, we shall use, for qualitative purposes, the projections of this curve in the planes $x_{10}x_{30}$, and $x_{10}x_{40}$ only. Evidently, the degenerate family $p/q=2/3$ will be presented in the above plane by the straight line $x_{10}=1$, according to the normalization mentioned before.

In Fig. 5 we present qualitative results for the families A,B,C obtained for the masses (13). In all cases, we have taken

$$x_{10}>0, \quad x_{30}>0, \quad x_{40}>0 \quad \text{at } t=0 \quad (\text{also } x_{10}<x_{30}<x_{40}).$$

The osculating eccentricities of the orbits of P_1, P_3 and P_4 are very small and for all orbits of all these families the ratio of rela- tive frequencies $(\omega_3-\omega_1)/(\omega_4-\omega_1)$ is found to be equal to $2/3$ to an accuracy of three decimal places. We have also found that the satel- lites I, II, III are in their pericenter (P) or apocenter (A) at $t=0$, along the families A, B, C, as shown below. The branches of these families are designated by the resonance T_1/T_3 (see Fig. 5a).

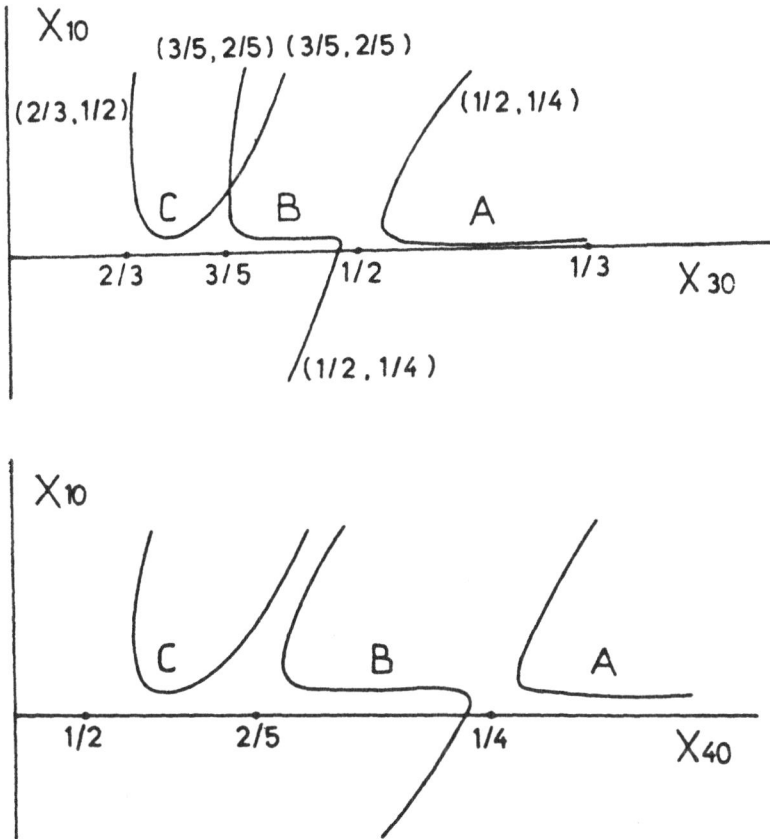

Fig. 5a,b: The families A,B,C (schematically). The lower branch
(almost parallel to the line $x_{10}=1$) of family A in the plane
$x_{10}x_{40}$ (Fig. 5b) extends to infinity while the lower branch
of A in the plane $x_{10}x_{30}$ (Fig.5a) stops at the point corresponding to the resonance 1/3. The ratios in the parentheses
near each branch of the families in Fig.5a denote the ratios
T_1/T_3, T_1/T_4, respectively.

Family	branch	I	II	III
A	1/3	P	P	P
A	1/2	P	P	A
B	1/2	A	A	P
B	3/5	P	P	A
C	3/5	P	A	P
C	2/3	P	P	A

The lower branch of family B corresponds to a resonance 1/2, 1/4 between P_1-P_3 and P_1-P_4, respectively and has many similarities with the motion of the three inner satellites of Jupiter. We have selected the periodic orbit with initial conditions

$$x_{10}=0.99429726, \quad x_{30}=1.59200305, \quad x_{40}=2.44984472$$
$$\dot{y}_{30}=-0.80992184, \qquad \dot{y}_{40}=-1.80974958 \tag{15}$$

as the closest orbit to the motion of Jupiter's satellites. The osculating elements of the orbits of the three satellites vary between the following limits:

$$0.9978 \leq a_1 \leq 0.9979, \qquad 0.01680 \leq e_1 \leq 0.01690,$$
$$1.5509 \leq a_3 \leq 1.5517, \qquad 0.02602 \leq e_3 \leq 0.02629,$$
$$2.4585 \leq a_4 \leq 2.4596, \qquad 0.00364 \leq e_4 \leq 0.00395.$$

At t=0 the bodies P_1 and P_3 are at apocenter and the body P_4 at pericenter.

The motion given by (15) has many similarities with the actual motion of Jupiter's satellites but does not coincide with it. This is so because the relative positions of P_1, P_3, P_4 with respect to P_2 are not those of the actual case. A periodic orbit representing closely the actual case has been obtained in the same way as (15), by the continuation of a degenerate orbit corresponding to $T_1/T_3=1/2$ and $T_1/T_4=1/4$, by increasing the masses to the values given by (13), if we take $x_{10}>0$, $x_{40}>0$ and $x_{30}<0$ (instead of $x_{30}>0$ we had in (15)). This orbit is given by

$$x_{10}=1.000594320, \quad x_{30}=-1.595000148, \quad x_{40}=2.572200612,$$
$$\dot{y}_{30}=0.801696725, \quad \dot{y}_{40}=-1.948632207$$

and its osculating elements vary between the limits

$$1.002595 \leq a_1 \leq 1.002693, \qquad 0.001982 \leq e_1 \leq 0.002012,$$
$$1.601014 \leq a_3 \leq 1.601357, \qquad 0.003780 \leq e_3 \leq 0.003949,$$
$$2.572558 \leq a_4 \leq 2.573604, \qquad 0.000154 \leq e_4 \leq 0.000368.$$

The satellites I and III are in conjuction and II in opposition when they are all in their perijoves, and this corresponds to the actual case. A periodic orbit for the three inner satellites of Jupiter has been found by de Sitter (Brower and Clemence, 1961, p.82, Hagihara, 1961, p.123) and was used as an intemediary orbit to obtain the ephemeris of the satellites.

As a general remark we can say that a large part of the families correspond to almost resonant motion of the three small bodies. This resonant motion corresponds to the part of the families in Figure 5 which is not parallel to the line $x_{10}=1$. The osculating eccentricities of the orbits of the planets along these branches are greater than zero, but in most cases remain small, of the other of 10^{-2}. This value is in agreement with the eccentricities in the solar system, in particular the Sun-Jupiter-Saturn system. (Compare this with the remark made in section 2(c) for the eccentricities of the Sun-Jupiter-Saturn system, considered there as a 3-body motion).

4. DISCUSSION

The method developed in this paper could provide us with useful information on the generation and evolution of planetary systems. This is so because by studying families of planetary-type orbits for several mass-ratios we can obtain an overall view of the problem. In this way we find which configurations are unstable, so that they are excluded as possible configurations for planetary systems existing in nature.

In the families of planetary-type orbits we have obtained both for three and for four bodies, we note that a large part of the obtained families correspond to resonant motion. Moreover, the appearance of resonances increases as the number of bodies increases. For N=3 the resonant orbits are those corresponding to the part of the families which are not parallel to the straight line $x_{10}=1$ and also some isolated orbits in the part which is almost parallel to $x_{10}=1$ (Figure 1). The former resonant orbits are associated with nonzero eccentricities of the orbits of the two planets (though in most cases the eccentricities are not large). The addition of a fourth body has as a consequence the increase of resonant cases, no matter how small the mass of the fourth body is. In fact, all orbits obtained from a degenerate family for N=4, corresponding to a fixed ratio p/q, when the masses of the planets assume nonzero values, are resonant in the sense that the ratio of the relative angular velocities in the rotating frame xOy $(\omega_3-\omega_1)/(\omega_4-\omega_1)$ is almost constant, equal to p/q, for all members of the family. Besides, the addition of a fourth body results in the appearance of more resonances, in the absolute orbits of the planets, not originally present in the (simple) periodic orbits of the three-body system. This can be seen from the comparison of Figs.1 and 5 (and Table III). The resonances in the ratios of the absolute motions of the three planets in the 4-body system appear in all the branches of the families in Fig.5 which are not parallel to the line $x_{10}=1$. For example, for N=3 we have the resonances 1/2, 2/3, 3/4 and when a fourth body is added, we can see from Figure 5, for p/q=2/3, that the resonances 1/4, 2/5, 3/5 are also present. In the same way, the addition of a fifth body will increase further the resonant cases.

Taking into account all the above, we can attempt an interpretation of the appearance of commensurabilities in our Solar System. Indeed, if we assume that the motion of the Solar System, or at least its most

important components, must be near a periodic orbit, then it is very
likely that many commensurabilities will exist, as a large part of the
families of periodic planetary-type orbits in the N-body problem are
resonant orbits.

As far as stability is concerned, we may note that if we restrict
ourselves to N=3, i.e. we study the Sun-Jupiter-Saturn system only and
ignore the other planets, then there exists an infinity of stable configurations for the actual masses of these bodies. It would be of interest to study whether the addition of more bodies make the system more
stable or, on the contrary, limit the stable configurations of the Solar
System. The stability of the 4-body systems has not yet been studied
and will appear elsewhere (Hadjidemetriou and Michalodimitrakis, 1976).

There seems to be some confusion on the role which the resonances
among the planets play in the stability of the Solar System (e.g. Moser
1973). The stability analysis of periodic planetary-type orbits of 3
bodies has shown that the commensurabilities do not play a very important
role in the stability of the system. For example, the commensurable
orbits 1/2 in family I_1 are stable for $m_1=m_3=0.001$, $m_2=0.998$ and the
commensurable orbits 1/2 in family I_2 for these masses are unstable.
Also, the commensurable orbits 1/2 of family I_1 for the masses (9) are
unstable. Thus, the presence of small divisors (e.g. Hagihara, 1961,
p.111) does not seem to play an important role in stability as, for the
same commensurability, the stability depends on the relative dimensions
of the orbits of the planets (i.e. on the position on the family) and
on the relative masses of the planets.

We would also like to comment on the meaning of (linear) instability. Usually, instability is associated with escape of at least one body,
and this seems to be the rule in most cases in other problems. In the
planetary orbits however for N=3 we could not establish such a close
connection between instability and escape. For small masses of the
planets, a perturbed orbit to an unstable periodic orbit did not lead to
escape but to random, bounded, motion. Of course, one can always argue
that escape will eventually happen if the computations are carried further
in time, but this remains an open question. We do have a case for N=3
where escape takes place after some hundred revolutions (Hadjidemetriou
1976a), but the masses of the planets were rather large ($m_1=m_3=0.05$,
$m_2=0.90$).

Finally, we may note that the method of periodic orbits may provide
a new approach to the accurate computation of the planets and satellites
of the Solar System. Instead of using the two-body approach as a reference orbit to compute the perturbations we may use for this purpose a
periodic motion for several members of the Solar System, and calculate
the perturbations for the actual motion starting from this periodic
orbit. The periodic motion can be obtained to a high degree of accuracy
numerically and analytically in the form of Fourier series. In this
way we avoid the small divisors which complicate the classical approach
to the solution of the planetary problem, as all the resonances will be

included in the reference periodic orbit itself, whose numerical computation does not depend on any resonance present and the gravitational
effect of one planet on the other is completely taken into account.

REFERENCES

1.   Brower, D. and Clemence, G.M.: 1961, Orbits and Masses of Planets
      and Satellites, in Kuiper G.P. and Middlehurst, B.M (eds.), The
      Solar System, The University of Chicago Press.
2.   Delibaltas, P.: 1976, Astrophys. Space Sience 45, 207.
3.   Hadjidemetriou, J.D.: 1975a, Celes. Mech. 12, 155.
4.   Hadjidemetriou, J.D.: 1975b, Celes. Mech. 12, 255.
5.   Hadjidemetriou, J.D.: 1976a, Astrophysics Space, Science 40, 201.
6.   Hadjidemetriou, J.D.: 1976b, in V. Szebehely and B.O. Tapley (eds).
      Long-Time Predictions in Dynamics, D. Reidel Publ. Co., 223.
7.   Hadjidemetriou, J.D.: 1976c, The Existence of Families of Periodic
      Orbits in the N-Body Problem, Celes. Mech. (to appear).
8.   Hadjidemetriou, J. D. and Christides, Th.: 1975, Celes. Mech. 12,
      175.
9.   Hadjidemetriou, J. D. and Michalodimitrakis, M.: Families of
      Periodic Planetary-Type Orbits in the N-Body Problem and their
      Stability (in preparation).
10.   Hagihara, Y.: 1961, The Stability of the Solar System, in Kuiper,
      G.P. and Middlehurst, B.M. (eds.). The Solar System, The University
      of Chicago Press.
11.   Moser, J.: 1973, Stable and Random Motions in Dynamical Systems,
      Princeton Univ. Press, ch. I.
12.   Peek, B.M.: 1958, The Planet Jupiter, Faber and Faber, London,
      p.256.

PERTURBATIONS OF CRITICAL MASS IN THE RESTRICTED THREE-BODY PROBLEM

R.K. Sharma and P.V. Subba Rao
Vikram Sarabhai Space Center

ABSTRACT

Based upon the authors' result (Astron. Astrophys. 43, 381-3, 1975) that oblateness of the more massive primary decreases the value of the critical mass, this note is concerned primarily with a few observations on Szebehely's result (Astron. J. 72, 7-9, 1967) that the Coriolis force is a stabilizing force. It has been deduced that a small change ε in the Coriolis force from unity results in a perturbation of the critical mass by a factor of $-8\varepsilon/81\sqrt{69}$.

GRAVITATIONAL RESTRICTED THREE–BODY PROBLEM : EXISTENCE OF RETROGRADE SATELLITES AT LARGE DISTANCE

Daniel Benest
Observatoire de .Nice

ABSTRACT

In the frame of the gravitational restricted three–body problem, we study by numerical simulation the retrograde satellites –and their sta- –bility– at large distance. In the circular plane case, the stable sat- –ellites mostly surround (in the phase–space X_0-V_0) the characteristic of a family of single–periodic orbits where this family is stable, and librates (in the physical space) around a curve which corresponds to the nearest (in the phase–space X_0-V_0) periodic orbit. An analytical analysis of this libration is made for Hill's case. The beginning of the study of the three–dimensional orbits is presented.

INTRODUCTION

Within the frame of the gravitational restricted three–body problem with point masses, we study by numerical simulation the motion of a sa- –tellite S (of infinitesimal mass) particularly at large distance of its primary P (of normalised mass μ), while P and the other massive body B (of normalised mass $1-\mu$) have keplerian orbits around each other. We use rotating–pulsating axes with origin in P, and B is fixed on the X axis with the abscissa −1 (fig. 1). We note e the eccentricity of the relative orbit of P around B, and T the true anomaly of P on its orbit; as usual, we use T instead of the physical time as independent variable. Then, the equations of motion for S are:

$$\frac{dX}{dT} = U,$$

$$\frac{dY}{dT} = V,$$

$$\frac{dZ}{dT} = W,$$

$$\left.\begin{array}{c} \\ \\ \\ \\ \end{array}\right\} (1)$$

V. Szebehely (ed.), Dynamics of Planets and Satellites and Theories of Their Motion, 285-303.

and

$$\frac{dU}{dT} = 2V + (X+1-\mu - K_1(X+1) - K_2 X)/(1+e \cos T),$$

$$\frac{dV}{dT} = -2U + Y(1-K_1-K_2)/(1+e \cos T),$$

$$\frac{dW}{dT} = -Z(K_1+K_2+e \cos T)/(1+e \cos T),$$

$$\text{with } K_1 = (1-\mu)((X+1)^2+Y^2+Z^2)^{-3/2},$$

$$K_2 = \mu(X^2+Y^2+Z^2)^{-3/2};$$

$$\left.\begin{array}{c} \\ \\ \\ \\ \\ \\ \end{array}\right\} \quad (2)$$

or, for Hill's case:

$$\frac{dU}{dT} = 2V + X(3-K)/(1+e \cos T),$$

$$\frac{dV}{dT} = -2U - KY/(1+e \cos T),$$

$$\frac{dW}{dT} = -Z(1+K+e \cos T)/(1+e \cos T),$$

$$\text{with } K = (X^2+Y^2+Z^2)^{-3/2}.$$

$$\left.\begin{array}{c} \\ \\ \\ \\ \\ \\ \end{array}\right\} \quad (3)$$

The term $e \cos T$ vanishes in the circular case (i.e. $e=0$) and terms in Z and W vanish in the plane case (i.e. when S must stay in the same plane as P and B). These equations are integrated with a fourth-order Runge-Kutta method, where the classical coefficients have been replaced by those given by Ralston.

Figure 2 shows a satellite orbit which approaches near to B. This lead us to extend the usual definition of a satellite: we shall call satellite of P a body whose mean motion around P, averaged over a suffi- -ciently long time, is zero in the rotating frame while its mean motion around B is different from zero. The stability of a periodic orbit can be defined in a more analytical way through the value of several indices easily computed by considerations on orbits close to the periodic one.

For simplicity, S starts always from the right part of the X axis, and perpendicularly to it, so that $X_0 > 0$, $Y_0 = 0$, $Z_0 = 0$ and $U_0 = 0$ (see fig. 1). Moreover, it has been shown that only retrograde orbits can be found stable at large distance, and numerical results indicate that the three-dimensional orbits are symmetrical with respect to the X-Y plane; then we limit ourselves to $V_0 < 0$ and $W_0 > 0$. Therefore, for given e and μ, an orbit can be represented by a point in the phase-space $(X_0, |V_0|, \theta_0)$, where we can examine the subspace of initial con- -ditions for families of periodic orbits (called their characteristic) and for stable orbits in general.

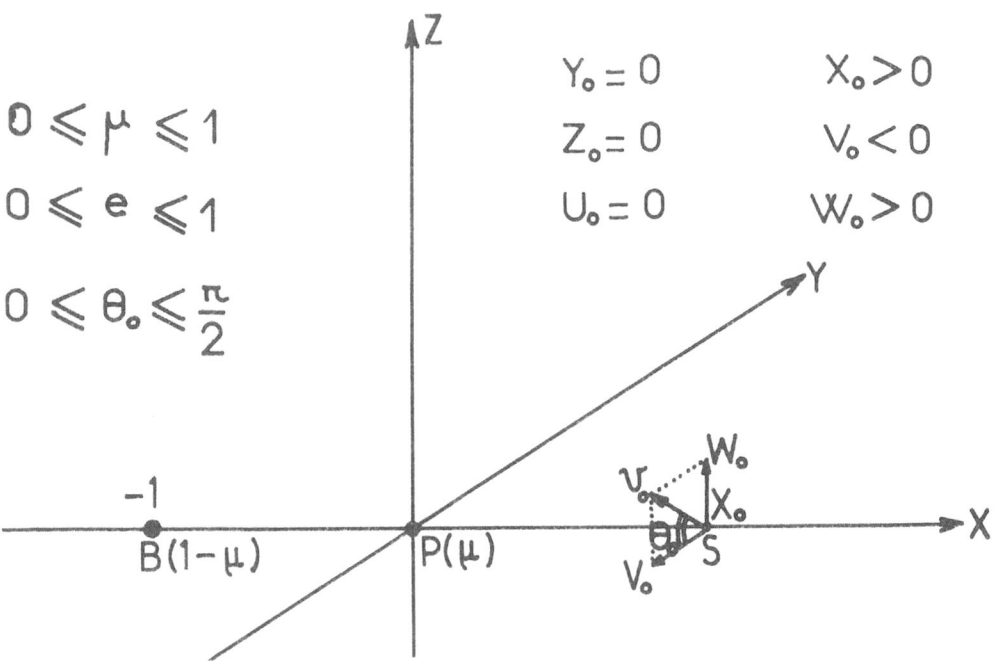

$$0 \leqslant \mu \leqslant 1$$

$$0 \leqslant e \leqslant 1$$

$$0 \leqslant \theta_o \leqslant \frac{\pi}{2}$$

$$Y_o = 0 \qquad X_o > 0$$

$$Z_o = 0 \qquad V_o < 0$$

$$U_o = 0 \qquad W_o > 0$$

Figure 1. Parameters and initial conditions.

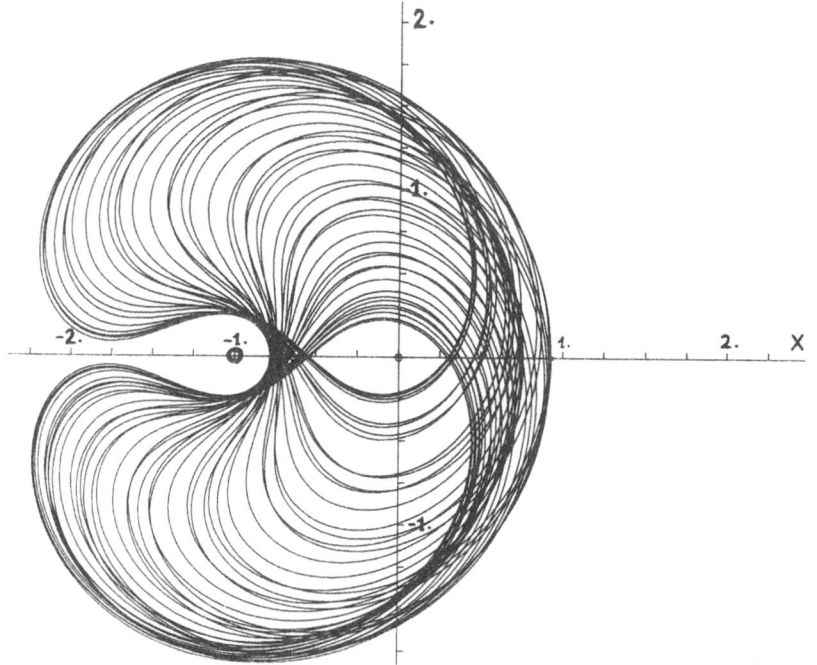

Figure 2. An example of large satellite orbit.

1. STABILITY OF THE ORBITS: EVOLUTION WITH μ

Note: this section is a synthesis of four papers (1974, 1975, 1976b, 1976c).

The fundamental role of the periodic orbits in the restricted problem is well-known since a long time. In the circular plane case,
Dr Hénon has shown that the knowledge of the monoparametric family of
single-periodic orbits symmetrical with respect to the X axis (called
family f, see fig. 3) is essential for the study of the satellite orbits
in general. Moreover, in 1974, we have found that only one other family
has some importance. For these symmetrical orbits, the stability indices
reduce to one, noted a, such that there is stability for $|a| < 1$ and
$a \neq -0.5$. First we consider the circular plane case, where the phase-
space reduces to the plane X_0-V_0.

1.1. Periodic orbits in the circular plane case

Figure 3 shows the general shape of the orbits of family f, together with a fictitious example of the characteristic of the family. As
the orbits grow from little ones in the vicinity of P, the point P_0 runs
along the characteristic up to the point E, corresponding to an ejection
orbit beyond which the orbits are no more satellite ones. Moreover, during the run of P_0, the variation of X_0 is monotonic, but for $\mu > 0.8$
in the vicinity of E.

Figure 4 shows the evolution of the characteristic of family f when
μ increases from 0 to 1. Only a sample of our results is represented
here, family f and its stability have been computed for 62 different
values of μ.

When the variation of X_0 is monotonic, we can establish a map of
stability in the plane $X_0-\mu$ (see fig. 5). The thick line represents the
ejection orbits, and the dashed thick line indicate where the variation
of X_0 becomes non-monotonic. Orbits corresponding to a $\approx$ -0.5 are in
dashed lines, and dash-dot lines indicate the extrema of a, especially
interresting in the hatched regions ($|a| > 1$) because they indicate there
the most unstable orbits. For $\mu < 0.0477...$, the orbits are continuously
stable until ejection. From 0.0477... to $\mu = 1$, the orbits are continu-
-ously stable until they reach the first ocurrence of a=-1, with one or
two intervals of stability furtherout.

For $\mu = 0.0477...$, an unstable segment appears on the characteristic
between two points where a=-1, called second sort critical points; these
points correspond to intersections with a double-periodic family, called
family φ. Figure 6 shows the general shape of the orbits of family φ,
together with a fictitious example of the characteristic of the family.
As these orbits have two perpendicular intersections with the positive
part of the X axis, an orbit is represented by two points in the X_0-V_0
plane. As the orbits grow, the two points P_1 and P_2 separate from P_{0_1}
and run along the characteristic to meet again in P_{0_2}, where the two
branches of the orbit of family φ blend into one orbit of family f, but
run two times. Moreover, during the run of P_1 and P_2, the variation of
the quantity $(X_1+X_2)/2$ is monotonic from X_{0_1} to X_{0_2}, at least up to
$\mu \approx 0.13$; therefore this quantity can be used as the single quantity X_0

for family f.

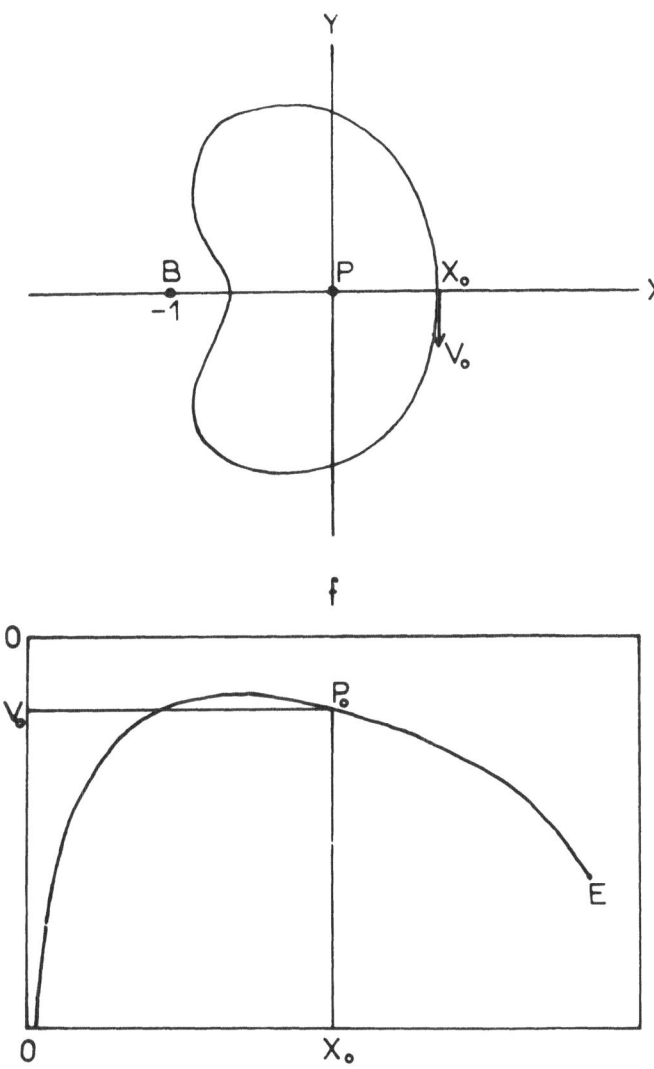

Figure 3. Family f: general shape of the orbit (top) and of the characteristic (bottom).

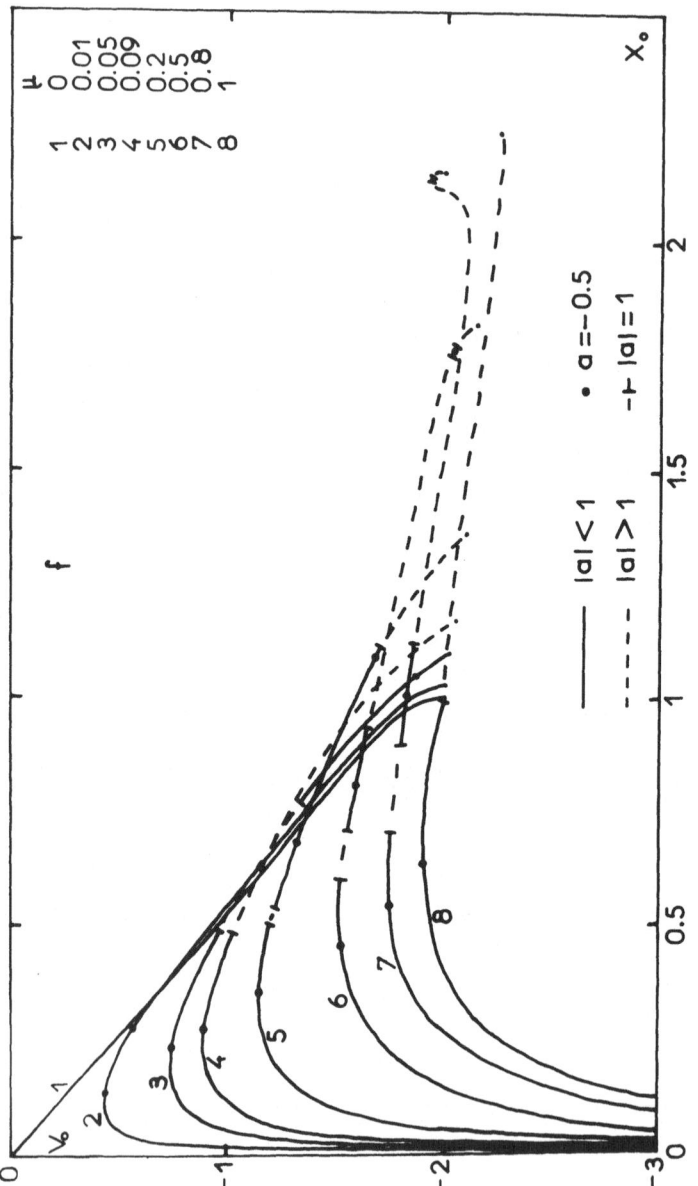

Figure 4. Family f: evolution of the characteristic.

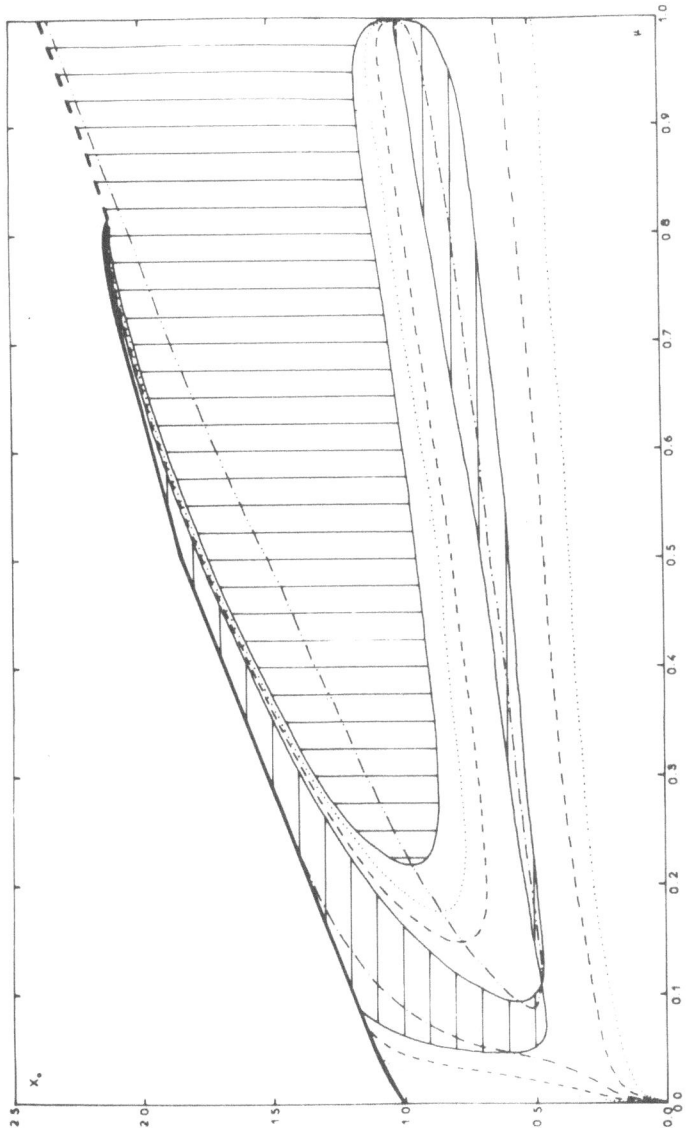

Figure 5. Family f: stability map; .....: a=0; - - - -: a=-0.5; ———: |a|=1; ≡: a<-1;
- - - -: minima of a; |||||: a>+1; - - -: maxima of a; ▅▅▅: ejection; ■ ■ ■: the variation
of a becomes non-monotonic.

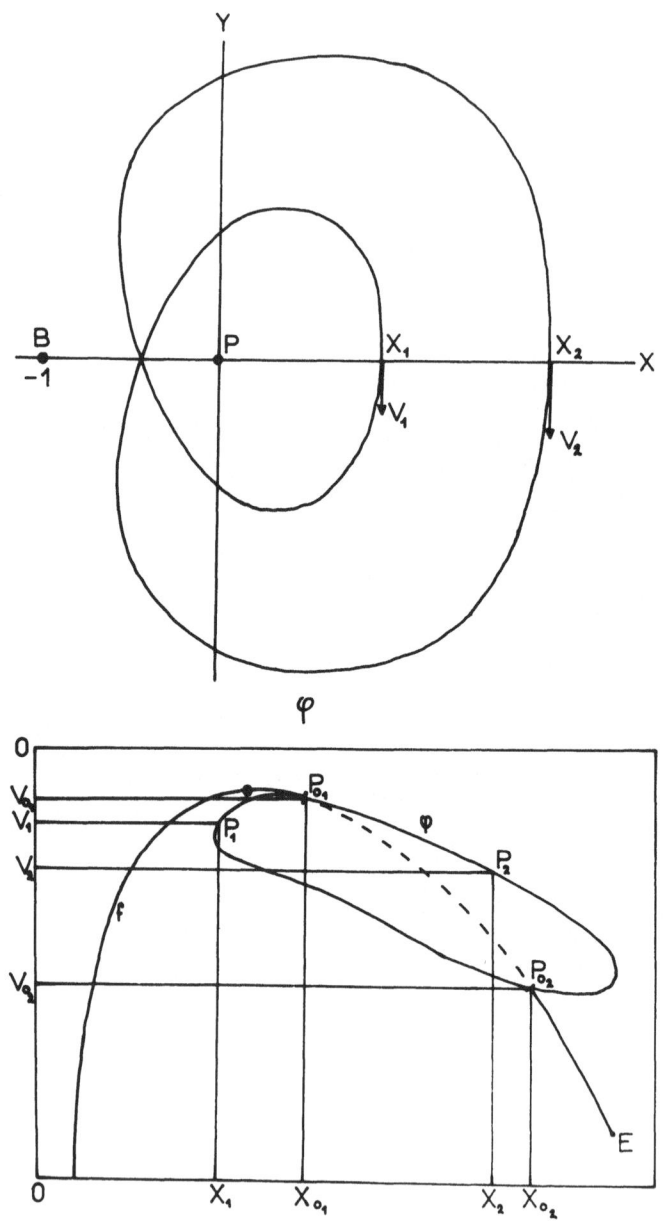

Figure 6. Family φ : general shape of the orbits(top)
and of the characteristic (bottom).

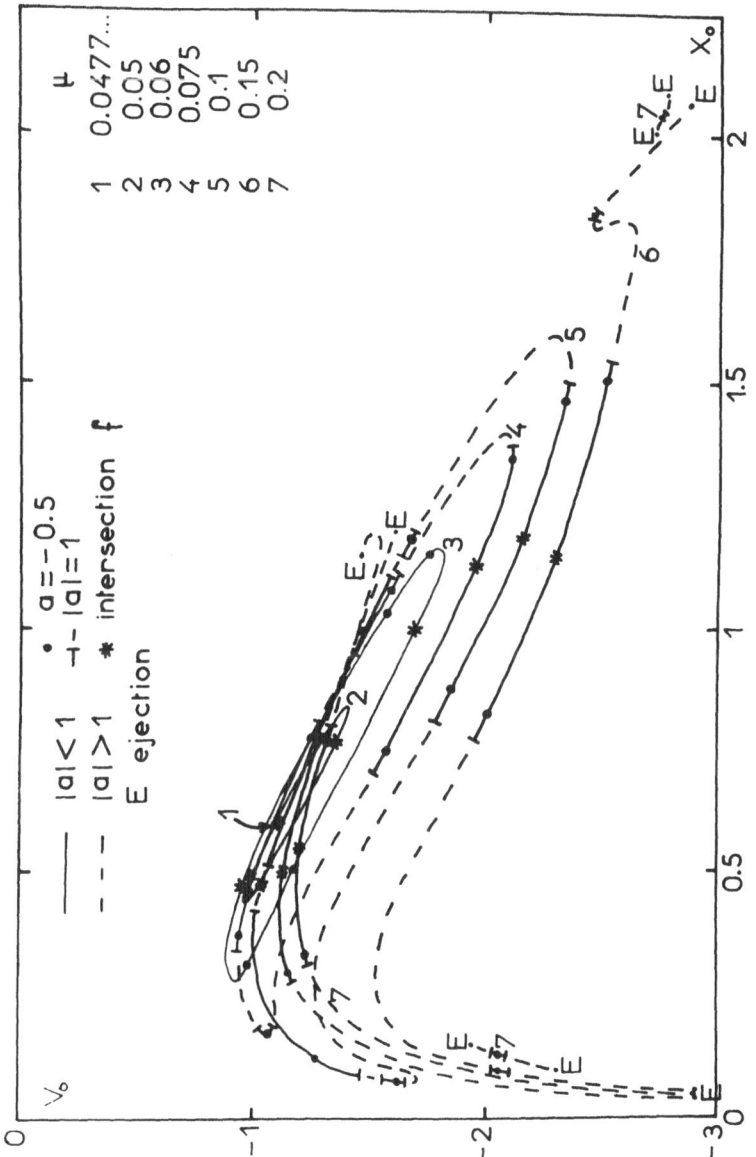

Figure 7. Family φ : evolution of the characteristic.

Figure 7 shows the evolution of the characteristic of family φ when μ increases from 0.0477... to 0.2, where the family has become almost unstable. Only a sample of our results is represented here, family φ and its stability have been computed for 26 different values of μ.

As for family f, we can establish a map of stability, with now $(X_1+X_2)/2$ as ordinate (fig. 8). From 0.0477... to $\mu \simeq 0.063$, the orbits are continuously stable. For $\mu \gtrsim 0.063$, there are two or more intervals of stability which decrease as μ grows. The second sort critical points of family φ are intersections with 4-periodic families, which are almost unstable.

1.2. Vertical stability of plane periodic orbits in the circular case

Nevertheless, an actual orbit can be called stable only if it is stable with respect to three-dimensional perturbations. Fortunately, perturbations in the plane are not coupled with those perpendicular to the plane. Thus we can obtain the three-dimensional stability by combining results on horizontal stability with a study on vertical stability, for which we can also define a vertical stability index a_v.

Figure 9 presents the vertical stability map of family f. the notations are the same as for the horizontal stability. We can combine horizontal and vertical stability to obtain the three-dimensional stability (fig. 10). Figure 11 presents the vertical stability -on the left- and the three-dimensional stability -on the right- for family φ.

As for horizontal stability, the vertical critical points (where $a_v = 1$) correspond to intersections with families of three-dimensional orbits. We are now planing to explore the three-dimensional families which intersect families f and φ.

1.3. Non-periodic orbits in the circular plane case

Now we turn to the non-periodic plane orbits. Figure 12 shows a fictitious example of the shape of the subspace of the initial conditions for stable orbits, which we shall call the Non-Periodic Stability zone. This zone is composed by a large continental region, approximately limited by the Lagrange points, and a peninsula more or less elongated, with sometimes one or more islands; there can be also lakes of instability enclosed in the zone. Inside the continental region, takes place the part of the characteristic of family f up to the first occurrence of a=-0.5. At this point, the peninsula is attached and surrounds the remaining part of the characteristic where and only where it is stable. From 0.0477... to $\mu \simeq 0.15$, the family φ can more or less neutralize the effects of the instability of family f. Figure 13 shows the evolution of the Non-Periodic Stability zone when μ increases up to 1.

Figure 14 presents some orbits along a section of the zone for $\mu =0.054$ and $V_0=-0.95$. We see here that, for large non-periodic orbits, the motion can be decomposed approximately into a fast "reference motion", looking like the nearest periodic orbit, and a slow libration around P of the centre of the corresponding curve.

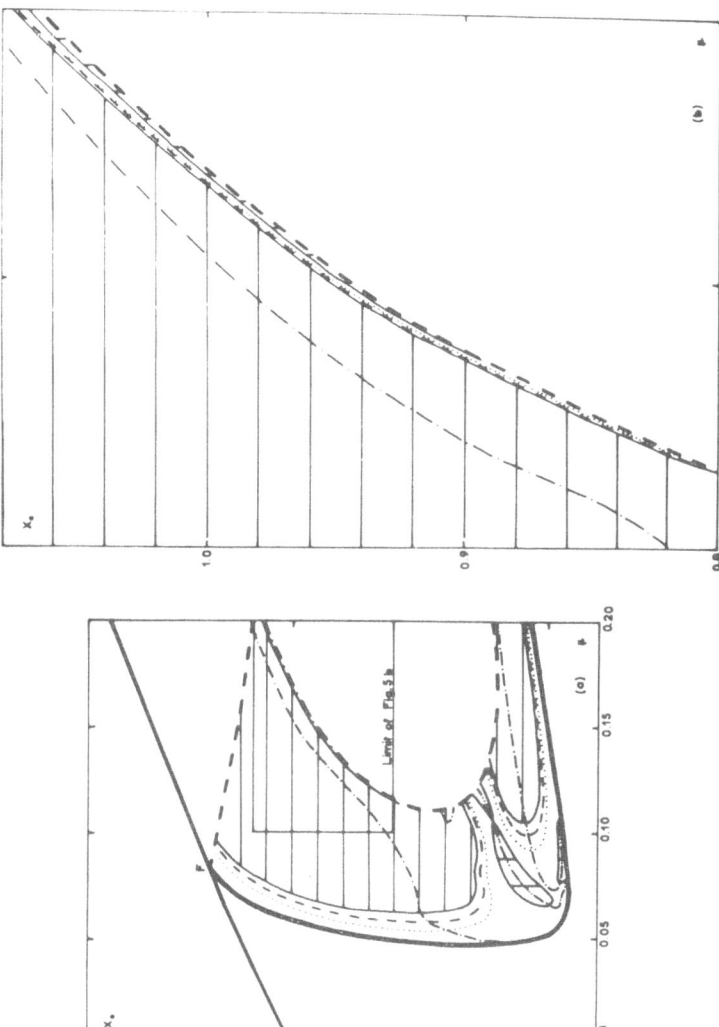

Figure 8. Family φ : stability map.

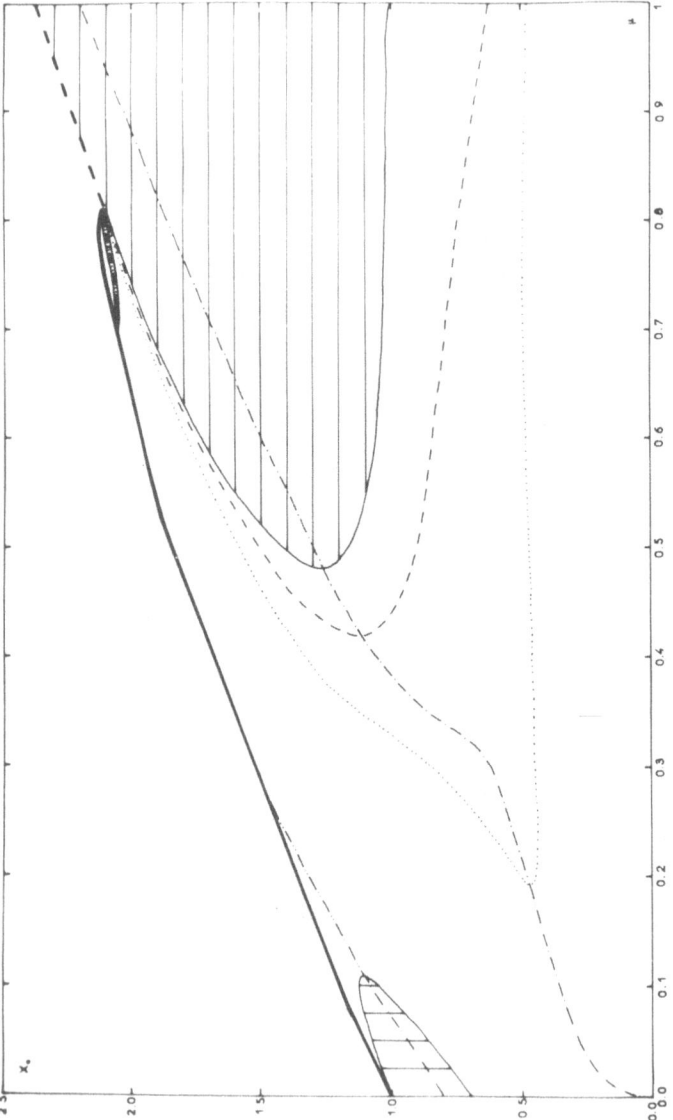

Figure 9. Family f: vertical stability map.

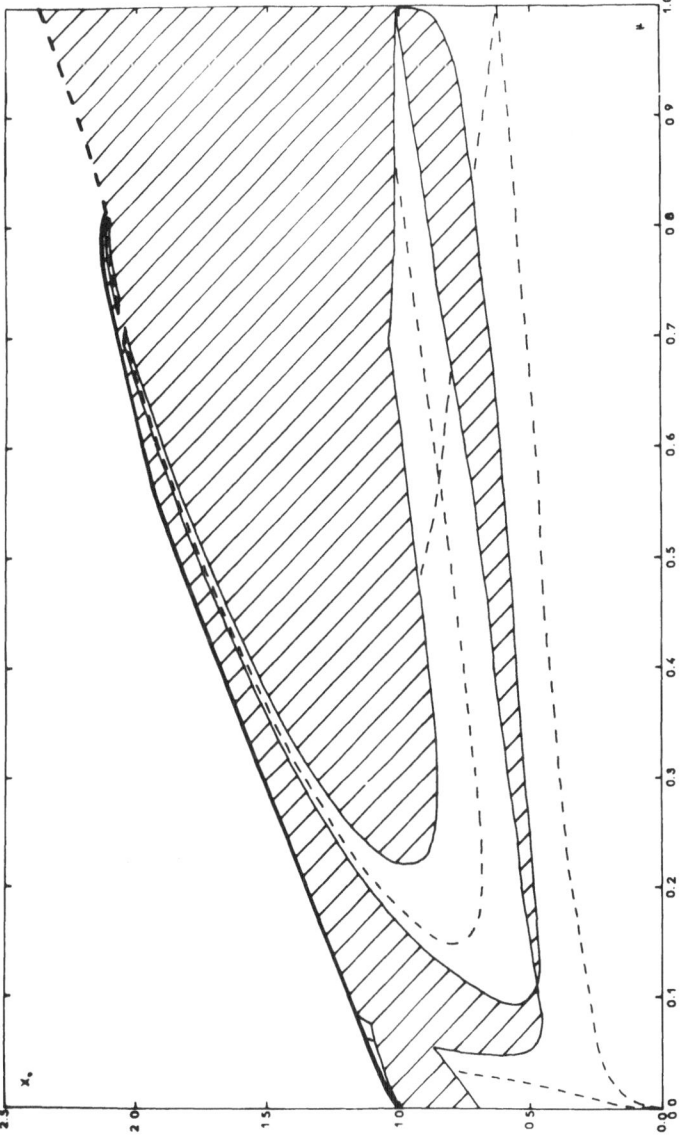

Figure 10. Family f: three—dimensional stability map.

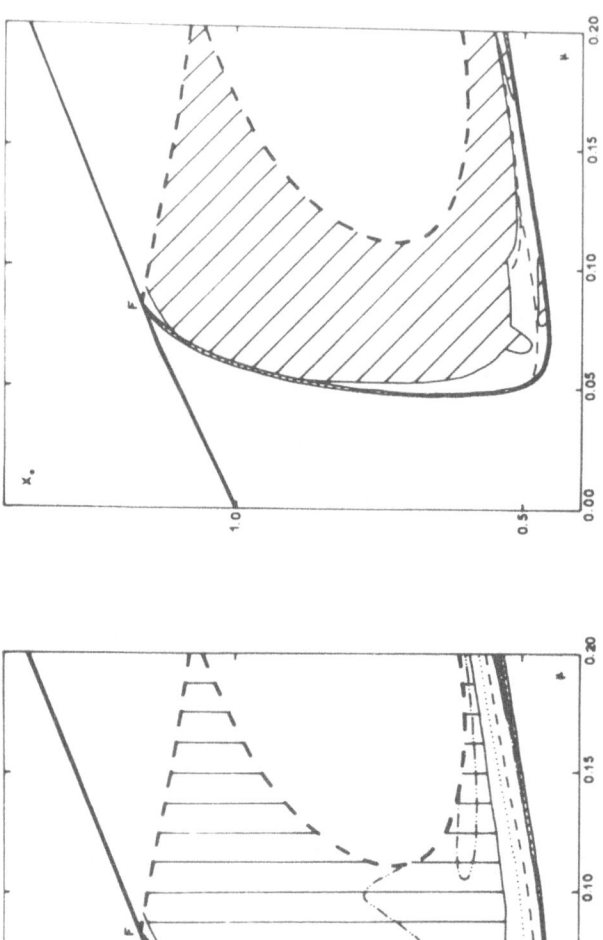

Figure 11. Family φ : vertical (left) and three-dimensional (right) stability maps.

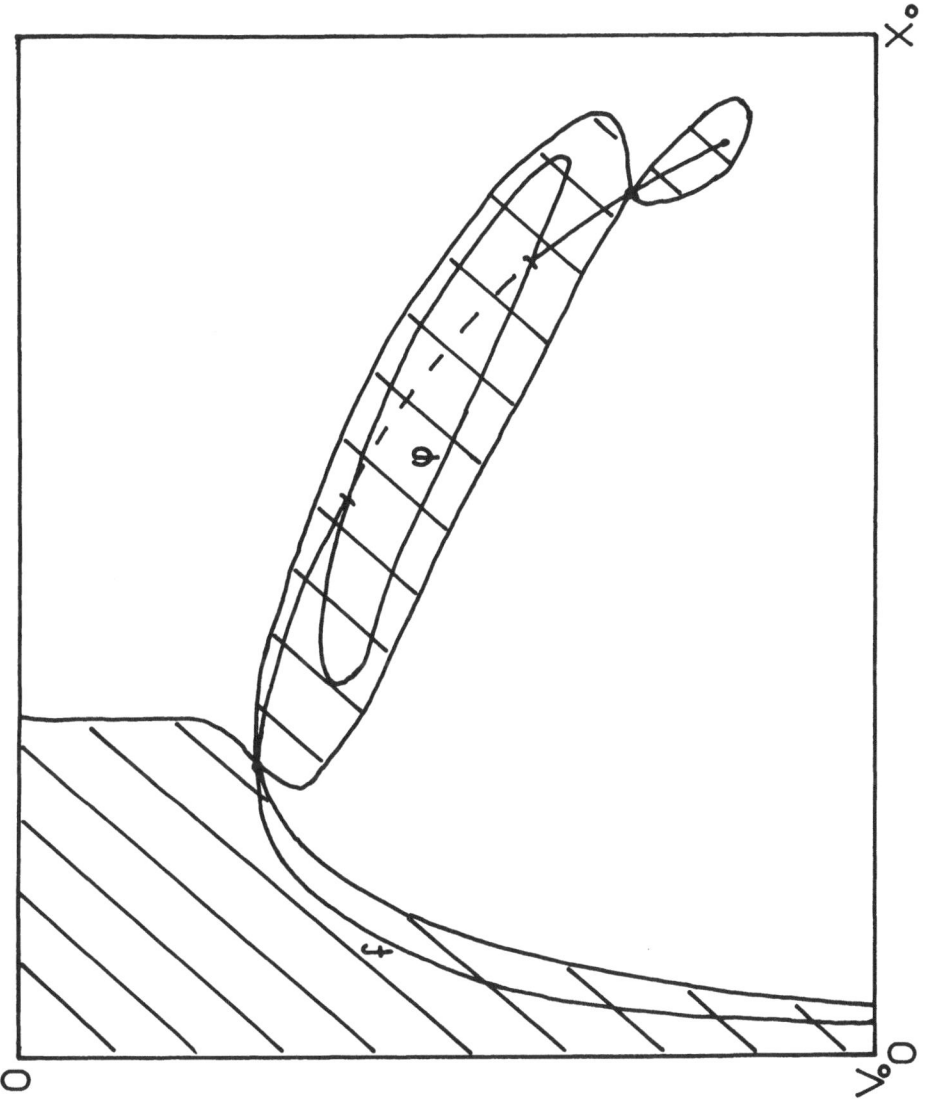

Figure 12. Non-Periodic Stability zone: general shape.

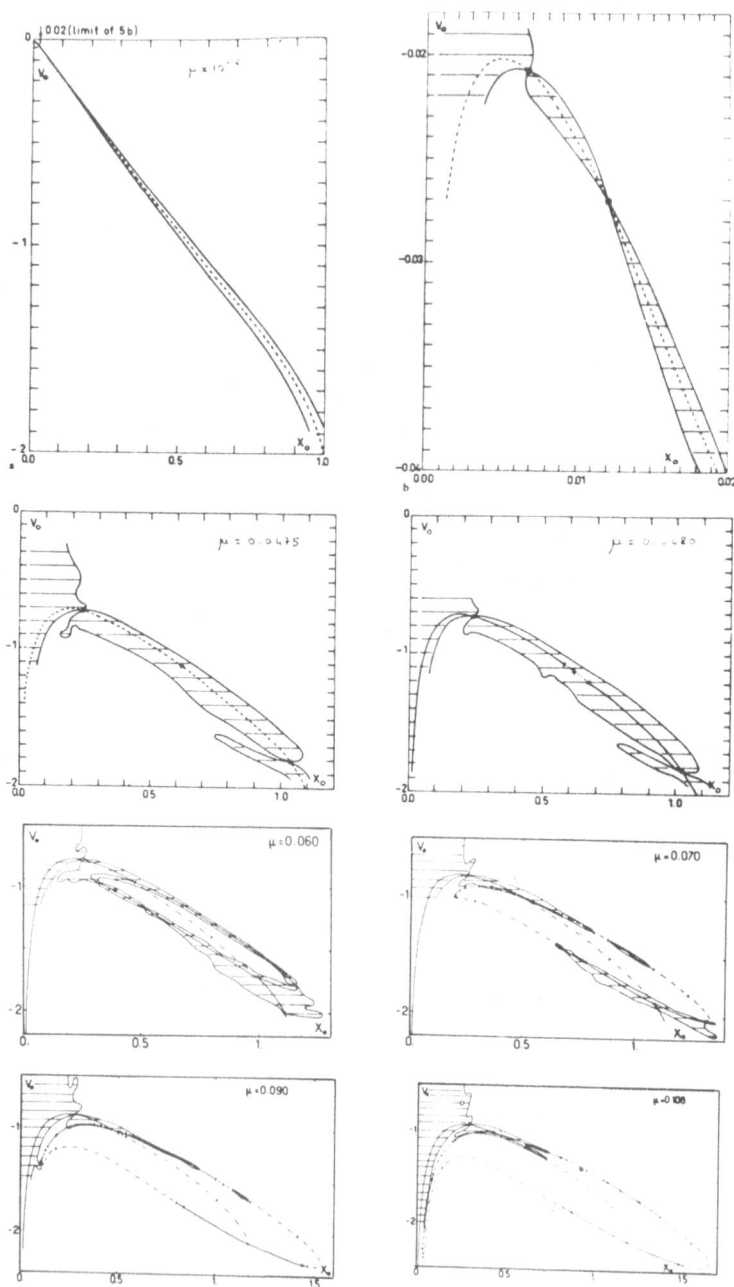

Figure 13. Non—Periodic Stability zone: evolution with μ

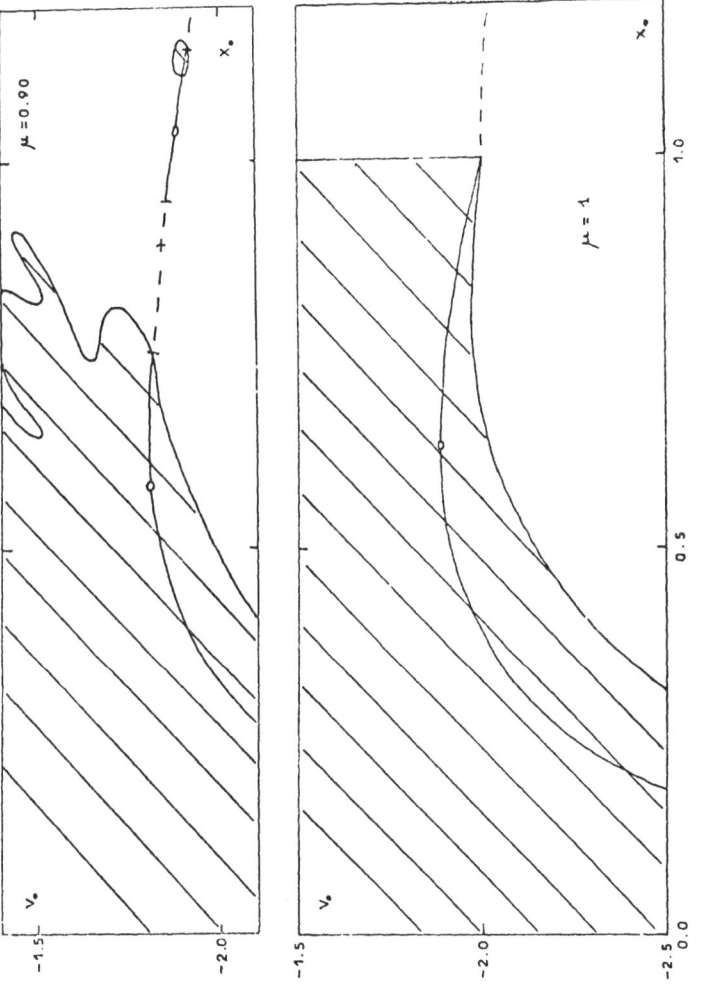

Figure 13.   continued

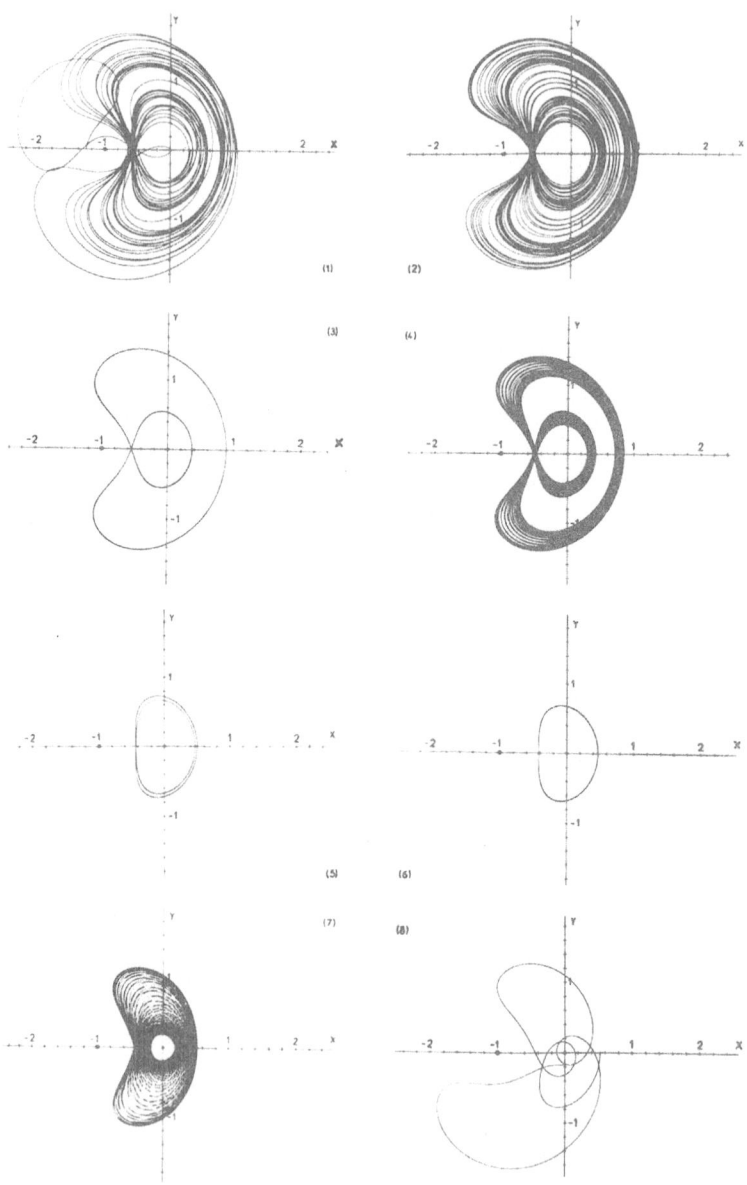

Figure 14. Orbits;   =0.054; V₀=-0.95; X₀ from 0.3 to 0.525.

2. ANALYSIS OF THE LIBRATION IN HILL'S CASE

Note: this section is detailed in $(1976a)$ whose abstract follows.

Numerical explorations of the restricted problem have shown that, for stable large non-periodic retrograde satellite orbits, the motion can be decomposed into a fast "reference motion" and a slow libration around B_2. We study here this libration in the circular plane Hill's case, for which the "reference motion" is elliptic. We establish the equations of motion for the coordinates of the centre of the ellipse. We find two integrals of motion: the first is the semi-major axis of the ellipse; the second is essentially Jacobi's integral, translated into the new coordinates. We give a formula for the period of the libration and we find its limiting value for small libration amplitudes. A numerical verification gives very good agreement for all these results.

REFERENCES

Benest,D., 1974,Astron. Astrophys. 32,39
  "   1975,   "   45,353
  "   1976a,Cel. Mech. 13,203
  "   b,Astron. Astrophys. to be published
  "   c,   "   "

DISPLACEMENT OF THE LAGRANGE EQUILIBRIUM POINTS IN THE RESTRICTED
THREE BODY PROBLEM WITH RIGID BODY SATELLITE

W.J. Robinson
University of Bradford

Abstract
    In the restricted problem of three point masses, the positions of
the equilibrium points are well known and are tabulated. When the
satellite is a rigid body, these values no longer correspond to the
equilibrium points. This paper seeks to determine the magnitudes of
the discrepancies.

1.   Introduction
    In the restricted problem of three point masses the positions of
the three collinear equilibrium points have been extensively calculated
(Szebehely, 1969). It has always been an attractive feature that the
other two points make equilateral triangles with the primaries.

    In this paper the possibility is considered that the satellite may
be a rigid body rather than a point mass. Since the satellite is always
regarded as extremely small, this assumption is unlikely to have noticeable effects on the overall picture. However, being small, it may have
an appreciable effect on the positions of the equilibrium points as
viewed from the satellite.

    It is assumed that the centre of mass of the satellite is situated
at an equilibrium point and that the attitude of the satellite is one
of stable equilibrium as described in an earlier paper by the author
(Robinson, 1974).

2.   Description of the system
    The primaries are point bodies with masses m_1 and m_2 located at
points A_1 and A_2. They rotate with constant angular velocity $\underline{\omega}$ about
their common mass centre O under the action of their mutual gravita-
tional attractions. The distance $A_1 A_2$ is constant. Choosing the units
of mass, length and time so that $m_1 + m_2 = 1$, $A_1 A_2 = 1$ and $|\underline{\omega}| = \omega = 1$
respectively, the gravitational constant also takes the value 1.

    A rectangular coordinate frame OXYZ is chosen so that OZ is the
axis of rotation and A_1 has coordinates $(m_2, 0, 0)$.

V. Szebehely (ed.), Dynamics of Planets and Satellites and Theories of Their Motion, 305-314.

The satellite has mass m, its centre of mass is located at G and

$$\overrightarrow{A_1 G} = \underline{r}_1, \qquad \overrightarrow{OG} = \underline{r}, \qquad \overrightarrow{A_2 G} = \underline{r}_2.$$

The mass m is assumed to be so small in comparison with the masses of the primaries that their motion is unaffected by its presence.

When the satellite is a point mass, its equation of motion is

$$\ddot{\underline{r}} + 2\underline{\omega} \times \dot{\underline{r}} + \underline{\omega} \times (\underline{\omega} \times \underline{r}) = -\Lambda \frac{m_1}{r_1^2} \hat{\underline{r}}_1. \tag{2.1}$$

In this equation, $\dot{\underline{r}}$ and $\ddot{\underline{r}}$ are the time derivatives relative to the rotating frame OXYZ. $\hat{\underline{r}}_1$ is the unit vector in the direction of $\underline{r}_1$ and $\underline{r}_1 = r_1 \hat{\underline{r}}_1$. The symbol Λ is always followed by an expression referring to m_1. The symbol means that a similar expression referring to m_2 has been omitted. Thus the symmetry of the expressions in terms of m_1 and m_2 is used to shorten the written equations. In the simplest case $\Lambda m_1 = 1$.

In the case where the satellite is a rigid body, let P be the position of an element of the body of mass μ. With

$$\overrightarrow{A_1 P} = \underline{\rho}_1, \qquad \overrightarrow{OP} = \underline{\rho}, \qquad \overrightarrow{A_2 P} = \underline{\rho}_2 \text{ and } \overrightarrow{GP} = \underline{\sigma}$$

it follows that

$$\Sigma \mu \underline{\rho}_1 = m \underline{r}_1, \qquad \Sigma \mu \underline{\rho} = m \underline{r} \text{ and } \Sigma \mu \underline{\sigma} = \underline{0}$$

and the equation of motion of the satellite is

$$m\{\ddot{\underline{r}} + 2\underline{\omega} \times \dot{\underline{r}} + \underline{\omega} \times (\underline{\omega} \times \underline{r})\} = -\Lambda m_1 \Sigma \frac{\mu \hat{\underline{\rho}}_1}{\rho_1^2} \tag{2.2}$$

where Σ indicates summation over all the elements of the satellite.

Since $\underline{\rho}_1 = \underline{r}_1 + \underline{\sigma}$ and $\frac{\sigma}{r_1}$ is usually an extremely small quantity, a polynomial $Q_{n,m}(\alpha_1)$ may be defined for non-negative integral values of n and m by the equation

$$\frac{1}{\rho_1^n} = \frac{1}{r_1^n} \sum_{m=0}^{\infty} \left(-\frac{\sigma}{r_1}\right)^m Q_{n,m}(\alpha_1)$$

where $\alpha_1 = (\hat{\underline{\sigma}} \cdot \hat{\underline{r}}_1)$ is the cosine of an angle. When $m < 0$, $Q_{n,m}(\alpha)$ is defined as having the value 0. Some of the properties of $Q_{n,m}(\alpha)$ are listed in Appendix 1.

Equation (2.2) can now be expressed in the form

$$m\{\ddot{\underline{r}} + 2\underline{\omega} \times \dot{\underline{r}} + \underline{\omega} \times (\underline{\omega} \times \underline{r})\}$$
$$= -\Lambda \frac{m_1}{r_1^2} \sum_{s=0}^{\infty} \left(-\frac{1}{r_1}\right)^s \{\hat{\underline{r}}_1 S_s(\alpha_1) - \underline{T}_s(\alpha_1)\} \tag{2.3}$$

where $S_s(\alpha_1)$ and $\underline{T}_s(\alpha_1)$ are scalar and vector moments of the satellite about G which are defined by

$$S_s(\alpha_1) = \Sigma \mu \sigma^s Q_{3,s}(\alpha_1) \tag{2.4}$$

$$\underline{T}_s(\alpha_1) = \Sigma \mu \underline{\sigma} \sigma^{s-1} Q_{3,s-1}(\alpha_1). \tag{2.5}$$

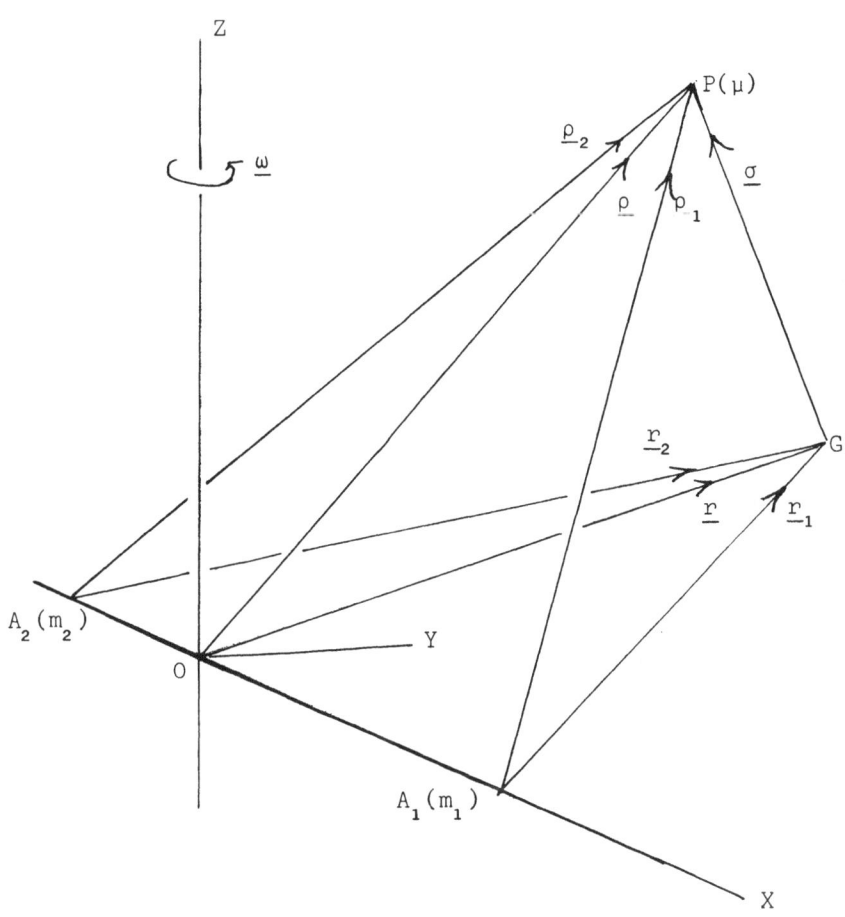

Diagram illustrating the notation of Section 2.

It can be seen that if a is a length commensurate with the linear dimensions of the satellite then $S_s(\alpha_1)$ and $\underline{T}_s(\alpha_1)$ have magnitudes of the order of ma^s. Some of the lower order moments are calculated in Appendices 2 and 3.

3.   The equilibrium points
      The positions of the equilibrium points are found by placing $\ddot{\underline{r}} = \dot{\underline{r}} = \underline{0}$ in the equation of motion and then solving the resulting equations.

In the case of the point mass satellite the five equilibrium points are usually denoted by L_1, L_2, L_3, L_4 and L_5, the first three being the collinear points the remaining two being the equilateral points.

If L_0 is one of these points and

$$\overrightarrow{A_1 L_0} = \underline{r}_{01}, \qquad \overrightarrow{OL_0} = \underline{r}_0$$

it follows that

$$\underline{\omega} \times (\underline{\omega} \times \underline{r}_0) = - \Lambda \frac{m_1}{r_{01}^2} \hat{\underline{r}}_{01}. \tag{3.1}$$

The corresponding equation for a rigid body satellite is

$$m \, \underline{\omega} \times (\underline{\omega} \times \underline{r}) = - m \Lambda \frac{m_1}{r_1^2} \hat{\underline{r}}_1$$

$$- \Lambda \frac{m_1}{r_1^2} \sum_{s=2}^{\infty} \left(- \frac{1}{r_1}\right)^s \left\{ \hat{\underline{r}}_1 S_s(\alpha_1) - \underline{T}_s(\alpha_1) \right\} \tag{3.2}$$

the values of $S_0(\alpha_1)$, $S_1(\alpha_1)$, $\underline{T}_0(\alpha_1)$ and $\underline{T}_1(\alpha_1)$ having been substituted in Equation (2.3).

The convergence of the infinite series depends on the moment of inertia terms rather than on the inverse powers of r_1.

It is clear that Equations (3.1) and (3.2) have the same solutions only when all the moments of the satellite about G vanish. Since the differences between the two equations are very small, it is to be expected that their solutions will differ by similarly small amounts. If G is at the point L, the equilibrium point of Equation (3.2) corresponding to the point L_0, let $\overrightarrow{L_0 L} = \underline{\varepsilon}$. It follows that $\underline{r}_1 = \underline{r}_{01} + \underline{\varepsilon}$ and

$$\frac{1}{r_1^n} = \frac{1}{r_{01}^n} \sum_{m=0}^{\infty} \left(- \frac{\varepsilon}{r_{01}}\right)^m Q_{n,m}(\beta_{01}) \tag{3.3}$$

where $\beta_{01} = \hat{\underline{r}}_{01} \cdot \hat{\underline{\varepsilon}}$, the expansion being possible on the assumption

that $\dfrac{\varepsilon}{r_{01}}$ is sufficiently small.

Regarding $\underline{r}_{01}$ as a known quantity (Szebehely, 1967), the object is to determine the displacement vector $\underline{\varepsilon}$, Equation (3.1) being regarded as a first approximation to Equation (3.2).

Under these circumstances the following expansions are necessary.

$$\alpha_1 = \sum_{s=0}^{\infty} \left(- \frac{\varepsilon}{r_{01}}\right)^s \alpha_{s1} \tag{3.4}$$

$$S_s(\alpha_1) = \sum_{p=0}^{\infty} \left(-\frac{\varepsilon}{r_{01}}\right)^p S_{sp1} \tag{3.5}$$

$$\underline{T}_s(\alpha_1) = \sum_{p=0}^{\infty} \left(-\frac{\varepsilon}{r_{01}}\right)^p \underline{T}_{sp1} \tag{3.6}$$

where α_{s1}, S_{sp1} and $\underline{T}_{sp1}$ are functions of ε, r_{01} and the moments of inertia. These coefficients are discussed in Appendix 4.

Equations (3.1) and (3.2) now reduce to

$$m\,\underline{\omega} \times (\underline{\omega} \times \underline{\varepsilon}) = -\Lambda \frac{m_1}{r_{01}^2} \sum_{s=1}^{\infty} \left(-\frac{\varepsilon}{r_{01}}\right)^s \underline{V}_{0s01}$$

$$- \Lambda \frac{m_1}{r_{01}^2} \sum_{s=2}^{\infty} \sum_{t=0}^{\infty} \sum_{p=0}^{\infty} \frac{\varepsilon^{t+p}}{(-r_{01})^{s+t+p}} \underline{V}_{stp1} \tag{3.7}$$

where

$$\underline{V}_{stp1} = \left\{ \hat{\underline{r}}_{01} Q_{3+s,t}(\beta_{01}) - \hat{\underline{\varepsilon}}\, Q_{3+s,\,t-1}(\beta_{01}) \right\} S_{sp1}$$

$$- Q_{2+s,t}(\beta_{01})\, \underline{T}_{sp1} \tag{3.8}$$

The properties of $\underline{V}_{stp1}$ are listed in Appendix 5. The suffices have the following significances. s is the order of the moments contained in $\underline{V}_{stp1}$ and $(t+p)$ is the power of the associated term in ε and $s+t+p$ the power of $\frac{1}{r_{01}}$.

With m and a, a characteristic dimension of the satellite, as the very small parameters which determine the relative magnitudes of the terms in Equation (3.7), the linearized form of the equation becomes

$$m\,\underline{\omega} \times (\underline{\omega} \times \underline{\varepsilon}) - \Lambda \frac{m_1}{r_{01}^3}\, \varepsilon\, \underline{V}_{0101}$$

$$= -\Lambda \frac{m_1}{r_{01}^2} \left(-\frac{1}{r_{01}}\right)^p \underline{V}_{p001} \tag{3.9}$$

where p is the smallest positive integer exceeding 1 which gives a non-zero solution.

4. The collinear equilibrium points
    At the collinear points $\hat{\underline{r}}_{01} = [\lambda_1, 0, 0]$ and $\hat{\underline{r}}_{02} = [\lambda_2, 0, 0]$ where λ_1, λ_2 have the values -1, -1 at L_1; -1, +1 at L_2 and +1, +1 at L_3.

    It follows that $\underline{\varepsilon} = \xi\underline{i} + \eta\underline{j} + \zeta\underline{k}$ is given by

$$m\xi\left(1 + 2\Lambda \frac{m_1}{r_{01}^3}\right) = \Lambda \frac{m_1}{r_{01}^{a+2}} (-1)^a \underline{V}_{a001} \cdot \underline{i}$$

$$m\eta\left(1 - \Lambda \frac{m_1}{r_{01}^3}\right) = \Lambda \frac{m_1}{r_{01}^{b+2}} (-1)^b \underline{V}_{b001} \cdot \underline{j}$$

$$m\zeta \Lambda \frac{m_1}{r_{01}^3} = \Lambda \frac{m_1}{r_{01}^{c+2}} (-1)^{c+1} \underline{V}_{c001} \cdot \underline{k}$$

where $\underline{i}$, $\underline{j}$ and $\underline{k}$ are the unit vectors along the axes of OXYZ and a, b and c are the smallest integers greater than 1 which give non-zero terms.

It has been shown (Robinson, 1974), that if the centre of mass is held at a collinear equilibrium point, the principal axes of the satellite align themselves parallel to the axes of OXYZ when the satellite is at relative rest. If A, B and C are the principal moments about OX, OY and OZ respectively the satellite reaches a stable attitude with C > B > A. For some bodies there is a second possibility with B > A > C.

If the satellite is such that 2A ≠ B + C, then

$$\underline{V}_{2001} = \frac{3}{2}\lambda_1(-2A + B + C)\,[1, 0, 0]$$

so that

$$m\xi\left(1 + 2\Lambda \frac{m_1}{r_{01}^3}\right) = \frac{3}{2}(-2A + B + C)\,\Lambda\,\frac{m_1\lambda_1}{r_{01}^4}$$

To obtain η and ζ, the third moments have to be considered. If these do not vanish

$$m\eta\left(1 - \Lambda \frac{m_1}{r_{01}^3}\right) = \frac{3}{2}\Sigma\mu y(4x^2 - y^2 - z^2)\,\Lambda\,\frac{m_1}{r_{01}^5}$$

$$m\zeta \Lambda \frac{m_1}{r_{01}^3} = \frac{3}{2}\Sigma\mu z(4x^2 - y^2 - z^2)\,\Lambda\,\frac{m_1}{r_{01}^5}$$

If a is a length commensurate with the linear dimensions of the satellite it can be seen that ξ is of the same order at a^2 and η and ζ of the same order as a^3.

5. The equilateral equilibrium points
In this case $r_{01} = r_{02} = 1$ and

$$\hat{\underline{r}}_{01} = \frac{1}{2} [-1, \sqrt{3}, 0] \quad \text{and} \quad \hat{\underline{r}}_{02} = \frac{1}{2} [1, \sqrt{3}, 0].$$

Referring again to the stable stationary position of the satellite (Robinson, 1974) when G is at L , the principal axes Ox and Oy are turned through an angle $-\kappa$ about Oz where $\cos 2\kappa = \frac{1}{N}$ and $N = \sqrt{1 + 12m_0^2}$ with $2m_0 = m_2 - m_1$.

Equation (3.9) now becomes

$$\begin{bmatrix} \xi + \sqrt{12} \ m_0 \eta \\ \sqrt{12} \ m_0 \xi + 3\eta \\ -\frac{4}{3} \zeta \end{bmatrix} = \frac{4}{3m} (-1)^p \ \Lambda \ m_1 \ \underline{V}_{p001} \tag{5.1}$$

ξ and η can be determined when p = 2 in some cases, but to find ζ, p must take the value 3 at least.

With these values

$$\xi = \frac{m_0 (68 - 11N^2)}{4Nm(4 - N^2)} (A - B)$$

$$\eta = \frac{6 - N^2}{4\sqrt{3} \ Nm(4 - N^2)} [4NC + 3(A-B) - 2N(A-B)] + \frac{N(B-A)}{2\sqrt{3}(4-N^2)m}$$

$$\zeta = \frac{1}{m} \Lambda \ m_1 \underline{k} \cdot \underline{V}_{3001}$$

Again it can be seen that ξ and η are of order a^2 while ζ is of order a^3.

Conclusions

It has been shown that the displacements of the equilibrium points are extremely small, which was to be expected. If a is the length of the satellite, remembering that $A_1 A_2 = 1$, then ξ is of the same order as a^2. η is of the same order in the equilateral case, but of order a^3 in the collinear case. The coordinate ζ is, at its greatest, of order a^3 in either case.

It can also be mentioned that since all bodies certainly have one stable attitude at each equilibrium point, some may have two stable attitudes. The conditions for the second case are given in the paper referred to earlier (Robinson, 1974). The outcome is that some bodies may actually have ten equilibrium points, while others have only five, or some intermediate number. Of course, in those cases where the points occur in pairs, the members of such pairs are very near each other.

Appendix 1.   The polynomials $Q_{mn}(\alpha)$

From the definition

$$(1 + 2\alpha x + x^2)^{-n/2} = \sum_{s=0}^{\infty} (-x)^s \, Q_{ns}(\alpha)$$

it follows that

$$Q'_{ns}(\alpha) = \frac{dQ_{ns}(\alpha)}{d\alpha} = n \, Q_{n+1,s-1}(\alpha)$$

and

$$Q_{n0}(\alpha) = 1$$
$$Q_{n1}(\alpha) = n\alpha$$
$$Q_{n2}(\alpha) = \tfrac{1}{2}n\{(n + 2)\alpha^2 - 1\}$$
$$Q_{n3}(\alpha) = \tfrac{1}{6}n(n + 2)\{(n + 4)\alpha^3 - 3\alpha\}.$$

Note that

$$Q_{1s}(\alpha) = P_s(\alpha)$$

which are the familiar Legendre Polynomials.

Appendix 2.   The scalar moments $S_s(\alpha_1)$

$$S_s(\alpha_1) = \Sigma \, \mu\sigma^s Q_{3s}(\alpha_1)$$

where

$$\alpha_1 = \underline{\hat{\sigma}} \cdot \underline{\hat{r}}_1$$

$$S_0(\alpha_1) = \Sigma\mu = m$$
$$S_1(\alpha_1) = 3\Sigma\mu(\underline{\sigma} \cdot \underline{\hat{r}}_1) = 0$$
$$S_2(\alpha_1) = \frac{15}{2} \Sigma\mu(\underline{\sigma} \cdot \underline{\hat{r}}_1)^2 - \frac{3}{2} \Sigma \, \mu\sigma^2$$
$$S_3(\alpha_1) = \frac{35}{2} \Sigma\mu(\underline{\sigma} \cdot \underline{\hat{r}}_1)^3 - \frac{15}{2} \Sigma \, \mu\sigma^2(\underline{\sigma} \cdot \underline{\hat{r}}_1)$$

Appendix 3.   The vector moments $\underline{T}_s(\alpha_1)$

$$\underline{T}_s(\alpha_1) = \Sigma \, \mu\underline{\sigma} \, \sigma^{s-1} Q_{3s-1}(\alpha_1).$$

Since the above definition holds for $s \geqslant 1$ the additional definition is made

$$\underline{T}_0(\alpha_1) = \underline{0}.$$

From the definition

$$\underline{T}_1(\alpha_1) = \Sigma \, \mu \, \underline{\sigma} = \underline{0}$$
$$\underline{T}_2(\alpha_1) = 3 \Sigma \, \mu \, \underline{\sigma}(\underline{\sigma} \cdot \underline{\hat{r}}_1)$$
$$\underline{T}_3(\alpha_1) = \frac{15}{2} \Sigma \, \mu \, \underline{\sigma}(\underline{\sigma} \cdot \underline{\hat{r}}_1)^2 - \frac{3}{2} \Sigma \, \mu \, \underline{\sigma} \, \sigma^2.$$

Appendix 4
Using Equation (3.3) it can be shown that

$$\alpha_1 = \sum_{s=0}^{\infty} \left(-\frac{\varepsilon}{r_{01}}\right)^s \alpha_{s1}$$

where

$$\alpha_{s1} = \alpha_{01} Q_{1,s}(\beta_{01}) - \gamma \, Q_{1,s-1}(\beta_{01}),$$

$$\alpha_{01} = \hat{\underline{r}}_{01} \cdot \underline{\sigma} \quad \text{and} \quad \gamma = \hat{\underline{\varepsilon}} \cdot \hat{\underline{\sigma}}.$$

Making use of the relation

$$Q'_{m,n}(\alpha) = m \, Q_{m+2,n-1}(\alpha)$$

and Taylor's Theorem

$$Q_{m,n}(\alpha_1) = \sum_{s=0}^{n} \frac{2^s}{s!} \frac{\Gamma\left(\frac{m}{2}+s\right)}{\Gamma\left(\frac{m}{2}\right)} \left\{ \sum_{t=1}^{\infty} \left(-\frac{\varepsilon}{r_{01}}\right)^t \alpha_{t1} \right\}^s Q_{m+2s,n-s}(\alpha_{01})$$

$$= \sum_{s=0}^{\infty} \left(-\frac{\varepsilon}{r_{01}}\right)^s q_{mns1}.$$

From the definition

$$S_s(\alpha_1) = \sum_{\mu} \mu \, \sigma^s \, Q_{3\cdot s}(\alpha_1)$$

it follows that

$$S_s(\alpha_1) = \sum_{\mu} \mu \, \sigma^s \sum_{t=0}^{\infty} \left(-\frac{\varepsilon}{r_{01}}\right)^t q_{3s t 1}$$

$$= \sum_{t=0}^{\infty} \left(-\frac{\varepsilon}{r_{01}}\right)^t S_{s t 1}$$

where

$$S_{s t 1} = \sum_{\mu} \mu \, \sigma^s \, q_{tmn1}$$

is a scalar moment of the s-th order.
In a similar manner

$$\underline{T}_s(\alpha_1) = \sum_{t=0}^{\infty} \left(-\frac{\varepsilon}{r_{01}}\right)^t \underline{T}_{s t 1}.$$

Since $S_0(\alpha_1) = m$, $S_1(\alpha_1) = 0$, $\underline{T}_0 = \underline{T}_1 = \underline{0}$

$S_{001} = m$, $S_{0t1} = 0 \ (t \neq 0)$, $S_{1t1} = 0$ for all t.

$\underline{T}_{0t1} = \underline{T}_{1t1} = \underline{0}$ for all t.

Also $S_s(\alpha_{01}) = S_{s01}$, $\underline{T}_s(\alpha_{01}) = \underline{T}_{s01}$.

Appendix 5.   The function $\underline{V}_{st p1}$

$$\underline{V}_{st p1} = \{\hat{\underline{r}}_{01} Q_{3+s,t}(\beta_{01}) - \hat{\underline{\varepsilon}} \, Q_{3+s,t-1}(\beta_{01})\} \, S_{s p1}$$

$$- Q_{2+s,t}(\alpha_{01}) \, \underline{T}_{s p1}.$$

If s, t and p are non-zero positive integers it follows that

$$\underline{V}_{\infty01} = m\,\hat{\underline{r}}_{01}$$

$$\underline{V}_{s001} = \hat{\underline{r}}_{01}S_s(\alpha_{01}) - \underline{T}_s(\alpha_{01})$$

$$\underline{V}_{0t01} = m\{\hat{\underline{r}}_{01}Q_{3,t}(\beta_{01}) - \hat{\underline{\varepsilon}}\,Q_{3,t-1}(\beta_{01})\}$$

$$\underline{V}_{00p1} = \underline{0}$$

$$\underline{V}_{0tp1} = \underline{0}$$

$$\underline{V}_{s0p1} = \hat{\underline{r}}_{01}S_{sp1} - \underline{T}_{sp1}$$

$$\underline{V}_{st01} = \{\hat{\underline{r}}_{01}Q_{3+s,t}(\beta_{01}) - \hat{\underline{\varepsilon}}\,Q_{3+s,t-1}(\beta_{01})\}\,S_s(\alpha_{01})$$
$$\qquad\qquad - Q_{2+s,t}(\beta_{01})\,\underline{T}_s(\alpha_{01})$$

$$\underline{V}_{1tp1} = \underline{0}\,.$$

References

Robinson, W.J.: 1974, Celes. Mech. <u>10</u>, 17.
Szebehely, V.: 1967, Theory of Orbits, Academic Press, New York and
    London.

A NEW KIND OF PERIODIC ORBIT :
THE THREE-DIMENSIONAL ASYMMETRIC

V. V. Markellos
University of Glasgow

A great deal of human and computer effort has been directed in
recent decades to the determination of the periodic orbits of the
restricted three-body problem and the study of their properties for
well known reasons of significance and feasibility.

In most cases it is the plane and symmetric (with respect to the
synodical line of the primaries) orbits that are investigated the
main reasons for this being simplicity of the algorithms involved and
computer time economy. Few results exist on the three-dimensional
symmetric periodic orbits and even fewer on the plane but asymmetric
ones.

It is felt that our understanding of the structure of the periodic
solutions of the restricted problem can be improved if more attention
is paid to the asymmetric orbits. Not only are these orbits necessary
in order to obtain a complete picture of the set of periodic orbits
but they seem to be highly relevant in the study of the motions of the
members of the Solar system, as they are characteristically associated
with *commensurabilities*.

Clearly the most general orbits, and the most relevant ones in the
study of the Dynamics of Planets and Satellites, are the *Asymmetric
Three-Dimensional*. This note is an announcement of their discovery.

They have been detected by examination of the stability of plane
asymmetric periodic orbits with respect to perturbations which tend
to change the plane motion into three-dimensional. They have been
determined numerically by application of a suitable predictor-corrector
algorithm. Details are given in another paper.

Here we illustrate a family of asymmetric three-dimensional periodic
orbits 'around' both primaries. The orbits shown are retrograde in the
rotating, but direct in the fixed, system of coordinates. Their periods
are nearly commensurate to the basic period of the system (twice as long).
The value used for the mass parameter of the problem was $\mu = 0.45$.

V. Szebehely (ed.), Dynamics of Planets and Satellites and Theories of Their Motion, 315-317.
All Rights Reserved. Copyright © 1978 by D. Reidel Publishing Company, Dordrecht, Holland.

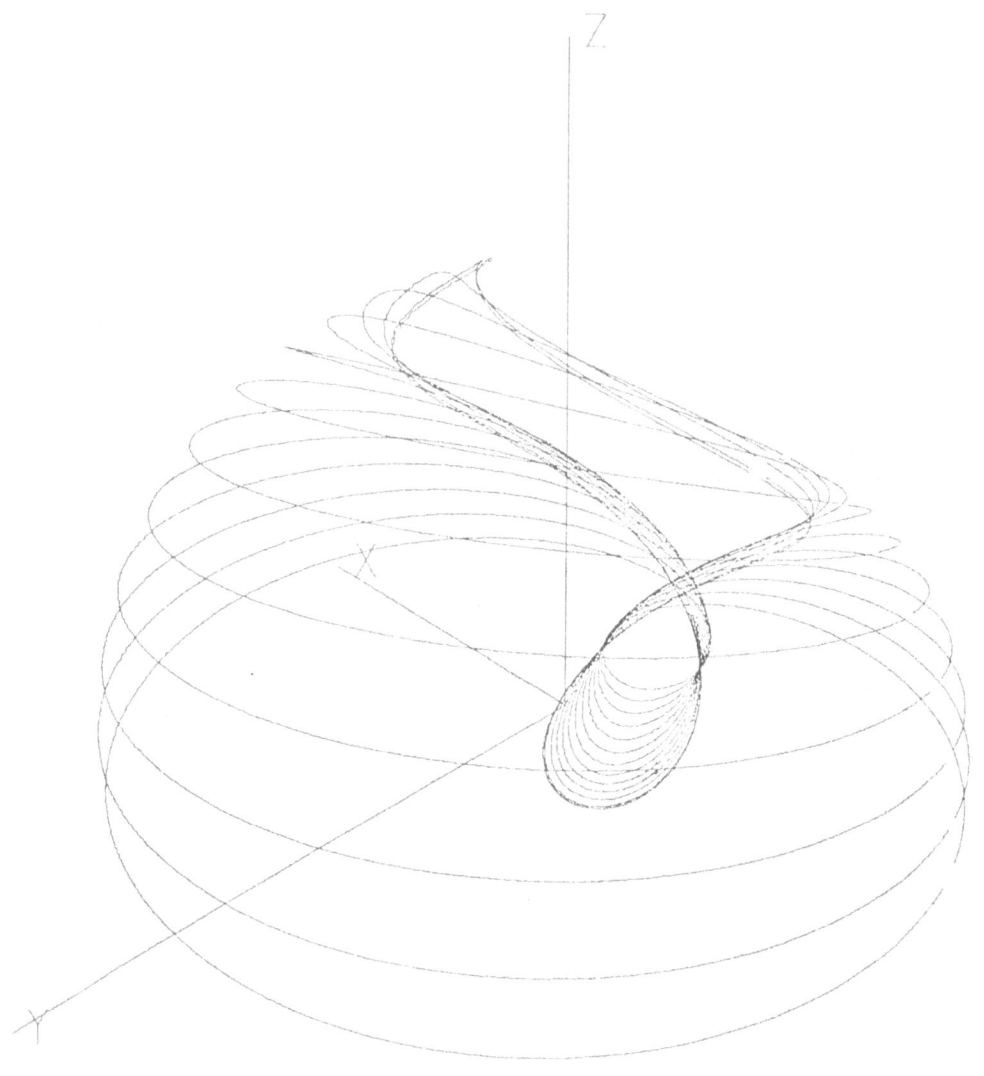

Fig.1 Computer plot of the family A2(ℓ_2) of three-
        dimensional asymmetric periodic orbits.   The
        image is created by conical projection of the
        orbits on a plane perpendicular to the line
        joining the point of "view" (x= -3, y= 3, z= 3)
        to the centre of mass of the primaries (origin).
        The axes are drawn to distances three times
        the mutual (unit) distance of the primaries.

This is the second of two families $(A1(\ell_2)$ and $A2(\ell_2))$ that have been determined of this new kind. Both families bifurcate from family $a(\ell_2)$ of plane asymmetric periodic orbits. The latter family has been discussed by Kristianson (1933) and Message (1969). It is a branch of Strömgren's family ℓ of plane symmetric periodic orbits around both primaries.

References

Kristianson,K.,1933.Astr.Nachr.,250,249
Markellos,V.V.,1977.Mon.Not.R.astr.Soc.,180,103
Message,P.J.,1969.In Periodic orbits, stability, and resonances,
          ed. G.E.O.Giacaglia,D.Reidel,Dordrecht,Holland

ON ASYMMETRIC PERIODIC SOLUTIONS OF THE PLANE RESTRICTED PROBLEM OF
THREE BODIES, AND BIFURCATIONS OF FAMILIES

P.J. Message
Liverpool University
D.B. Taylor
Glasgow University

Previous work on the plane circular restricted problem of three
bodies (Message 1958, 1959, 1970, and Fragakis 1973) has shown the
existence, in association with each of the commensurabilities 2:1 and
3:1 of the orbital periods, of a pair of families of asymmetric
periodic solutions, branching from the stable series of symmetric
periodic solutions of Poincaré's second sort associated with that
commensurability. (Each solution of either family is the mirror image,
in the line of the two finite bodies, of a member of the other family
of solutions associated with the commensurability.) The stability is
transferred at the bifurcation to the two series of asymmetric orbits,
each of which is therefore stable. Recent numerical integrations
carried out by one of us (P.J.M.) have found such asymmetric periodic
orbits associated also with the 4:1 commensurability, and quantities
describing orbits of one of the two series are given in Table 1, show-
ing the run of such orbits up to a second bifurcation with the same
series of symmetric periodic orbits from which it sprang. Quantities
describing some members of this series of symmetric orbits are given
in Table 2. It is seen that stability is transferred back to the
symmetric series at the second bifurcation. (The unit of distance is
the distance between the two finite bodies, the unit of speed is the
speed of their relative motion, and the initial conditions given
$(x_0, \dot{x}_0, \dot{y}_0)$ are for a crossing of the line of the two finite bodies,
this line being taken as axis of "x" in a rotating Cartesian frame in
the usual way. The mean values of the major semi-axis and eccentricity
are denoted by $\bar{a}$ and $\bar{e}$, respectively, C is Jacobi's constant, and $\bar{y}_2$
is the mean value of the critical argument $y_2 = 4\lambda - \lambda' - 3\omega$. The
mass ratio used is 0.000954927, T is the period of the solution in
units of the period of the motion of the two finite bodies, and $2\pi\, c/T$
is the non-zero characteristic exponent.)

It was shown earlier (Message 1970) that a bifurcation of a series
of asymmetric orbits from a series of symmetric orbits of the second
sort implies a zero of $\dfrac{\partial^2 H^*}{\partial y_2^2}$, where H^* is the long-period part of the

319

V. Szebehely (ed.), Dynamics of Planets and Satellites and Theories of Their Motion, 319-323.
All Rights Reserved. Copyright © 1978 by D. Reidel Publishing Company, Dordrecht, Holland.

x_0	$\dot{x}_0$	$\dot{y}_0$	C	T	$\bar{a}$	$\bar{e}$	$y_2/(2\pi)$	$\cosh(2\pi c)$
3.02441	0.00186	-2.51016	-1.75377	4.00007	2.5225	0.2002	0.4541	0.99996
3.07723	0.02208	-2.57461	-1.74513	4.00002	2.5225	0.2252	0.3403	0.99783
3.13406	0.03133	-2.64379	-1.73496	3.99996	2.5226	0.2514	0.3104	0.99394
3.38456	0.05364	-2.94724	-1.67858	3.99956	2.5231	0.3617	0.2348	0.94319
3.63377	0.06320	-3.25382	-1.59915	3.99914	2.5236	0.4708	0.1947	0.80572
3.88892	0.07828	-3.55530	-1.49587	3.99911	2.5235	0.5763	0.1690	0.60221
4.09096	0.08154	-3.79879	-1.39372	3.99954	2.5229	0.6581	0.1564	0.41258
4.19809	0.08083	-3.92788	-1.33280	3.99989	2.5225	0.6995	0.1646	0.29455
4.33799	0.07645	-4.12082	-1.23200	4.00041	2.5219	0.7589	0.1821	0.09419
4.50418	0.06919	-4.29817	-1.12631	4.00076	2.5214	0.8113	0.2095	-0.06989
4.61277	0.06188	-4.43128	-1.03583	4.00087	2.5211	0.8495	0.2420	-0.11806
4.73631	0.05132	-4.58516	-0.91481	4.00079	2.5207	0.8925	0.2821	-0.01065
4.82849	0.04119	-4.70317	-0.80353	4.00057	2.5204	0.9245	0.3235	0.23279
4.92632	0.02567	-4.83421	-0.65220	4.00019	2.5201	0.9583	0.3842	0.66697
4.99326	0.00006	-4.93263	-0.50117	3.99993	2.5200	0.9817	0.4784	0.96223

Table 1

Initial conditions and other data for asymmetric periodic solutions associated with the 4:1 commensurability of periods

x_o	$\dot{y}_o$	C	T	$\bar{a}$	$\bar{e}$	$\cosh(2\pi c)$
2.56473	-1.94591	-1.78565	4.00015	2.5226	0.0179	1.00000
2.71867	-2.13659	-1.78098	4.00007	2.5227	0.0789	0.99985
2.91225	-2.37380	-1.76657	4.00004	2.5227	0.1556	0.99939
2.98435	-2.46153	-1.75873	4.00005	2.5226	0.1842	0.99960
3.06438	-2.55858	-1.74842	4.00008	2.5225	0.2160	0.99960
3.30311	-2.84634	-1.70724	4.00027	2.5222	0.3106	1.00066
3.78319	-3.42012	-1.57200	4.00064	2.5208	0.5009	1.01487
4.45657	-4.22863	-1.21423	4.00014	2.5200	0.7683	1.11917
4.56761	-4.36504	-1.12367	4.00000	2.5200	0.8125	1.30153
4.74135	-4.58334	-0.94761	3.99983	2.5199	0.8816	1.30798
4.97828	-4.90843	-0.54617	3.99988	2.5200	0.9757	1.23146
4.98689	-4.92219	-0.52111	3.99990	2.5200	0.9791	1.01915
5.00482	-4.95174	-0.46104	3.99997	2.5200	0.9862	0.98854
5.02001	-4.98065	-0.39602	4.00006	2.5201	0.9922	0.90331
5.03784	-5.02543	-0.26094	4.00032	2.5202	0.9992	0.78708

Note: 0.41582

Table 2

Initial conditions and other data for symmetric periodic orbits associated with the 4:1 commen-surability of periods

Table 3

Values of the mean eccentricity at zeros of $\dfrac{\partial^2 H^*}{\partial y_2^2}$ in the limit $m' \to 0$

Commensurability	mean eccentricities	
2:1	0.03652,	0.95927
3:1	0.12211,	0.97178
4:1	0.20112,	0.97792
5:1	0.26711,	0.98162
6:1	0.32184,	0.98413
7:1	0.36767,	0.98594
8:1	0.40657,	0.98732

Hamiltonian function, and $y_2 = (p+q)\lambda - p\lambda' - q\omega$ is the critical argument. Recalculations of $\dfrac{\partial^2 H^*}{\partial y_2^2}$ by one of us (D.B.T.) have led to the finding of an error in the programme originally used in the calculation of this quantity, of such a type that the values previously given for the mean eccentricities at which such bifurcations can be expected need correction for those commensurabilities with $q \neq 1$. Corrected values of the mean eccentricity at which zeros occur, in the limit as the mass ratio tends to zero, for commensurabilities with $p = 1$ and $q = 1, 2, 3, \ldots, 7$, are given in Table 3. No zeros are now found for the cases with $p = 2$, $q = 1, 3, 5, 7$, with $p = 3$, $q = 1, 2, 4, 5$, with $p = 4$, $q = 1, 3, 5$, with $p = 5$, $q = 1, 2, 3, 4$, with $p = 6$, $q = 1$, with $p = 7$, $q = 1, 2$, and with $p = 8$, $q = 1$, so that no asymmetric periodic orbits are now indicated at these commensurabilities.

REFERENCES

Fragakis, C.N., 1973, Astrophysics and Space Science, vol. 22, pp. 421-440, and vol. 23, pp. 17-42.

Message, P.J., 1958, Astronomical Journal, vol. 63, pp. 443-448.

Message, P.J., 1959, Astronomical Journal, vol. 66, pp. 226-236.

Message, P.J., 1970, in "Periodic Orbits, Stability, and Resonances", (ed. G.E.O. Giacaglia), pp. 19-32.

CONSTRUCTION DE SOLUTIONS PERIODIQUES DU PROBLEME RESTREINT
ELLIPTIQUE PAR LA METHODE DE HALE

Anne Sergysels-Lamy, Roland Sergysels
Service de Mécanique Analytique, Université Libre
de Bruxelles, Belgique

1. INTRODUCTION

Les équations du mouvement du troisième corps, dans les variables sans dimensions définies par Szebehely (1967, page
591) ont été données dans un article précédent (Sergysels-
Lamy, 1975). Plusieurs changements de variables successifs
permettent de ramener ces équations à la forme standard au
sens de Hale :

$$\frac{dx}{d\theta} = \varepsilon \ X(x,\theta,\varepsilon) \tag{1}$$

où x et X sont des vecteurs à quatre composantes, et X est
périodique en la variable indépendante θ, anomalie vraie de
l'orbite du troisième corps dans le système synodique.

Il est alors possible de démontrer, en utilisant le rapport
de masse des primaires comme petit paramètre ε, l'existence
d'une classe de solutions de longue période du système (1),
à l'aide de la méthode de Hale (1963). La démonstration
d'existence reste valable pour un rapport de masse quelconque, à condition de considérer le mouvement au voisinage d'un
des primaires.

2. RESULTATS NUMERIQUES

L'adaptation de la méthode de Hale au calcul numérique des
solutions périodiques d'un système sous forme standard a été
effectuée par un des auteurs (Sergysels, 1975).

La méthode comprend deux parties.

i) L'intervalle [0,T] , où T est la période du système (1),

V. Szebehely (ed.), Dynamics of Planets and Satellites and Theories of Their Motion, 325-331.
All Rights Reserved. Copyright © 1978 by D. Reidel Publishing Company, Dordrecht, Holland.

est divisé en 2n pas :

$$\theta_i = \frac{(i-1)T}{2n} \qquad\qquad i = 1,2,\ldots, 2n+1 \qquad\qquad (2)$$

La suite numérique $x_n(\theta_i)$ est alors construite de la manière suivante, les intégrales étant évaluées par des méthodes classiques (règle du trapèze, règle de Simpson,...).
Le premier élément de la suite est le vecteur constant a :

$$x_o(\theta_i) = a. \qquad\qquad (3)$$

Connaissant l'élément x_n, l'élément x_{n+1} est obtenu par
l'algorithme :

$$PX(x_n) = \frac{1}{T} \int_o^T d\theta\ X\ [x_n(\theta_i,a,\varepsilon),\theta_i,\varepsilon] \qquad\qquad (4)$$

$$F[x_n(\theta_i,a,\varepsilon),\theta_i,\varepsilon] = \int_o^{\theta_i} d\theta\ \{X[x_n(\theta_j,a,\varepsilon),\theta_j,\varepsilon]-PX(x_n)\} \qquad (5)$$

$$PF(x_n) = \frac{1}{T} \int_o^T d\theta\ F\ [x_n(\theta_i,a,\varepsilon),\theta_i,\varepsilon] \qquad\qquad (6)$$

$$x_{n+1}(\theta_i,a,\varepsilon) = a + \varepsilon F[x_n(\theta_i,a,\varepsilon),\theta_i,\varepsilon] - \varepsilon PF(x_n). \qquad\qquad (7)$$

ii) La suite $\{x_n\}$ converge pour ε suffisamment petit et sa
limite x^* est une solution T-périodique de moyenne a du
système (1) s'il existe un vecteur $a(\varepsilon)$ tel que :

$$\int_o^T d\theta\ X\ [x^*(\theta_i,a,\varepsilon),\theta_i,\varepsilon] = 0. \qquad\qquad (8)$$

La méthode analytique ne permet de calculer la valeur de a
qu'en $\varepsilon = 0$. Il est donc nécessaire de rechercher $a(\varepsilon)$,
soit en appliquant la méthode de Newton-Raphson aux équa-
tions de bifurcation (8), soit en effectuant un balayage
systématique des valeurs de a, ce qui n'est évidemment
pratique que si le vecteur a n'est pas de dimension trop
élevée.

Dans le cas du problème restreint elliptique le vecteur a
est de dimension quatre. Toutefois, pour effectuer la
démonstration d'existence nous avons dû utiliser la propriété E de Hale (1963) (page 43) :

$$S\ X\ (Sx,-\theta,\varepsilon) = -X(x,\theta,\varepsilon) \qquad\qquad (9)$$

où S est la matrice diag (1,-1,1,-1).

Nous nous restreignons alors à la recherche d'orbites symé-

triques par rapport à la ligne des syzygies, pour lesquelles
nous avons la propriété :

Sa = a.   (10)

Les composantes a_2 et a_4 du vecteur a sont donc nulles.
Comme l'application de la méthode de Newton-Raphson exige
une itération supplémentaire et allonge donc le programme
nous avons recherché la valeur des composantes non nulles
par balayage, jusqu'à ce que les membres de gauche des
équations (8) soient inférieurs à 10^{-14}.

Tous les calculs ont été effectués dans le cas où les deux
primaires sont de masses égales et nous avons choisi une
période T de 40π.

Les figures 1 et 2 représentent une orbite rétrograde,
respectivement dans le système pulsant et dans le système
physique, l'excentricité de l'orbite des primaires étant .6.
Dans le système pulsant, le primaire perturbateur se trouve
à 20 unités sur l'axe $\bar{\xi}$. Dans le système physique, il oscille entre 5.625 et 22.5 unités sur l'axe x.

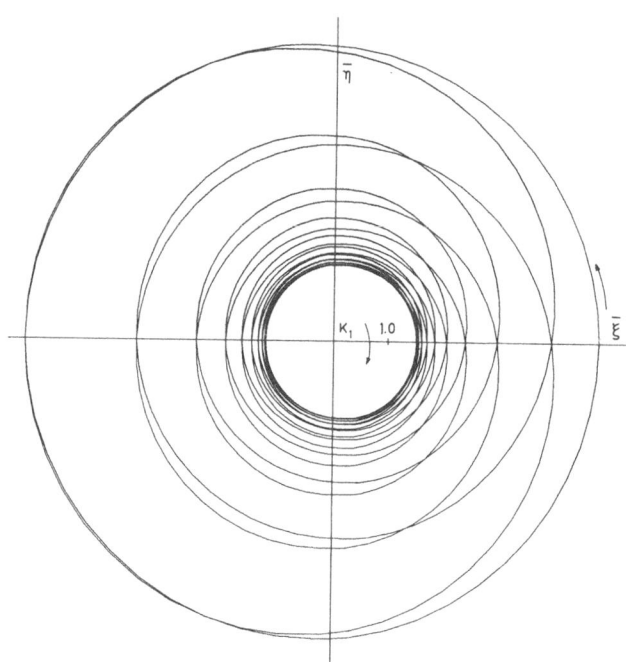

Fig.1.   Orbite rétrograde, système pulsant, e=.6.

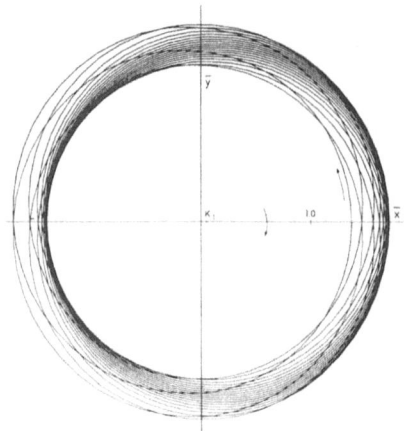

Fig. 2.   Orbite rétrograde, système physique, e=.6.

Les figures 3 et 4 représentent une orbite directe, l'excentricité des primaires étant .4.   Dans le système pulsant, le
primaire perturbateur est situé à 20 unités sur l'axe ξ̄.
Dans le système physique, il oscille entre 8.57142 et 20 unités sur l'axe x̄.

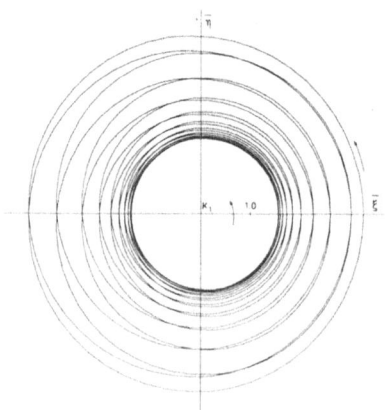

Fig. 3.   Orbite directe, système pulsant, e=.4.

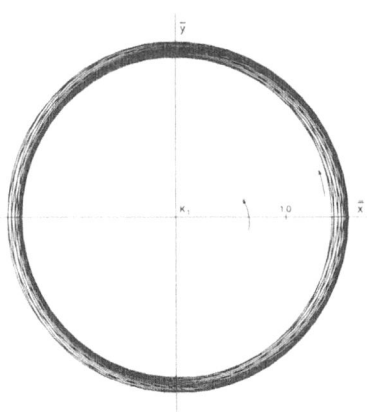

Fig. 4. Orbite directe, système physique, e=.4.

Les figures 5 et 6 représentent une orbite directe, l'excentricité des primaires étant .001. Bien que sa période soit
de 40π, les lobes sont pratiquement superposés, donnant à
cette orbite une période apparente de 2π.

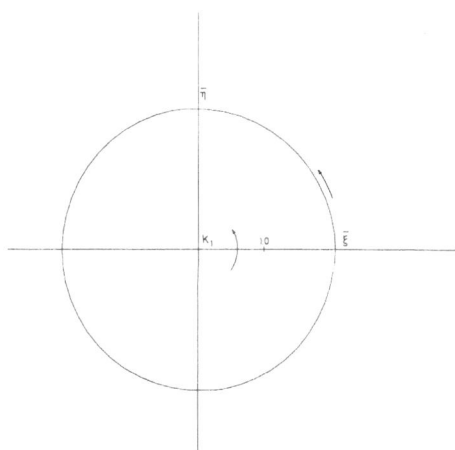

Fig. 5 Orbite directe, système pulsant, e=.001.

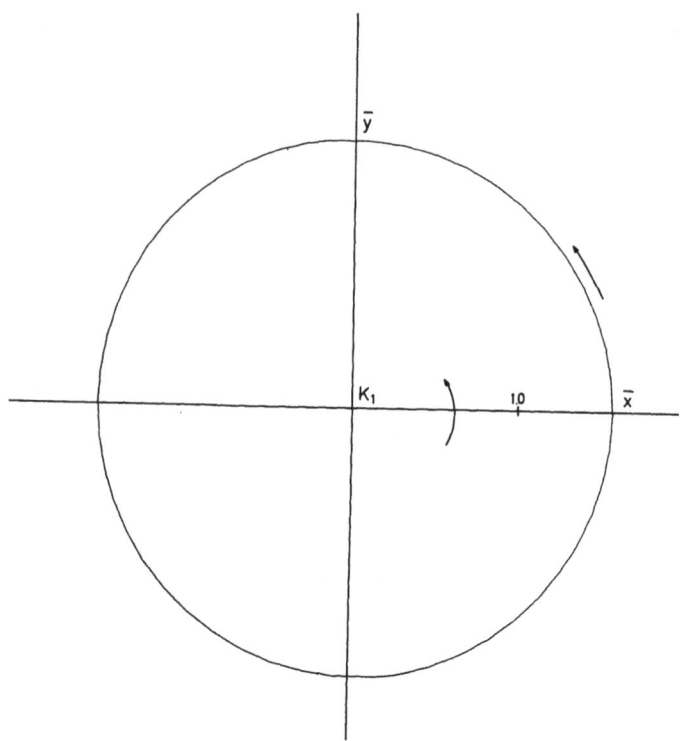

Fig. 6.   Orbite des primaires   Orbite génératrice

L'évolution de l'excentricité de l'orbite génératrice, en
fonction de l'excentricité de l'orbite des primaires est ré-
sumée dans le tableau 1.

	Oribite des primaires	Orbite génératrice
Orbites directes	.001	.000081
	.2	.0219
	.3	.041
	.4	.073
Orbites rétrogrades	.001	.000079
	.2	.0189
	.4	.070
	.6	.229

Tableau 1.

Il semble donc que la classe d'orbites que nous avons étu-
diée constitue lè prolongement des solutions périodiques de
première sorte du problème restreint circulaire.

'Construction of Periodic Orbits of the Elliptic Restricted Problem by
Hale's Method' by A. Sergysels-Lamy

ABSTRACT. This paper contains numerical results following the theoretical
demonstration already published in Celes. Mech. 11 (1975) 43; the numerical method is described in R. Sergysel's paper in Bull. Cl. Sci., Acad.
Roy. Belg. 61 (1975) 888.

REFERENCES

HALE, J.K. : 1963, Oscillations in Nonlinear Systems,
            Mc Graw-Hill, New-York.

SERGYSELS, R : 1975, Acad. Roy. Belg. Bull. Cl. Sc. 61, 888.

SERGYSELS-LAMY, A. : 1975, Celes. Mech. 11, 43.

SZEBEHELY, V. : 1967, Theory of Orbits, Academic Press,
            New-York.

ORBITAL STABILITY IN THE ELLIPTIC RESTRICTED THREE BODY PROBLEM

C.A. Williams and J.G. Watts
Dept. of Astronomy, University of South Florida, Tampa, Fla.
Fla. 33620

ABSTRACT

Based on the concept of orbital stability introduced by G. W. Hill, a method is presented to facilitate the determination of the orbital stability of solutions to the planar elliptic restricted problem of three bodies. The invariant relation introduced by Szebehely and Giacaglia (1964) contains an integral which is expanded here about a Keplerian solution to the problem. If the expansion converges, it can be used to determine the conditions for Hill stability. With it one can also define stability in a periodic sense.

Szebehely and Giacaglia (1964) give the equations of motion of the elliptic restricted three body problem in a coordinate system rotating and pulsating with respect to an inertial system. The coordinate system is chosen in such a way that as a consequence of the rotation and pulsation the two primary masses are located at fixed points on the horizontal ξ axis. In two dimensions, the equations of motion of this system are

$$\xi'' - 2\eta' = \omega_\xi$$

$$\eta'' + 2\xi' = \omega_\eta$$

(1)

The independent variable is the true anomaly f of the primaries. A prime indicates differentiation with respect to f. The potential function is defined by

$$\omega = (1 + \bar{e} \cos f)^{-1} \Omega,$$

(2)

where $\qquad \Omega = \tfrac{1}{2}(1-\mu)\rho_1^2 + \tfrac{1}{2}\mu\rho_2^2 + \dfrac{1-\mu}{\rho_1} + \dfrac{\mu}{\rho_2}$,

/. Szebehely (ed.), Dynamics of Planets and Satellites and Theories of Their Motion, 333-337.

with $\rho_1^2 = (\xi - \mu)^2 + \eta^2$ and $\rho_2^2 = (\xi - \mu+1)^2 + \eta^2$.

The eccentricity of the primary orbit is $\overline{e}$, the mass of the primary at
the point $(\mu,0)$ is $1-\mu$ and the mass of the other primary is μ. ρ_1 and
ρ_2 are the distances of the particle from the masses $1-\mu$ and μ
respectively.

Since this system is nonconservative, the equations (1) do not
furnish a Jacobi integral as do the equations of motion in the circular
restricted problem. (The latter can be derived from Eqs. (1) and (2)
by putting $\overline{e} = 0$, $f = t$.) However, the following invariant relation
can be formally derived in the same manner as can the Jacobi integral
in the circular problem:

$$(\xi')^2 + (\eta')^2 = 2\omega - C - 2\int_{f_o}^{f} \frac{\partial\omega}{\partial f} \, df. \tag{3}$$

Since in the circular case $\partial\omega/\partial t \equiv 0$, Eq. (3) would in that case
represent an integral. Szebehely (1967) remarks that Eq. (3) may be
used to obtain time varying curves of zero velocity in the case that
$\overline{e} \neq 0$.

In order to discuss orbital stability, we need to describe the
time variation of the curves of zero velocity. If we can establish
that a particular solution of Eq. (1) has closed curves of zero
velocity for some domain $f_1 \leq f \leq f_2$, then we will have established
orbital stability for the system in the same sense as that defined by
G. W. Hill (1905) for the lunar theory. Hill stability excludes the
possibility of escape from a primary (i.e., ρ_1 and ρ_2 cannot approach
infinity), although collisions may still occur. If we can establish
Hill stability for $f_2 \to \infty$, we will have established orbital stability
in a non-trivial sense.

In his above quoted book (p. 596), Szebehely suggests an expansion
of the invariant relation Eq. (3) with respect to powers of $\overline{e}$. The
integral term is included in this expansion. To do this, we note that

$$\frac{\partial\omega}{\partial f} = \frac{\Omega}{(1 + \overline{e} \cos f)^2} \frac{\overline{e} \sin f}{} \tag{4}$$

Ω in Eq. (4) is to be evaluated along a solution of Eq. (1); the
integral is therefore a function of f.

Define $I(f) = \int_{f_o}^{f} \frac{\Omega(f) \sin f}{(1 + \overline{e} \cos f)^2} \, df$. \tag{5}

The curves of zero velocity, if they exist, can be obtained from the equation

$$2\Omega(\xi,\eta) = \left[C + 2\bar{e}\ I(f)\right]\ (1 + \bar{e}\cos f) \equiv K(f). \tag{6}$$

As the particle follows its orbit, we can define a continuous set of zero velocity curves for the motion if the equation $2\Omega = K$ has a real locus. In order to evaluate I, it is assumed that the motion of the particle can be represented by perturbations of Keplerian motion. This will be true in three cases: Case I, ρ_1 is very small (C is large); Case II, ρ_2 is very small (C is large); Case III, ρ_1 and ρ_2 are not small but C is very large.

Cases I and II describe the motion of a particle moving close to one of the primaries and perturbed by the other.

Case III applies to the system in which a negligibly small outer planet moves in the gravitational field of the Sun and one larger planet whose orbit is completely inside the small planet's orbit.

The first step in the evaluation of I is to expand the various terms in Ω into series in the mean anomaly of the particle's orbit. In this paper, only Case II was considered; however, Case I can be obtained from Case II by interchanging μ and $1-\mu$.

ρ_2 is related to the radius vector of the particle's orbit about mass μ. To illustrate the method, consider the expansion of the term μ/ρ_2 in the function Ω. The expansion of a/r, where r is the radius vector in elliptic motion, is well known. In this problem, the radius vector r must be transformed, since the coordinate system defined by the axes ξ and η is rotating and pulsating. The rotation leaves r invariant, but because of the pulsation

$$\rho_2 = r(1 + \bar{e}\cos f)/(1 - \bar{e}^2) = r\bar{a}/\bar{r},\ \text{where}\ \bar{a} \equiv 1. \tag{7}$$

Therefore, we have the relation $\mu/\rho_2 = (\mu/a)\ (a/r)\ (\bar{r}/\bar{a})$, where the expansion of a/r is a Poisson series in the particle's mean anomaly and eccentricity $\bar{e}$. The form of the expansion of this term will be

$$\frac{\mu}{\rho_2} = \frac{\mu}{a}\ \sum_{j_1=0}^{\infty}\ \sum_{j_2=-\infty}^{+\infty} A_{j_1 j_2}\ \cos\ (j_1\ell + j_2\bar{\ell})\ ,$$

where $A_{j_1 j_2}$ is a function of e and $\bar{e}$. A similar method is applied to the expansion of the other terms in Ω.

After the expansion is performed, it is necessary to include the expansion of $\sin f(1 + \bar{e}\cos f)^{-2}$ before integrating Eq. (4). It will be necessary to adopt $\bar{\ell}$ as the independent variable instead of f, which

requires one more step in the expansion process. Fully expanded, the
integrand in Eq. (4) has the form

$$\frac{1}{\bar{e}} \frac{\partial \omega}{\partial f} = \sum_{j_1=0}^{\infty} \sum_{\substack{j_2,j_3 \\ -\infty}}^{\infty} A_{j_1 j_2 j_3} \sin(j_1 \ell + j_2 \bar{\ell} + j_3 w) \tag{8}$$

where A is a function of μ, a, e, and $\bar{e}$; and w is the argument of
pericenter of the particle's orbit. One may set $\bar{w} = 0$ without any
loss of generality. The integration of (8) is performed, assuming a,
e, and w are constants. This will give a first order approximation to
I. For a small perturbation this will give sufficient accuracy for the
determination of the range of I(f).

For Case II, $\mu/\rho_2 \simeq \mu/a$. If $a \leq \mu$, this will be the dominant
term in the expansion. Another large term is $(1-\mu)(\rho_1^2/2+1/\rho_1) \simeq 3(1-\mu)/2$.
Both of these lead to the largest term in I and have the frequency $\bar{n}$.
(In dimensionless variables, $\bar{n} = 1$.) μ/ρ_2 contributes another large
term with frequency $2\bar{n}$. The remaining terms are factored by powers of
$\bar{e}$, a^2, e, or μ and are considerably smaller. The frequency of a term
in the expansion is given by $pn + q\bar{n}$ where p and q are integers. If
a resonance condition exists, i.e., if $n/\bar{n} = -q/p$, the integral will
not converge and this representation is not applicable. If we exclude
a resonance condition and if the orbit is quasi-periodic, then the
Poisson series representation of the integral should be a good one.
Using Eq. (8), we can write

$$I = \sum_{j_1,j_2,j_3} - \frac{A_{j_1 j_2 j_3}}{j_1 n + j_2 \bar{n}} \cos(j_1 \ell + j_2 \bar{\ell} + j_3 w) \Bigg|_{t_o}^{t} \tag{9}$$

If this series converges, then

$$|I| \leq M, \text{ where} \tag{10}$$

$$M = 2 \sum_{j_1 j_2 j_3} \left| \frac{A_{j_1 j_2 j_3}}{j_1 n + j_2 \bar{n}} \right|$$

If Eq. (9) holds, then we can substitute these results into Eq. (6) to
obtain $(1 - \bar{e})(C - 2\bar{e}M) \leq K(f) \leq (1 + \bar{e})(C + 2\bar{e}M)$ if $C - 2\bar{e}M > 0$.
The lower limit will be $(1 + \bar{e})(C - 2\bar{e}M)$ if $C - 2\bar{e}M < 0$. Considering
Eq. (2) for Ω, we can derive the well known result that curves of zero
velocity exist when $2\Omega \geq 3$. Whether or not these curves are closed
depends on the value of μ. In the notation of Szebehely and Williams
(1964), the curves of zero velocity are closed when $2\Omega > C_2$, for Case II.

This implies that the motion of the particle is restricted to the vicinity of the mass μ. We will have Hill stability for all time if

$$(1 - \bar{e}) (C - 2\overline{e}M) \geq C_2$$

It is also possible to define a slightly different type of stability using this method. Assume that there exists at least one value of f, say f*, where the integral in Eq. (10) takes on a value I* such that

$$(C + 2\overline{e}I^*) (1 + \bar{e} \cos f^*) = K(f^*) > C_2 .$$

Then, if the integral given by Eq. (10) is quasi-periodic, there will be another value of f, say f** > f*, where again (K(f**) > C_2. This condition will occur periodically if Eq. (10) converges. One could say that regardless of the motion of the particle, the particle returns periodically to the vicinity of the mass μ for all time. The existence of this type of stability depends entirely on the behavior of the integral given by Eqn. (10) and the initial conditions of the system.

This theory can be applied to the motion of satellites of planets for which the solar perturbations are dominant. If the satellite is so close to the planet that the perturbations are very small, then these results are almost trivial. A more interesting problem would be to study the motion of satellites farther from the planet where the possibility exists that K(f) < C_2 for some value of f. In this case, however, a representation of the motion different from a Keplerian one, may be necessary.

REFERENCES

Hill, G. W. 1905, Collected Mathematical Works of G. W. Hill, 1,
      Carnegie Inst. of Washington, Washington, D. C.
Szebehely, V. and Giacaglia, G. 1964, A. J.
Szebehely, V. 1967, Theory of Orbits, Acad. Press.
Szebehely, V. and Williams, C. A. 1964, A. J. 69, 460.

RESONANCE IN THE RESTRICTED PROBLEM OF THREE BODIES WITH SHORT-PERIOD
PERTURBATIONS IN THE ELLIPTIC CASE

K. B. BHATNAGAR and BEENA GUPTA
Department of Mathematics, University of Dehli, Delhi, India

ABSTRACT. This is a generalization of the paper by Bhatnagar and Beena
Gupta 'Resonance in the restricted problem of three bodies with short-
periodic perturbations'. The motion of an asteroid moving in the gravitational field of Jupiter is considered. In the original paper it was
assumed that Jupiter is moving in a circular orbit around the Sun. In
the present paper we consider the orbit to be elliptic. The series
occurring in the problem are expanded in powers of a small parameter ε,
which represents the ratio of the mass of Jupiter to that of the Sun.
The perturbations in the osculating elements are obtained up to $0(\varepsilon)$.

1. EQUATIONS OF MOTION

Let us suppose that Jupiter moves in an unperturbed elliptic orbit
with the Sun at one of its foci. Take the orbital plane of Jupiter as
the (x,y) plane. Let e' be the eccentricity of Jupiter's orbit, a' its
semi-major axis, λ' its mean longitude, ℓ' its mean anomaly and n' its
mean motion. The corresponding elements of the asteroid are denoted by
e, a, λ, ℓ, and n.

The equations of motion of the asteroid with negligible mass are:

$$\frac{dx}{dt} = \frac{\partial H}{\partial \dot{x}} , \qquad\qquad \frac{d\dot{x}}{dt} = -\frac{\partial H}{\partial x} ,$$

$$\frac{dy}{dt} = \frac{\partial H}{\partial \dot{y}} , \qquad\qquad \frac{d\dot{y}}{dt} = -\frac{\partial H}{\partial y} ,$$

$$\frac{dz}{dt} = \frac{\partial H}{\partial \dot{z}} , \qquad\qquad \frac{d\dot{z}}{dt} = -\frac{\partial H}{\partial z} ,$$

where the Hamiltonian, $H = H_0 + H_1$, is given by

$$H_0 = 1/2 \ (\dot{x}^2 + \dot{y}^2 + \dot{z}^2) - \frac{\mu}{r},$$

V. Szebehely (ed.), Dynamics of Planets and Satellites and Theories of Their Motion, 339-353.
All Rights Reserved. Copyright © 1978 by D. Reidel Publishing Company, Dordrecht, Holland.

$$H_1 = \varepsilon \, \mu (\frac{1}{\Delta} - \frac{xx' + yy'}{r'^3}) .$$

In these equations (x,y,z) are the coordinates of the asteroid, $(x',y',0)$ are the coordinates of Jupiter, Δ is the Jupiter-asteroid distance, r' the Sun-Jupiter distance, r the Sun-asteroid distance and $\mu = k^2 \varepsilon$.

Let us introduce the change of variables

$$(x,y,z,\dot{x},\dot{y},\dot{z}) \rightarrow (L,G,H,\ell,g,\tilde{h})$$

defined by the canonical transformations

$$\ell = - \frac{\partial W'}{\partial L}, \qquad g = - \frac{\partial W'}{\partial G}, \qquad \tilde{h} = - \frac{\partial W'}{\partial H},$$

$$\dot{x} = \frac{\partial W'}{\partial x}, \qquad \dot{y} = \frac{\partial W'}{\partial y}, \qquad \dot{z} = \frac{\partial W'}{\partial z}.$$

Here W' is a generating function; and L,G,H,ℓ,g and h are the Delauny variables given by

$$L = \sqrt{\mu a}, \qquad G = \sqrt{1-e^2}, \qquad H = G \, \text{Cos} \, i,$$

$$\ell = \ell, \qquad g = \omega \qquad , \qquad \tilde{h} = \Omega,$$

where l is the inclination of the orbital plane of the asteroid with respect to the reference plane, ω the argument of the perihelion and Ω the longitude of the ascending node.

The equations of motion become

$$\frac{dL}{dt} = \frac{\partial \tilde{F}}{\partial \ell}, \qquad\qquad \frac{d\ell}{dt} = - \frac{\partial F}{\partial L},$$

$$\frac{dG}{dt} = \frac{\partial \tilde{F}}{\partial g}, \qquad\qquad \frac{dg}{dt} = - \frac{\partial \tilde{F}}{\partial G},$$

$$\frac{dH}{dt} = \frac{\partial \tilde{F}}{\partial \tilde{h}}, \qquad\qquad \frac{d\tilde{h}}{dt} = - \frac{\partial \tilde{F}}{\partial H}$$

with $\qquad \tilde{F} = \mu^2/2L^2 + F_1, \qquad F_1 = \varepsilon k^2 [\frac{1}{\Delta} - \frac{xx' + yy'}{\gamma'^3}] .$

k is the Gaussian constant.

The equations of motion can be written as (Brouwer and Glemence, 1961),

$$\frac{dL}{dt} = \frac{\partial F}{\partial \ell}, \qquad\qquad \frac{d\ell}{dt} = - \frac{\partial F}{\partial L},$$

$$\frac{dG}{dt} = \frac{\partial F}{\partial g}, \qquad \frac{dg}{dt} = -\frac{\partial F}{\partial G},$$

$$\frac{dH}{dt} = \frac{\partial F}{\partial h}, \qquad \frac{dh}{dt} = -\frac{\partial F}{\partial H}, \qquad (1)$$

$$\frac{dK}{dt} = \frac{\partial F}{\partial k}, \qquad \frac{dk}{dt} = -\frac{\partial F}{\partial K}.$$

with

$$F = F_0 + F_1,$$

$$F_0 = \frac{\mu^2}{2L^2} - n'K,$$

$$F_1 = \varepsilon \; \Sigma \; C_{p_1, p_2, p_3, p_4}^{m_2, m_3, m_4} \; (\text{Sin } \frac{i}{2})^{2m_3} e^{m_2} e'^{m_4} \cos(p_1 \ell + p_2 g + p_3 h + p_4 k).$$

The coefficients C's are functions of a and a' of degree-1. And the D'Alembert's characteristics give

$$m_2 = |j_2| + 2k_2 = |p_1 - p_2| + 2k_2,$$

$$2m_3 = |j_3| + 2k_3 = |p_2 - p_3| + 2k_3, \qquad (2)$$

$$m_4 = |j_4| + 2k_4 = |p_3 + p_4| + 2k_4,$$

where k_2, k_3 and k_4 are positive integers of zero. Let us introduce the new variables

$$(x_1, x_2, x_3, x_4; \; y_1, y_2, y_3, y_4)$$

defined by the following canonical transformations:

$$x_1 = L + \frac{p}{q} K, \qquad y_1 = \ell,$$

$$x_2 = -\frac{1}{q} K, \qquad y_2 = p\ell + q\omega + q(\Omega - \lambda'),$$

$$x_3 = G + K, \qquad y_3 = \omega, \qquad (3)$$

$$x_4 = -H - K, \qquad y_4 = \omega'.$$

The system of Equation (1) reduces to

$$\frac{dx_i}{dt} = \frac{\partial K'}{\partial y_i}; \qquad \frac{dy_i}{dt} = \frac{\partial K'}{\partial x_i}. \qquad (i = 1,2,3,4) \qquad (4)$$

with $K' = K_0 + K_1$,

$$K_0 = \frac{\mu^2}{2(x_1 + px_2)^2} + qn'x_2,$$ (5)

and

$$K_1 = \varepsilon R,$$

where

$$R = \Sigma \ f(a,e,i,a',e') \ \cos(p_1\ell + p_2\omega - p_3\omega' + p_4\ell').$$ (6)

and restrictions on p_1, p_2, p_3 and p_4 are given by the Equations (2).

2. SHORT-PERIOD PERTURBATIONS

Let us eliminate the short-periodic terms, i.e. the terms which contain mean anomaly in their argument. The elimination is achieved through the well known Von Zeipel method. Here we assume canonical transformations (x,y) to (ξ,η) defined by the generating function $W(\xi,y,\varepsilon)$ such that the new Hamiltonian $\phi(\xi,\eta,\varepsilon)$ is free from the angular variable η_1. Also we assume the two series

$$W = W_0 + W_{\frac{1}{2}} + W_1 + W_{3/2} + \ldots \ ,$$

$$\phi = \phi_0 + \phi_{\frac{1}{2}} + \phi_1 + \phi_{3/2} + \ldots \ ,$$

where

$$W_j = O(\varepsilon^j) \text{ and } \phi_j = O(\varepsilon^j).$$

We consider the problem by assuming that

$$|pn - qn'| \leq n \ \varepsilon^{1/2},$$ (7)

where p and q are mutually prime integers. Since W does not contain time explicitly, the Hamilton-Jacobi equation will be

$$\phi(\xi; \ \frac{\partial W}{\partial \xi_2}, \ \frac{\partial W}{\partial \xi_3}, \ \frac{\partial W}{\partial \xi_4}, \ \varepsilon) = K(\frac{\partial W}{\partial y}; \ y; \ \varepsilon).$$ (8)

Here ξ means ξ_1, ξ_2, ξ_3 and ξ_4 and y means y_1, y_2, y_3 and y_4.

Following the procedure of Giacaglia (1969) we shall have

$$\phi_0 = K_0,$$

$$\phi_{\frac{1}{2}} = W_{\frac{1}{2}} = 0,$$

$$\phi_1 = \frac{1}{2\pi q} \int_0^{2\pi q} R \; dy_1,$$

$$W_1 (\xi,y,\varepsilon) = -(\frac{\partial K_0}{\partial \xi_1})^{-1} \int (K_1 - \phi_1) \; dy_1, \tag{9}$$

$$\phi_{3/2} = 0,$$

$$W_{3/2} = - (\frac{\partial K_0}{\partial \xi_1})^{-1} \int (\frac{\partial K_0}{\partial \xi_2}) (\frac{\partial W_1}{\partial y_2}) \; dy_1.$$

Thus we have established the two series of W and ϕ up to $O(\varepsilon^{3/2})$. Since we are considering terms only up to $O(\varepsilon^{3/2})$, we neglect terms of $O(\varepsilon^2)$. It may be noted that the series are of the same form as in the circular case except that the value of K differs from one case to another.

Thus in this case, i.e. up to $O(\varepsilon^{3/2})$, the Hamiltonian become

$$\phi^{(3/2)} = \phi_0 + \phi_1, \tag{10}$$

where

$$\phi_0 = \frac{\mu^2}{2(\xi_1 + p\xi_2)^2} + qn'\xi_2. \tag{11}$$

and

$$\phi_1 = \varepsilon \; \Sigma C(a^*,e^*,i^*,a'^*,e'^*) \cos [-p_1 n_2 + (p_2 + p_1 q) n_3 -$$

$$- (p_3 + p_1 q) n_4]. \tag{12}$$

Also up to this order we have

$$\dot{\xi}_1 = \frac{\partial \phi^{(3/2)}}{\partial n_1} = 0.$$

Hence

$$\xi_1 = \text{const.}$$

or

$$L^* + (p/q) K^* = \text{const.}$$

The short-period perturbations are given by the generating function W in an implicit form as

$$x_j = \varepsilon_j + \frac{\partial W_1}{\partial y_j} + \frac{\partial W_{3/2}}{\partial y_j} = \xi_j + \varepsilon \; \Delta \; x_j,$$

$$\eta_j = y_j + \frac{\partial W_1}{\partial \xi_j} + \frac{\partial W_{3/2}}{\partial \xi_j} = y_j + \varepsilon \, \Delta \, y_j, \tag{13}$$

where Δx_j and Δy_j are short-periodic terms.

3. ELIMINATION OF THE CRITICAL ARGUMENT

At the critical point the motion is stationary and this occurs when $pn = qn'$. Now we will further decrease the degrees of freedom by introducing a new transformation given by a generating function S.

Here ϕ is a function of $(\xi; \eta_2, \eta_3, \eta_4, \varepsilon)$. Let us change the Hamiltonian ϕ to $F(X;Y;\varepsilon)$ by introducing a generating function S such that the new Hamiltonian F is independent of Y_2.
Let us introduce the new variables

$$(X_1, X_2, X_3, X_4; -, Y_3, Y_4, \varepsilon)$$

with the transformation defined by the equation

$$\xi_j = \frac{\partial S}{\partial \eta_j}; \; Y_j = \frac{\partial S}{\partial X_j}. \; (j = 1,2,3,4)$$

We also assume that

$$S = S_0 + S_{\frac{1}{2}} + S_1 + S_{3/2} + \ldots$$

$$F = F_0 + F_{\frac{1}{2}} + F_1 + F_{3/2} + \ldots$$

and

$$S_0 = X_1 \eta_1 + X_2 \eta_2 + X_3 \eta_3 + X_4 \eta_4,$$

where

$$S_j = 0(\varepsilon^j) \text{ and } F_j = 0(\varepsilon^j).$$

In general, the stationary solution will exist for the mean motion of the orbit and it will correspond to exact mean reasonance, i.e. at the point,

$$\xi_2 = \frac{\partial \phi}{\partial \eta_2} = 0,$$

$$\dot{\eta}_2 = -\frac{\partial \phi}{\partial \xi_2} = 0, \tag{14}$$

and

$$pn** - qn' = 0. \tag{15}$$

Here the double asterisks denote the averaged value over n_1 and n_2.
To obtain the series for S and F we will solve the Hamilton-Jacobi
equation by successive approximations.
    a) If we take the case of zero-order approximation then we will get

$$F_0(X_1,X_2) = \phi_0(X_1,X_2) = \frac{\mu^2}{2} (X_1 + pX_2)^{-2} + qn'X_2 \tag{61}$$

which is constant.
    b) Approximation of order $(\varepsilon^{\frac{1}{2}})$
Taking the approximation upto $0(\varepsilon^{\frac{1}{2}})$ we have

$$F_{\frac{1}{2}} = 0. \tag{17}$$

    c) Approximation of order (ε)
Taking the approximation upto $0(\varepsilon^{\frac{1}{2}})$ we have

$$\phi = \phi_0 + \phi_1$$

$$F = F_0 + F_1$$

and

$$S = S_0 + S_{\frac{1}{2}} + S_1.$$

Also from Equation (8) and taking transformations up to this order
we have the Hamilton-Jacobi equation

$$\phi(X + \frac{\partial S_{\frac{1}{2}}}{\partial n} + \frac{\partial S_1}{\partial n} ; n_2, n_3, n_4, \varepsilon) = F(X;n_3 + \frac{\partial S_{\frac{1}{2}}}{\partial x_3} + \frac{\partial S_1}{\partial x_3},$$

$$n_4 + \frac{\partial S_{\frac{1}{2}}}{\partial x_4} + \frac{\partial S_1}{\partial x_4}, \varepsilon).$$

Expanding this equation in Taylor's series and considering them up to
$0(\varepsilon)$ we have

$$F_1(X;n_3,n_4,\varepsilon) = \phi_1(X;n_2,n_3,n_4,\varepsilon) + \frac{1}{2} (\frac{\partial S_{\frac{1}{2}}}{\partial n_2})^2 \frac{\partial^2 \phi_0}{\partial x_2^2} + \frac{\partial S_{\frac{1}{2}}}{\partial n_2} \frac{\partial \phi_0}{\partial x_2}.$$

In this equation both F_1 and $S_{\frac{1}{2}}$ are unknown quantities. For determining these two we consider the approximate relations:

$$\xi_2 = X_2 + \frac{\partial S_{\frac{1}{2}}}{\partial n_2},$$

$$\tag{18}$$

$$Y_2 = n_2 + [\partial S_{\frac{1}{2}} / \partial X_2].$$

We know that X_2 is constant at any event. And by considering the Euqation (14) we see that ξ_2 is constant. Because from Equation (14) we see that up to $0(\varepsilon)$, $\partial\phi^{(1)}/\partial n_2 = 0$ is the necessary condition for the solution and therefore for satisfying Equation (18) we see that $S_{\frac{1}{2}}$ should be identically zero for the stable stationary solution.

Let $n_2 = n_2^0 \ (\xi; n_3, n_4, \varepsilon)$ be the point of minimum of $\phi(\xi; n; \varepsilon)$ such that

$$\left|\frac{\partial\phi_1}{\partial n_2}\right| \ n_2 \neq n_2^0 = 0. \tag{19}$$

The point will exist because ϕ_1 is periodic in n_2 with period π. Now to make the condition ($S_{\frac{1}{2}} = 0$) sufficient for the stable stationary solution we take

$$F_1(X; n_3, n_4, \varepsilon) = \phi_1(X; n_2^0 \ (X, n_3, n_4, \varepsilon), \ n_3, n_4, \varepsilon), \tag{20}$$

where ϕ_1 is given by Equation (12).

And the general equation defining $S_{\frac{1}{2}}$ is given by

$$\frac{\partial S_{\frac{1}{2}}}{\partial n_2} = \frac{L^{**}}{3p^2 n^{**}} \ [-qn'-pn^{**} \pm \{(qn'-pn^{**})^2 - \frac{6p^2 n^{**}}{L^{**}} \ U_1\}^{\frac{1}{2}}], \tag{21}$$

where

$$U_1(X; n_2, n_3, n_4, \varepsilon) = \phi_1(X; n_2, n_3, n_4, \varepsilon) - \phi_q(X; n_2^0 (X, n_3, n_4, \varepsilon) n_3, n_4, \varepsilon)$$

At the stationary solution the condition $S_{\frac{1}{2}} = 0$ is satisfied by Equation (21), Also from this equation we see that in general the motion will be of circulation, asymptotic or libration in n_2 if

$$\frac{6p^2 n^{**}}{L^{**}} \ U_1 \ \underset{>}{\overset{<}{=}} \ (qn' - pn^{**})^2$$

provided n_2 is taken to be maximum. U_1 is minimum at the libration centre ($n_2 = n_2^0$) where it is zero and it is maximum at the end points of the oscillation.

The amplitude of libration is given by the equation

$$U_1(X; n_2, n_3, n_4, \varepsilon) = \frac{L^{**}}{6p^2 n^{**}} \ (qn' - pn^{**})^2,$$

and is obtained as

$$n_2 = \bar{n}_2 \ (X; n_3, n_4, \varepsilon).$$

which is of order (ε) in this case.

Finally up to $0(\varepsilon^{3/2})$ the Hamiltonian is given by

$$F = \frac{\mu^2}{2} \ (X_1 + pX_2)^{-2} + qn'X_2 + F_1(X, Y_3, Y_4, \varepsilon).$$

which is a system with two degrees of freedom.

Also the parameters of the trajectory are given by the following equations:

$$a^* = a^{**} = const = (\frac{p^2 \mu}{q^2 n'})^{1/3}$$

$$K^{**} = const$$

$$\dot{Y}_1 = n^{**} - \frac{\partial F_1}{\partial X_1} = n^{**} - \varepsilon \quad R'(X;Y_3,Y_4),$$

$$\dot{Y}_2 = pn^{**} - qn' - \frac{\partial F_1}{\partial X_2} = pn^* - qn' - \varepsilon \ R''(X;Y_3,Y_4) \qquad (22)$$

$$\dot{X}_3 = \frac{\partial F_1}{\partial Y_3} = \varepsilon \ h'(X,Y_3,Y_4),$$

$$\dot{X}_4 = \frac{\partial F_1}{\partial Y_4} = \varepsilon \ h''(X.Y_3,Y_4),$$

$$\dot{Y}_3 = - \frac{\partial F_1}{\partial X_3} = \varepsilon \ F'(X,Y_3,Y_4,t),$$

$$\dot{Y}_4 = - \frac{\partial F_1}{\partial X_4} = \varepsilon \ F'(X;Y_3,Y_4,t).$$

The period of Y_1 is $2\pi/n^{**}$ which is short, and of Y_2 is given by $2\pi/(pn^{**}-qn')$ which is long and that of Y_3,X_3,Y_4 and X_4 is very long and given by $2\pi/n^{**} \ \varepsilon$.

d) Approximation of $0(\varepsilon^{3/2})$

Taking the approximation of $0(\varepsilon^{3/2})$ we will get

$$F_{3/2} = P_{3/2}(X;n_2^0(X;n_3,n_4,\varepsilon), \ n_3,n_4,\varepsilon),$$

where

$$P_{3/2}(X;n_3,n_4,\varepsilon) = \frac{\partial S_{\frac{1}{2}}}{\partial n_2} \frac{\partial \phi_1}{\partial X_2} + \frac{\partial S_{\frac{1}{2}}}{\partial n_3} \frac{\partial \phi_1}{\partial X_3} + \frac{\partial S_{\frac{1}{2}}}{\partial n_4} \frac{\partial \phi_1}{\partial X_4}$$

$$+ \frac{1}{6} (\frac{\partial S_{\frac{1}{2}}}{\partial n_2})^3 \frac{\partial^3 \phi_0}{\partial X_2^3} - \frac{\partial S_{\frac{1}{2}}}{\partial X_3} \frac{\partial F_1}{\partial n_3} - \frac{\partial S_{\frac{1}{2}}}{\partial X_4} \frac{\partial F_1}{\partial n_4}.$$

Therefore

$$F_{3/2} = |\frac{\partial S_{\frac{1}{2}}}{\partial n_3} \frac{\partial \phi_1}{\partial X_3} - \frac{\partial S_{\frac{1}{2}}}{\partial X_3} \frac{\partial F_1}{\partial n_3} + \frac{\partial S_{\frac{1}{2}}}{\partial n_4} \frac{\partial \phi_1}{\partial X_4} - \frac{\partial S_{\frac{1}{2}}}{\partial X_4} \frac{\partial F_1}{\partial n_4}|_{n_2=n_2^0} \qquad (23)$$

and n_2^0 in this case is given by the equation

$$\left|\frac{\partial\phi^{(3/2)}}{\partial\eta_2}\right|_{\eta_2 = \eta_2^0} = 0 \quad \text{or} \quad \left|\frac{\partial\phi_1}{\partial\eta_2}\right|_{\eta_2 = \eta_2^0} = 0,$$

for $\phi_\frac{1}{2}$ and $\phi_{3/2}$ are zero. Hence, the location of the libration centre is not changed.

Also S_1 is given by the equation

$$\frac{\partial S^{(1)}}{\partial\eta_2} = \frac{L^{**}}{3p^2n^{**}} \left[-(qn'-pn^{**}) \pm \{(qn'-np)^2 - 6\frac{p^2n^{**}}{L^{**}}(U_1+U_3)\right]^{\frac{1}{2}}}{2}$$

where   (24)

$$S^{(1)} = S_\frac{1}{2} + S_1,$$

and

$$U_{3/2}(X;\eta_2,\eta_3,\eta_4,\varepsilon) = P_{3/2}(X;\eta_2,\eta_3,\eta_4,\varepsilon) - F_{3/2}(X\eta_2,\eta_3,\eta_4,\varepsilon)\ldots$$
(25)

Since in general S_1 is real there are three possible motions in the variable η_2. The case of circulation, asymptotic motion and libration in η_2 occurs when

$$\{\max\}_{\eta_2}\ \frac{6p^2n^{**}}{L^{***}}(U_1 + U_{3/2}) \lessgtr (qn' - pn^{**})^2.$$

In the circulation and asymptotic cases S_1 is defined by choosing plus or minus sign. In the libration case the sign changes at the end points of oscillation where

$$\frac{6p^2n^{**}}{L^{**}}(U_1 + U_{3/2}) = (qn' - pn^{**})^2,$$

which also gives the amplitude of libration and can be found from

$$\eta_2 = \bar{\bar{\eta}}_2(X;\eta_3,\eta_4,\varepsilon).$$

Now up to $0(\varepsilon^{3/2})$ the system is reduced to two degrees of freedom with the Hamiltonian given by

$$F = F_0 + F_1 + F_{3/2}.$$

where F_0, G_1 and $F_{3/2}$ are given by the Equations (16), (20) and (23).
Two integrals of motion can be found from the equations

$$a^{**} = \text{const.}$$
$$\cdots$$ (26)
$$K^{**} = \text{const.}$$

and the other parameters of the trajectory can be found from the
following six equations:

$$\dot{Y}_1 = n^{**} - \frac{\partial F_1}{\partial X_1} - \frac{\partial F_{3/2}}{\partial X_1} = n^{**} - \varepsilon\, U(X;Y_3,Y_4,\varepsilon),$$

$$\dot{Y}_2 = pn^{**} - qn' - \frac{\partial F_1}{\partial X_2} - \frac{\partial F_{3/2}}{\partial X_2} = pn^{**} - qn' - \varepsilon U'(X;Y_3,Y_4,\varepsilon),$$

$$\dot{X}_3 = \frac{\partial F_1}{\partial Y_4} + \frac{\partial F_{3/2}}{\partial Y_3} = \varepsilon\, V(X;Y_3,Y_4,\varepsilon), \qquad (27)$$

$$\dot{X}_4 = \frac{\partial F_1}{\partial Y_4} + \frac{\partial F_{3/2}}{\partial Y_4} = \varepsilon\, V'(X;Y_3,Y_4,\varepsilon),$$

$$\dot{Y}_3 = -\frac{\partial F_1}{\partial X_3} - \frac{\partial F_{3/2}}{\partial X_3} = \varepsilon\, \bar{W}(X;Y_3,Y_4,\varepsilon,t),$$

$$\dot{Y}_4 = -\frac{\partial F_1}{\partial X_4} - \frac{\partial F_{3/2}}{\partial X_4} = \varepsilon\, \bar{\bar{W}}(X;Y_3,Y_4,\varepsilon,t).$$

The period of Y_1 is $2\pi/n^{**}$ which is short. The period of Y_2 is given
by $2\pi/(pn^{**} - qn')$ which is long and that of X_3,X_4,Y_3 and X_4 is very
long and given by $2\pi/n^{**}\,\varepsilon^{3/2}$.

4. PERTURBATIONS IN THE OSCULATING ELEMENTS UP TO $0(\varepsilon^{\frac{1}{2}})$

We see that up to $0(\varepsilon^0)$ there are no perturbations in the osculating
elements. Up to $0(\varepsilon^{\frac{1}{2}})$ the variations in the osculating elements can be
found out by considering the transformations:

$$\xi_j = X_j + \frac{\partial S_{\frac{1}{2}}}{\partial n_j},$$

$$\eta_j = Y_j + \frac{\partial S_{\frac{1}{2}}}{\partial x_j} \qquad (j = 1,2,3,4).$$

We shall first find the perturbations in Delaunay's variables and
then we shall obtain the variations in the osculating elements taking
terms up to $0(\varepsilon^{\frac{1}{2}})$. From Equations (3) we have.

$$L = x_1 + px_2, \qquad\qquad \ell = y_1,$$

$$G = x_3 + qx_2, \qquad\qquad \Omega - \lambda' = \frac{1}{q}\, y_2 - \frac{p}{q}\, y_1 - y_3,$$

$$H = qx_2 - x_4, \qquad\qquad \omega = y_3,$$

$$K = -qx_2, \qquad\qquad\quad \omega' = y_4.$$

Also we know that

$$L^* = \xi_1 + p\xi_2 = X_1 + \frac{\partial S_{\frac{1}{2}}}{\partial n_1} + px_2 + p\frac{\partial S_{\frac{1}{2}}}{\partial n_2},$$

and

$$\frac{\partial S_{\frac{1}{2}}}{\partial n_1} = 0.$$

Therefore

$$L^* = L^{**} + p\frac{\partial S_{\frac{1}{2}}}{\partial n_2},$$

Similarly

$$G^* = G^{**} + q\frac{\partial S_{\frac{1}{2}}}{\partial n_2} + \frac{\partial S_{\frac{1}{2}}}{\partial n_3},$$

$$H^* = H^{**} + q\frac{\partial S_{\frac{1}{2}}}{\partial n_2} - \frac{\partial S_{\frac{1}{2}}}{\partial n_4}, \qquad (28)$$

and

$$K^* = K^{**} - q\frac{\partial S_{\frac{1}{2}}}{\partial n_2}.$$

The variation of the mean semi-major axis is given by

$$a^* = \frac{L^{*2}}{\mu} = \frac{1}{\mu}[L^{**} + p\frac{\partial S_{\frac{1}{2}}}{\partial n_2}]^2.$$

Substituting the value of $\partial S_{\frac{1}{2}}/\partial n_1$ from Equation (21) in this equation we have

$$a^* = a_0^* \pm \Delta a^*, \qquad (29)$$

where

$$a^* = a^{**}(\frac{5}{3} - \frac{2}{3}\frac{qn'}{pn^{**}}), \qquad (30)$$

$$\Delta a^* = \frac{2}{3}a^{**}[(1 - \frac{qn'}{pn^{**}}) - \frac{1}{n^{**}L^{**}}U_1]^{1/2}. \qquad (31)$$

For a stationary solution, we have that

$$a^* = a_0^* = a^{**},$$

but, in general, the maximum variation from the mean value a_0^* is given by putting $n_2 = n_2^0$ in Equation (31), i.e.,

$$(\Delta a^*)_{max.} = \frac{2}{3}a^{**}(1 - \frac{qn'}{pn^{**}}).$$

Also from the second and third relation to Equation (28) we see that at exact resonance

$$G^* = G^{**} \text{ and } H^* = H^{**}.$$

The variation in eccentricity and inclination can be found if the system of equation $a^* = $ const. and $K^* = $ const. are completely integrated. Similarly we can find the variations in the angular variables as follows:

$$\ell^* = \ell^{**} - \frac{\partial S_{\frac{1}{2}}}{\partial L^{**}} - \frac{\partial S_{\frac{1}{2}}}{\partial G^{**}},$$

$$\omega^* = \omega^{**} + \frac{\partial S_{\frac{1}{2}}}{\partial G^{**}},$$

$$\Omega^* = \Omega^{**} - \frac{\partial S_{\frac{1}{2}}}{\partial K^{**}} - \frac{\partial S_{\frac{1}{2}}}{\partial G^{**}},\tag{32}$$

$$\omega'^* = \omega'^{**} - \frac{\partial S_{\frac{1}{2}}}{\partial H^{**}}.$$

5. PERTURBATIONS IN THE OSCULATING ELEMENTS UP TO $0(\varepsilon)$

In this case the transformations are

$$\xi_j = X_j + \frac{\partial S_{\frac{1}{2}}}{\partial \eta_j} + \frac{\partial S_1}{\partial \eta_j},$$

$$(j=1,2, ,)$$

$$\xi_j = Y_j + \frac{\partial S_{\frac{1}{2}}}{\partial X_j} + \frac{\partial S_1}{\partial X_j}.$$

Again, also in this case, we shall first find the perturbations in the Delaunay variables and from that we shall obtain the variation in the osculating elements taking terms up to $0(\varepsilon)$.

Following the same procedure as in Section 4, the variations in the Delauny variables are given by

$$L^* = L^{**} + p\,\frac{\partial S^{(1)}}{\partial \eta_2},$$

$$G^* = G^{**} + q\,\frac{\partial S^{(1)}}{\partial \eta_2} + \frac{\partial S^{(1)}}{\partial \eta_3},$$

$$H^* = H^{**} + q\,\frac{\partial S^{(1)}}{\partial \eta_2} - \frac{\partial S^{(1)}}{\partial \eta_4},\tag{33}$$

$$K^* = K^{**} - q\,\frac{\partial S^{(1)}}{\partial \eta_2}.$$

where

$$S^{(1)} = S_{\frac{1}{2}} + S_1.$$

The variation in the mean semi-major axis is given by

$$a^* = \frac{L^{*2}}{\mu} = \frac{1}{\mu} [L^{**} + p \frac{\partial S^{(1)}}{\partial \eta_2}]^2.$$

Simplifying this result we get

$$a^* = a_0^* \pm \Delta a^*, \tag{34}$$

where

$$a_0^* = a^{**} (\frac{5}{3} - \frac{2}{3} \frac{qn'}{pn^{**}}),$$

and

$$\Delta a^* = \frac{2}{3} a^{**} [(1 - \frac{qn'}{pn^{**}}) - \frac{6}{n^{**}L^{**}} (U_1 + U_{3/2})]^{1/2}. \tag{35}$$

At exact resonance, we have, as before,

$$a^* = a_0^* = a^{**}.$$

In general, the maximum variation from the mean semi-major axis a_0^* is obtained by putting $\eta_2 = \eta_2^0$ in Equation (35), i.e.,

$$(\Delta a^*)_{max.} = \frac{2}{3} a^* \left| 1 - \frac{qn'}{pn^{**}} \right|.$$

The variations in eccentricity and inclination can be found if integrals of Equation (26) are completely known. The variations in the angular variables are given by

$$\ell^* = \ell^{**} - \frac{\partial S^{(1)}}{\partial L^{**}} - \frac{\partial S^{(1)}}{\partial G^{**}},$$

$$\omega^* = \omega^{**} + \frac{\partial S^{(1)}}{\partial G^{**}},$$

$$\Omega^* = \Omega^{**} - \frac{\partial S^{(1)}}{\partial K^{**}} - \frac{\partial S^{(1)}}{\partial G^{**}}.$$

$$\omega'^* = \omega'^{**} - \frac{\partial S^{(1)}}{\partial H^{**}}.$$

Hence we can find perturbations in all the osculating elements.

REFERENCES

Brouwer, D. and Glemence, G.M.: 1961, *Methods of Celestial Mechanics*, Academic Press Inc., New York, pp. 582–84.
Giacaglia, G.E.O.: 1969, 'Resonance in the Restricted Problems of Three Bodies', Astron. J. 74, No. 10, H-1254–61.

Bhatnagar, K.B. and Beena Gupta: 'Resonance in the Restricted Problem
     of Three Bodies with Short-Periodic Perturbations', to appear in
     Porc. Nation. Sci. Acad. India.

PERIODIC ORBITS OF THE FIRST KIND IN THE RESTRICTED THREE-BODY PRO-
BLEM WHEN THE MORE MASSIVE PRIMARY IS AN OBLATE SPHEROID

R.K. Sharma
Vikram Sarabhai Space Center

ABSTRACT

The existence of periodic orbits of the first kind using
Delaunay's canonical variables is established through analytic
continuation in the planar restricted three-body problem when the
more massive primary is an oblate spheroid with its equatorial plane
coincident with the plane of motion.

TRIPLE COLLISION AS AN UNSTABLE EQUILIBRIUM

Jörg Waldvogel
Swiss Federal Institute of Technology (ETH), Zürich

ABSTRACT

The family of trajectories near a triple collision solution in the planar problem of three bodies is investigated by means of linearization in the neighborhood of the parabolic free fall. The local topological structure of this family is found to be that of a saddle point in the R^8. The corresponding stable manifold is the set of all triple collision solutions, whereas the instable manifold is formed by the parabolic solutions.

In important cases of nonzero total energy the family of close encounters is quantitatively described in terms of hypergeometric functions. By means of homothetic transformations the close triple encounters are then related to three-body motion with zero energy and zero angular momentum. In this way almost all close encounters near a homothetic solution can be treated by using a small number of parti-cular solutions of the three-body problem that may be calculated once for all.

In practical examples the first order theory presented here pre-dicts the escape velocity after a close triple encounter with a rela-tive accuracy comparable to the closeness of the encounter.

REFERENCES

Waldvogel, J.: 1976, "Triple Collision as an Unstable Equilibrium" submitted to the Académie Royale de Belgique, Brussels.

Waldvogel, J.: 1976, "The Three-Body Problem Near Triple Collision", to appear in Celestial Mechanics.

REGIONS OF ESCAPE ON THE VELOCITY ELLIPSOID FOR THE PLANAR THREE BODY
PROBLEM

George Bozis
Department of Theoretical Mechanics, University of Thessaloniki

ABSTRACT. The notion of the velocity ellipsoid for the planar three body
problem is given. Using the sufficient conditions for escape of one
member of a triple system, given by Standish (1971), a region is found
on the velocity ellipsoid for which escape is guaranteed.

1. INTRODUCTION

We shall consider planar systems of three bodies P_1, P_2, P_3 with masses

m_1, m_2, m_3 and with a certain value L of the total angular momentum

with respect to their center of mass G. To describe the motion we
shall use a non-inertial frame Oxy with its origin O as the center
of mass of the primaries P_1 and P_2, the x-axis permanently directed
from P_2 to P_1 and the y axis perpendicular to the x-axis. Let
x,y be the coordinates of the body P_3 and let $x_1 \geq 0$ be the ordinate of P_1. The location of the body P_2 on the x-axis is
easily found when x_1 and x,y are given. Therefore the problem is of
three degrees of freedom. The space $Oxyx_1$, may be taken as the configuration space for the dynamical system.

The distances r_1, r_2 of the body P_3 from the bodies P_1, P_2
respectively are given by

$$r_1 = \sqrt{(x-x_1)^2+y^2} \quad \text{and} \quad r_2 = \sqrt{(x-x_1 + \frac{x_1}{\mu})^2+y^2} \qquad (1)$$

where

$$\mu = \frac{m_2}{m_1+m_2}$$

and

$$m_1+m_2+m_3 = 1.$$

If ϑ is the angle formed by the moving axis Ox and any fixed
axis, say GX, on the plane of motion, the non-constant angular velocity

357

V. Szebehely (ed.), Dynamics of Planets and Satellites and Theories of Their Motion, 357-369.

of the non-inertial frame will be $\dot{\vartheta} = d\vartheta/dt$. The coordinates of the relative velocities are $\dot{x}, \dot{y}$ for the body P_3 and $\dot{x}_1$ for the body P_1.

The system provides the energy integral, given by

$$2E = \frac{(1-\mu)(1-m_3)}{\mu} \dot{x}_1^2 + \frac{R(1-m_3)}{\mu} \dot{\vartheta}^2 + m_3(1-m_3)(\dot{x}^2+\dot{y}^2) +$$

$$+ 2m_3(1-m_3)(x\dot{y}-\dot{x}y)\dot{\vartheta} - Q. \tag{2}$$

As to the angular momentum integral

$$L = m_3(1-m_3)(x\dot{y}-\dot{x}y) + \frac{R(1-m_3)}{\mu} \dot{\vartheta}, \tag{3}$$

this will be used to eliminate $\dot{\vartheta}$. Thus Equation (2) may be written as follows (Bozis, 1976)

$$(1-\mu)\dot{x}_1^2 + \mu m_3(\dot{x}^2+\dot{y}^2) - \frac{\mu^2 m_3^2}{R}(x\dot{y}-\dot{x}y)^2 = \frac{\mu}{1-m_3} \Phi \tag{4}$$

where

$$\Phi = 2E + Q - \frac{\mu L^2}{R(1-m_3)} \tag{5}$$

Both R and Q appearing in the above expressions are functions of the position coordinates x_1 and x,y and they are given by

$$R = (1-\mu)x_1^2 + \mu m_3(x^2+y^2) \tag{6}$$

and

$$Q = 2(1-m_3) \left\{ \frac{\mu^2(1-\mu)(1-m_3)}{x_1} + \frac{(1-\mu)m_3}{r_1} + \frac{\mu m_3}{r_2} \right\} \tag{7}$$

A simple calculation shows that the moment of inertia I of the three masses with respect to their center of mass G is given by

$$I = \frac{R(1-m_3)}{\mu}.$$

Also the quantity Q, defined by Equation (7), is related to the potential V of the system by the Equation

$$Q = -2V.$$

2. THE VELOCITY ELLIPSOID

The left member of Equation (4) is a definite positive quantity for all values of the variables involved. The inequality

$$\Phi(x,y,x_1) \geq 0$$

defines in the position space $0xyx_1$ regions where motion is allowed to take place (Bozis, 1976).

Therefore we conclude that, for definite values of E and L, to each point $P(x,y,x_1)$ of the position space there corresponds a "velocity ellipsoid" given by Equation (4) in the velocity space $0'\dot{x}\dot{y}\dot{x}_1$.

As P moves in the permissible region of the configuration space $0xyx_1$ its velocity vector may have any direction. Its magnitude, however, is such that, in the velocity space $0'\dot{x}\dot{y}\dot{x}_1$, the vector $\overrightarrow{0'V}$ parallel to the velocity of P, terminates on the surface of the velocity ellipsoid of the point P.

One of the main axes of the velocity ellipsoid is along the $0'\dot{x}_1$ axis. The other two main axes are on the plane $0'\dot{x}\dot{y}$. In order to orient the ellipsoid along its main axes let us use the new orthogonal system $0'\dot{\xi}\dot{n}\dot{x}_1$.

If $\varphi(0\leq\varphi<\frac{\pi}{2})$ is the angle formed by the positive semi-axes $0'\dot{x}$ and $0'\dot{\xi}$ on the plane $0'\dot{x}\dot{y}$ we have

$$tg2\varphi = \frac{2xy}{x^2-y^2} . \tag{8}$$

The equation of the velocity ellipsoid in the new system $0'\dot{\xi}\dot{n}\dot{x}_1$ is then

$$\frac{\dot{\xi}^2}{a^2} + \frac{\dot{n}^2}{b^2} + \frac{\dot{x}_1^2}{c^2} = 1$$

where the semi-axes a,b,c are given by the formulae

$$a = \sqrt{\frac{\Phi}{m_3(1-m_3)}}$$

$$b = \frac{1}{x_1}\sqrt{\frac{R\Phi}{m_3(1-m_3)(1-\mu)}} \tag{9}$$

$$c = \sqrt{\frac{\mu\Phi}{(1-\mu)(1-m_3)}}$$

We observe that for all points x,y,x_1 it is

$$a \leq b$$

and that

$$\frac{c}{a} = \sqrt{\frac{\mu m_3}{1-\mu}} = constant.$$

The quantities a,b and c all vanish for points x,y,x_1 on the surface of zero velocity $\Phi = 0$. The same quantities tend to infinity near collisions.

The value of a may be smaller or larger than the value of c since the quantity $\mu m_3/1-\mu$ ranges from 0 to ∞. It may be that a=c for values of μ and m_3 on the part of the hyperbola $\mu m_3=1-\mu$ inside the square $0 \leq \mu \leq 1$, $0 \leq m_3 \leq 1$.

It may also be that a=b≠0 and this actually happens only for points on the x_1-axis of the position space. It is therefore understood that for certain values of the parameters μ and m_3 and for points x_1, $x_3=0$, $y_3=0$ the velocity ellipsoid is reduced to a sphere. In this case the magnitude of the velocity, at these points, is completely defined if E and L are given.

In all other cases the magnitude of the velocity at a point x,y,x_1, depends on its direction and ranges between the maximum and the minimum value of the quantities a, b and c, evaluated at this point.

We shall now compare this statement with the information given by the well known Sundmann's inequality (e.g. Birkhoff, 1927). In our notation this inequality may be written as

$$(\frac{dI}{dt})^2 \leq 4I\Phi \tag{10}$$

Its meaning is that for each point (x,y, x_1) of the position space there exists a certain velocity which makes the quantity $(dI/dt)^2$ maximum. The quantity

$$\frac{dI}{dt} = \frac{1-m_3}{\mu}\{2(1-\mu)x_1\dot{x}_1 + 2\mu m_3(x\dot{x}+y\dot{y})\} \tag{11}$$

is a function of $\dot{x}$ and $\dot{y}$ only, since $\dot{x}_1 = \dot{x}_1(\dot{x}_3, \dot{y}_3)$ by Equation (4). The critical values of (11) are then found to be

$$\dot{x}_o = \pm \sqrt{\frac{\Phi}{I}}x, \quad \dot{y}_o = \pm \sqrt{\frac{\Phi}{I}}y \text{ and } \dot{x}_{1,o} = \pm \sqrt{\frac{\Phi}{I}}x_1 \tag{12}$$

and they define on the velocity ellipsoid two points symmetric with respect to the origin 0.

From another point of view Sundmann's inequality implies that for each point (x,y,x_1) of the position space the vector of the velocity $\overrightarrow{OV}$ in the velocity space $0'\dot{x}\dot{y}\dot{x}_1$ must be selected so that its terminal point V lies somewhere between the two parallel planes

$$(1-\mu)x_1\dot{x}_1 + \mu m_3(x\dot{x}+y\dot{y}) = -\frac{\mu}{1-m_3}\sqrt{I\Phi} \tag{13a}$$

and

$$(1-\mu)x_1\dot{x}_1 + \mu m_3(x\dot{x}+y\dot{y}) = \frac{\mu}{1-m_3}\sqrt{I\Phi} . \tag{13b}$$

One could probably think that the planes (13) intersect the surface (4) and that part of the surface is excluded. However, this is not the case.

It can easily be shown that "Sundmann's planes" (13a) and (13b) are always tangent to the velocity ellipsoid (4) at the points

$$P_o(\sqrt{\tfrac{\Phi}{I}}\,x, \quad \sqrt{\tfrac{\Phi}{I}}\,y, \quad \sqrt{\tfrac{\Phi}{I}}\,x_1)$$

and   (14)

$$P_o^*(-\sqrt{\tfrac{\Phi}{I}}\,x, \quad -\sqrt{\tfrac{\Phi}{I}}\,y, \quad -\sqrt{\tfrac{\Phi}{I}}\,x_1)$$

The conclusion is that, as far as the magnitude of the velocity is concerned, more information is given by the velocity ellipsoid than by Sundmann's planes.

3. CONDITIONS FOR ESCAPE

We shall prove in this section that in general to a certain part of the surface of the velocity ellipsoid there correspond escaping orbits.

We shall limit our study to negative values of the energy E. For such values of E the zero velocity surfaces allow for the following types of dissolution: (i). The body P_3 escapes to infinity leaving the bodies P_1 and P_2 in a close binary. (ii). The body P_3 goes to infinity in a close binary with either P_1 or P_2. In either case (for E<0) there is a minimum distance of the bodies bounded by the quantity (Birkhoff, 1927)

$$r_* = \frac{(1-m_3)\{\mu(1-\mu)(1-m_3)+m_3\}}{-E} \,.$$ (15)

Let us study case (i). It is exactly for this case that Standish (1971) has given sufficient conditions for escape. In our notation these conditions are the following:

$$\sqrt{x^2+y^2} > r_*$$

$$\frac{d}{dt}(\sqrt{x^2+y^2}) > 0$$

and

$$(\frac{d}{dt}\sqrt{x^2+y^2})^2 \geq 2\left[\frac{1}{\sqrt{x^2+y^2}} + \frac{\mu(1-\mu)r_*^2}{(\sqrt{x^2+y^2}-r_*)(x^2+y^2)}\right]$$

These are rewritten as follows:

$$x^2+y^2 > r_*^2$$ (16a)

$$x\dot{x}+y\dot{y} > 0$$ (16b)

$$x\dot{x}+y\dot{y}-S \geq 0$$ (16c)

where

$$S = \sqrt{2\left[\sqrt{x^2+y^2} + \frac{\mu(1-\mu)r_*^2}{\sqrt{x^2+y^2} - r_*}\right]} \qquad (17)$$

Only the positive square root in (17) is considered in view of the inequality (16b). Since (16c) is stronger than (16b) we only need consider (16c).

Equation

$$x\dot{x} + y\dot{y} - S = 0 \qquad (18)$$

represents, in the $0'\dot{x}\dot{y}\dot{x}_1$ space, a plane parallel to the $\dot{x}_1$ axis. The inequality (16c) holds to the other side of this plane from the side where the origin $0'$ is.

The question now is: Does the plane (18) intersect the velocity ellipsoid (4)? This happens indeed if, on the plane $\dot{x}_1=0$, the straight line (18) intersects the ellipse

$$\mu m_3(\dot{x}^2+\dot{y}^2) - \frac{\mu^2 m_3^2}{R}(x\dot{y}-\dot{x}y)^2 = \frac{\mu\Phi}{1-m_3} . \qquad (19)$$

Eliminating $\dot{y}$ between (18) and (19) we get, with $y\neq0$, the quadratic equation

$$\alpha\dot{x}^2 + \beta\dot{x} + \gamma = 0 \qquad (20)$$

where
$$\alpha = (1-\mu)\mu m_3 \frac{x^2+y^2}{Ry^2} x_1^2$$

$$\beta = -2(1-\mu)\mu m_3 \frac{Sx}{Ry^2} x_1^2$$

$$\gamma = -\frac{\mu\Phi}{1-m_3} + \mu m_3 \frac{S^2}{Ry^2} \{(1-\mu)x_1^2 + \mu m_3 y^2\}.$$

In order that the Equation (20) has two real roots we must have

$$\frac{\Phi}{m_3(1-m_3)} r^2 - S^2 \geq 0 \qquad (21)$$

where Φ is given by (5), S is given by (17) and
$$r^2 = x^2+y^2 \qquad (22)$$

We shall replace the inequality (21) by a weaker but much simpler inequality.

We have, in view of Equations (6), (7) and (22),

$$R \geq R_o = \mu m_3 r^2 \tag{23}$$

and

$$Q \geq Q_o = \frac{2\mu^2(1-\mu)(1-m_3)^2}{x_1} . \tag{24}$$

Therefore, in view of Equation (5),

$$\Phi \geq \Phi_o = 2E + Q_o - \frac{\mu L^2}{R_o(1-m_3)} . \tag{25}$$

We thus come to the conclusion that the inequality (21) is always true provided that

$$\frac{\Phi_o r^2}{m_3(1-m_3)} - S^2 \geq 0. \tag{26}$$

The inequality (26) is equivalent to the inequality.

$$0 < x_1 \leq f(r) \tag{27}$$

where

$$f(r) = \frac{2\mu^2(1-\mu)(1-m_3)^2 r^2}{m_3(1-m_3)S^2-(2Er^2- \frac{L^2}{m_3(1-m_3)})} \tag{28}$$

The meaning of the last inequality (27) is the following: Suppose that the initial conditions (i.e. coordinates of position and velocity) of the problem are given. Also suppose that the angular momentum L is given. The value of E is then known and from Equation (15) the value of r_* may be found. Then from Equation (17) and (28) the values of S and f(r) corresponding to the given initial conditions are found. Now if the inequalities

$$r > r_* \quad \text{and} \quad x_1 \leq f(r) \tag{29}$$

are satisfied we understand that the orbit corresponding to the given initial conditions may be escaping. This is because these two inequalities are sufficient conditions for the plane (18) and the ellipsoid (4) to intersect each other, thus forming on the velocity ellipsoid a patch to the points of which there correspond velocities leading to escaping orbits.

Therefore it will depend on the direction of the velocity $\dot{x},\dot{y},\dot{x}_1$ whether or not the orbit will be escaping.

Another way of interprating inequality (27) is the following:
Suppose that the third body is brought to a distance $r > r_*$ from the center
of mass of the primaries P_1 and P_2. The question is: With a certain
negative value of the energy E (by means of which r_* was determined) and of the angular momentum L are there any velocities which
make the body P_3 escape to infinity? The answer may definetely be af-
firmative provided that the distance x_1/μ between the primaries P_1
and P_2 is sufficiently small as to satisfy the inequality (27).

We now observe that
(i)   As $r \to r_*$ we have

$$\lim_{r \to r_*} S^2 = \infty \qquad \text{and} \qquad \lim_{r \to r_*} f(r) = 0$$

(ii) As $r \to \infty$ we have

$$\lim_{r \to \infty} f(r) = \frac{\mu^2 (1-\mu)(1-m_3)^2}{-E}$$

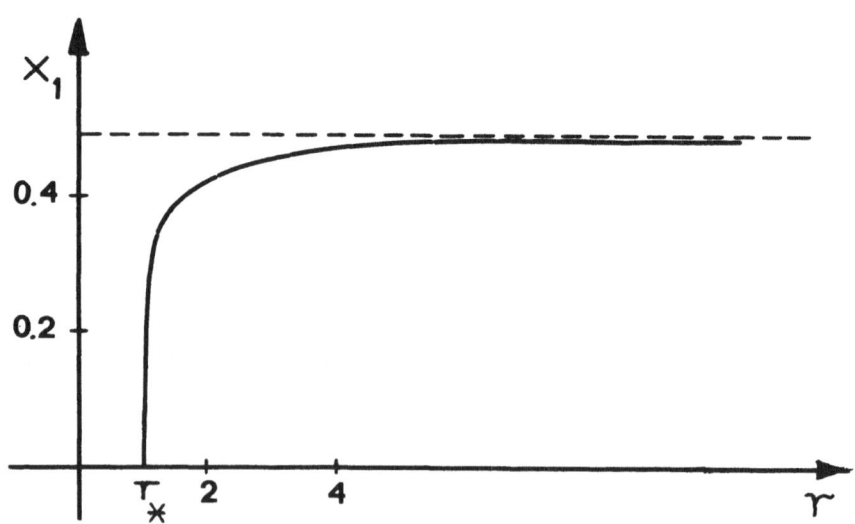

Fig. 1: The function $x_1 = f(r)$ as given by Equation (28), drawn for
$\mu = 0.50$, $m_3 = 0.01$, $E = -0.25$ and $L = 0.05$. For a certain $r > r_*$
and for $x_1 < f(r)$ there is always a region of escaping orbits
on the velocity ellipsoid.

It is worth noticing that the value of the $\lim_{r \to \infty} f(r)$ is the same with
the value k in the position space of the plane $x_1 = k = \mu^2(1-\mu)(1-m_3)^2/-E$
which is asymptotically tangent to the zero velocity surfaces
(Bozis, 1976).
(iii). A direct calculation gives

$$\frac{df(r)}{dr} = \frac{4}{r^3}\left[f(r)\right]^2 \mu^2 (1-\mu)(1-m_3)^2 \{m_3(1-m_3)r + \frac{L^2}{m_3(1-m_3)} +$$

$$+ m_3(1-m_3)\frac{\mu(1-\mu)r_*^2}{(r-r_*)^2}(3r-r_*)\}.$$

Since $r>r_*$, we have $df(r)/dr>0$, i.e. the function $f(r)$ is increasing in the interval $[r_*,\infty)$ (Figure 1).

For a given value of $r>r_*$ we have from Figure 1 the values of x_1 for which the inequality (27) is valid.

Inequality (27) may also be interpreted as follows: The Equation

$$x_1 - f(\sqrt{x^2+y^2}) = 0 \tag{30}$$

represents in the position space a surface of revolution around the $0x_1$ axis. Figure 1 may also serve to visualize the intersection of this surface with either of the planes $y=0$ or $x=0$. The surface of revolution is asymptoticaly tangent to the plane

$$x_1 = \frac{\mu^2(1-\mu)(1-m_3)^2}{-E}.$$

We thus come to the conclusion that for given values of μ, m_3, E and L escape of the third body is not guaranteed in the part of the position space underneath the zero velocity surfaces, above the surface of revolution (30) and outside the cylinder

$$x^2+y^2 = r_*^2 \tag{31}$$

Obviously, inside the cylinder (31) escape is not guaranteed since the inequality (16a) is not valid.

A final remark concerns the coefficients of Equation (20). These coefficients have no meaning for $y=0$. This case, however, may be studied separately. In fact the Equation of the straight line (18) becomes

$$x\dot{x} = S$$

whereas the Equation of the ellipse (19) becomes

$$\frac{\dot{x}^2}{a^2} + \frac{\dot{y}^2}{b^2} = 1$$

These two curves intersect each other if

$$\frac{S}{x} \leq a$$

or, since x=r, if

$$r \geq \frac{S}{a}$$

Again taking into account (23), (24), (25) we can easily prove
that the inequality r≥S/a is equivallent to the inequality (27). Thus
(27) covers the case y=0 as well.

4. NUMERICAL EXAMPLE

Figure 1 was drawn for the following values of the parameters:

$$\mu = 0.50, \quad m_3 = 0.01 \quad E = -0.25 \quad L = 0.05.$$

From Equation (15) we find

$$r_* = 1.0197.$$

We now select a distance $r=\sqrt{5} > r_*$ and, for this r, we find from
Equation (17)

$$S = 2.2135.$$

Then, either from the diagram of Figure 1 or from Equation (28) we find

$$f(r=\sqrt{5}) = 0.4374.$$

Let us now select x=1, y=2 (so that $x^2+y^2=5$) and $x_1=0.30$ (so
that the inequality (27) is satisfied).

For this position of the triple system (with given masses and E,
L) there will be a patch on the velocity ellipsoid for all points of
which we will have an escaping orbit for the body P_3.

In Figure 2 we give the projection of this patch on the $0'\dot{x}\dot{y}$ plane.
This is found as follows:

For the point
$$x = 1, \quad y = 2, \quad x_1 = 0.30$$

of the configuration space we calculate the values of R, Q and Φ from
Equations (6), (7) and (5) respectively. Then with the aid of Formulae
(9) and (8) we calculate the semi-axes a,b,c of the velocity ellipsoid
as well as its orientation.

We have

$$a = 5.5735, \qquad b = 6.9514, \qquad c = 0.5573$$

and
$$\text{tg}\varphi = 2.$$

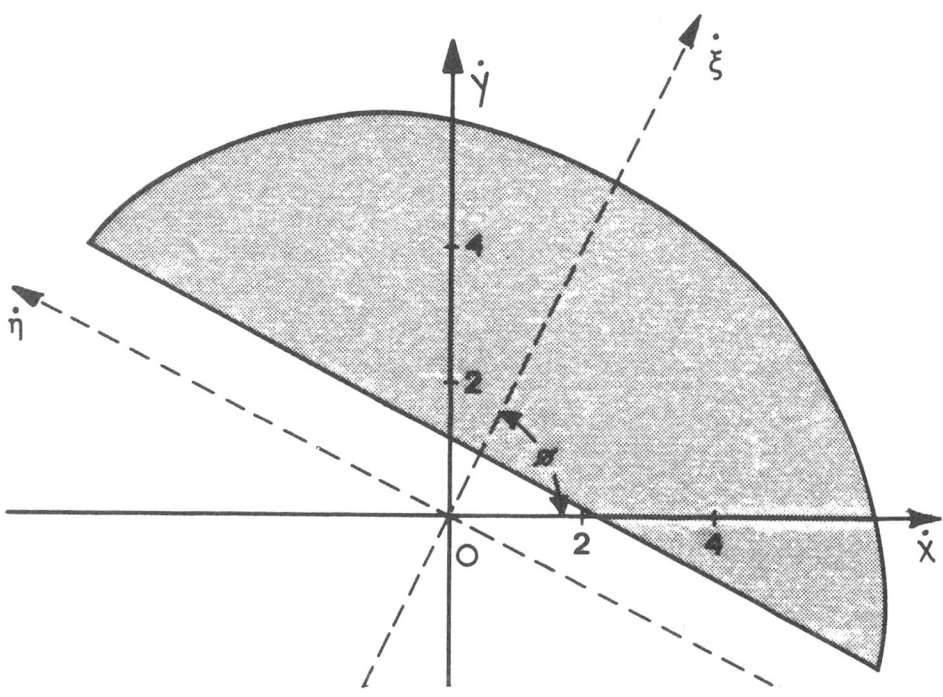

Fig.2? The shaded part of this Figure is the projection of the
escaping region of the velocity ellipsoid on the $0'\dot{x}\dot{y}$ plane
for $\mu=0.50$, $m_3=0.01$, $E=-0.25$, $L=0.05$. For a certain pair
$\dot{x}$, $\dot{y}$ of the shaded region the value of $\dot{x}_1$ is found from
Equation (4). The velocities $(\dot{x}, \dot{y}, \dot{x}_1)$ and $(\dot{x}, \dot{y}, -\dot{x}_1)$
are then escaping velocities.

We now draw the ellipse

$$\frac{\dot{\xi}^2}{a^2} + \frac{\dot{\eta}^2}{b^2} = 1.$$

Next we draw the straight line (18), i.e.

$$\dot{x} + 2\dot{y} = 2.2135$$

The shaded part of Figure 2 corrsponds to velocities for which the third
body by all means escapes to infinity. For any pair $\dot{x},\dot{y}$ of the shaded
part of Figure 2 the corresponding value of $\dot{x}_1$ must be found from
Equation (4). The velocity $(\dot{x},\dot{y},\dot{x}_1)$ as well as the velocity $(\dot{x},\dot{y},-\dot{x}_1)$
is then an escaping velocity.

5. COMMENTS AND CONCLUSIONS

The planar problem of three bodies is of three degrees of freedom in a conveniently selected non-inertial frame. As a consequence, we can think of a representative point P moving in a triaxial space $0xyx_1$. For given values of the masses of the three bodies and the parameters E and L there exist always, in the position space, regions where motion of the point $P(x,y,x_1)$ is allowed to take place.

On the other hand to each point $P(x,y,x_1)$ of the allowable position space there corresponds, in the velocity space $0\,\dot{x}\dot{y}\dot{x}_1$, a velocity ellipsoid, i.e. a surface with Equation (4). The vector of the velocity of the point P may have any direction. However the magnitude of the velocity is found from Equation (4) for any given direction. The range of the velocities which can be used from any point P with certain values of E, L can be found by calculating the semi-axes (9) of the velocity ellipsoid.

Let us look at this from another point of view: Suppose that μ, m_3, E and L are given and also a point $P(x,y,x_1)$ is given. The velocity ellipsoid corresponding to this point P is then determined with the aid of Equations (9) and (8). To each point of this ellipsoid there corresponds a certain orbit through the point P with a definite velocity. The question naturally arises: Do all the orbits through P have any common features? In particular how do they behave as to escaping? The answer is as follows: If the inequalities (29) are satisfied at the point P, then there exists a part on the surface of the velocity ellipsoid to all points of which there correspond escaping orbits. This part is always less than half of its surface. Its projection on the $0\,\dot{x}\dot{y}$ plane is defined, as in Fig.2, by the intersection of the ellipse (19) with the straight line (18), which is always parallel to one of the main axes of the ellipse. (See Figure 2).

In case that the inequalities (29) are not simultaneously satisfied there is no velocity for which escape of the third body may be guaranteed. This happens inside the cylinder (31) or outside this cylinder but between the surface of revolution (30) and the zero velocity surface.

If, for instance, in the numerical example worked in §4, we selected x_1=0.50 (instead of x_1=0.30) the second inequality (29) would not hold and no patch of escaping orbits on the velocity ellipsoid could be found.

Consider now a symmetric periodic orbit. Since at t=0 we have $x=x_0$, y=0, it follows from (8) that φ=0 and the velocity ellipsoid is

$$\frac{\dot{x}^2}{a^2} + \frac{\dot{y}^2}{b^2} + \frac{\dot{x}_1^2}{c^2} = 1.$$

Also, since at $t=0$ it is $\dot{x}_o=0$, $\dot{x}_{1,o}=0$ we have $\dot{y}_o=\pm b$.

On the other hand the straight line (8) is, in this case,

$$\dot{x} = \frac{S}{x_o} .$$

Obviously the initial velocity $(0,0,b)$ or $(0,0,-b)$ of the periodic orbit through the point $(x_o, 0, x_{10})$ corresponds to a point of the velocity ellipsoid outside the shaded region $\dot{x} \geq S/x_o$.

REFERENCES

1. Birkhoff, G. : 1927, Dynamical Systems, (published by the A.M.S., Providence, Rhode Island).
2. Bozis, G.: 1976, Astroph. and Space Science 43, 355.
3. Standish, M.: 1971, Celes. Mech 4, 44.

INDEX OF NAMES

INDEX OF SUBJECTS

ASTROPHYSICS AND SPACE SCIENCE LIBRARY

Edited by

J. E. Blamont, R. L. F. Boyd, L. Goldberg, C. de Jager, Z. Kopal, G. H. Ludwig, R. Lüst,
B. M. McCormac, H. E. Newell, L. I. Sedov, Z. Švestka, and W. de Graaff

1. C. de Jager (ed.), *The Solar Spectrum, Proceedings of the Symposium held at the University of Utrecht, 26–31 August, 1963.* 1965, XIV + 417 pp.
2. J. Ortner and H. Maseland (eds.), *Introduction to Solar Terrestrial Relations, Proceedings of the Summer School in Space Physics held in Alpbach, Austria, July 15–August 10, 1963 and Organized by the European Preparatory Commission for Space Research.* 1965, IX + 506 pp.
3. C. C. Chang and S. S. Huang (eds.), *Proceedings of the Plasma Space Science Symposium, held at the Catholic University of America, Washington, D.C., June 11–14, 1963.* 1965, IX + 377 pp.
4. Zdeněk Kopal, *An Introduction to the Study of the Moon.* 1966, XII + 464 pp.
5. B. M. McCormac (ed.), *Radiation Trapped in the Earth's Magnetic Field. Proceedings of the Advanced Study Institute, held at the Chr. Michelsen Institute, Bergen, Norway, August 16–September 3, 1965.* 1966, XII + 901 pp.
6. A. B. Underhill, *The Early Type Stars.* 1966, XII + 282 pp.
7. Jean Kovalevsky, *Introduction to Celestial Mechanics.* 1967, VIII + 427 pp.
8. Zdeněk Kopal and Constantine L. Goudas (eds.), *Measure of the Moon. Proceedings of the 2nd International Conference on Selenodesy and Lunar Topography, held in the University of Manchester, England, May 30–June 4, 1966.* 1967, XVIII + 479 pp.
9. J. G. Emming (ed.), *Electromagnetic Radiation in Space. Proceedings of the 3rd ESRO Summer School in Space Physics, held in Alpbach, Austria, from 19 July to 13 August, 1965.* 1968, VIII + 307 pp.
10. R. L. Carovillano, John F. McClay, and Henry R. Radoski (eds.), *Physics of the Magnetosphere, Based upon the Proceedings of the Conference held at Boston College, June 19–28, 1967.* 1968, X + 686 pp.
11. Syun-Ichi Akasofu, *Polar and Magnetospheric Substorms.* 1968, XVIII + 280 pp.
12. Peter M. Millman (ed.), *Meteorite Research. Proceedings of a Symposium on Meteorite Research, held in Vienna, Austria, 7–13 August, 1968.* 1969, XV + 941 pp.
13. Margherita Hack (ed.), *Mass Loss from Stars. Proceedings of the 2nd Trieste Colloquium on Astrophysics, 12–17 September, 1968.* 1969, XII + 345 pp.
14. N. D'Angelo (ed.), *Low-Frequency Waves and Irregularities in the Ionosphere. Proceedings of the 2nd ESRIN-ESLAB Symposium, held in Frascati, Italy, 23–27 September, 1968.* 1969, VII + 218 pp.
15. G. A. Partel (ed.), *Space Engineering. Proceedings of the 2nd International Conference on Space Engineering, held at the Fondazione Giorgio Cini, Isola di San Giorgio, Venice, Italy, May 7–10, 1969.* 1970, XI + 728 pp.
16. S. Fred Singer (ed.), *Manned Laboratories in Space. Second International Orbital Laboratory Symposium.* 1969, XIII + 133 pp.
17. B. M. McCormac (ed.), *Particles and Fields in the Magnetosphere. Symposium Organized by the Summer Advanced Study Institute, held at the University of California, Santa Barbara, Calif., August 4–15, 1969.* 1970, XI + 450 pp.
18. Jean-Claude Pecker, *Experimental Astronomy.* 1970, X + 105 pp.
19. V. Manno and D. E. Page (eds.), *Intercorrelated Satellite Observations related to Solar Events. Proceedings of the 3rd ESLAB/ESRIN Symposium held in Noordwijk, The Netherlands, September 16–19, 1969.* 1970, XVI + 627 pp.
20. L. Mansinha, D. E. Smylie, and A. E. Beck, *Earthquake Displacement Fields and the Rotation of the Earth, A NATO Advanced Study Institute Conference Organized by the Department of Geophysics, University of Western Ontario, London, Canada, June 22–28, 1969.* 1970, XI + 308 pp.
21. Jean-Claude Pecker, *Space Observatories.* 1970, XI + 120 pp.
22. L. N. Mavridis (ed.), *Structure and Evolution of the Galaxy. Proceedings of the NATO Advanced Study Institute, held in Athens, September 8–19, 1969.* 1971, VII + 312 pp.
23. A. Muller (ed.), *The Magellanic Clouds. A European Southern Observatory Presentation: Principal Prospects, Current Observational and Theoretical Approaches, and Prospects for Future Research, Based on the Symposium on the Magellanic Clouds, held in Santiago de Chile, March 1969, on the Occasion of the Dedication of the European Southern Observatory.* 1971, XII + 189 pp.

24. B. M. McCormac (ed.), *The Radiating Atmosphere. Proceedings of a Symposium Organized by the Summer Advanced Study Institute, held at Queen's University, Kingston, Ontario, August 3–14, 1970.* 1971, XI + 455 pp.
25. G. Fiocco (ed.), *Mesospheric Models and Related Experiments. Proceedings of the 4th ESRIN-ESLAB Symposium, held at Frascati, Italy, July 6–10, 1970.* 1971, VIII + 298 pp.
26. I. Atanasijević, *Selected Exercises in Galactic Astronomy.* 1971, XII + 144 pp.
27. C. J. Macris (ed.), *Physics of the Solar Corona. Proceedings of the NATO Advanced Study Institute on Physics of the Solar Corona, held at Cavouri-Vouliagmeni, Athens, Greece, 6–17 September 1970.* 1971, XII + 345 pp.
28. F. Delobeau, *The Environment of the Earth.* 1971, IX + 113 pp.
29. E. R. Dyer (general ed.), *Solar-Terrestrial Physics/1970. Proceedings of the International Symposium on Solar-Terrestrial Physics, held in Leningrad, U.S.S.R., 12–19 May 1970.* 1972, VIII + 938 pp.
30. V. Manno and J. Ring (eds.), *Infrared Detection Techniques for Space Research. Proceedings of the 5th ESLAB-ESRIN Symposium, held in Noordwijk, The Netherlands, June 8–11, 1971.* 1972, XII + 344 pp.
31. M. Lecar (ed.), *Gravitational N-Body Problem. Proceedings of IAU Colloquium No. 10, held in Cambridge, England, August 12–15, 1970.* 1972, XI + 441 pp.
32. B. M. McCormac (ed.), *Earth's Magnetospheric Processes. Proceedings of a Symposium Organized by the Summer Advanced Study Institute and Ninth ESRO Summer School, held in Cortina, Italy, August 30–September 10, 1971.* 1972, VIII + 417 pp.
33. Antonin Rükl, *Maps of Lunar Hemispheres.* 1972, V + 24 pp.
34. V. Kourganoff, *Introduction to the Physics of Stellar Interiors.* 1973, XI + 115 pp.
35. B. M. McCormac (ed.), *Physics and Chemistry of Upper Atmospheres. Proceedings of a Symposium Organized by the Summer Advanced Study Institute, held at the University of Orléans, France, July 31–August 11, 1972.* 1973, VIII + 389 pp.
36. J. D. Fernie (ed.), *Variable Stars in Globular Clusters and in Related Systems. Proceedings of the IAU Colloquium No. 21, held at the University of Toronto, Toronto, Canada, August 29–31, 1972.* 1973, IX + 234 pp.
37. R. J. L. Grard (ed.), *Photon and Particle Interaction with Surfaces in Space. Proceedings of the 6th ESLAB Symposium, held at Noordwijk, The Netherlands, 26–29 September, 1972.* 1973, XV + 577 pp.
38. Werner Israel (ed.), *Relativity, Astrophysics and Cosmology. Proceedings of the Summer School, held 14–26 August, 1972, at the BANFF Centre, BANFF, Alberta, Canada.* 1973, IX + 323 pp.
39. B. D. Tapley and V. Szebehely (eds.), *Recent Advances in Dynamical Astronomy. Proceedings of the NATO Advanced Study Institute in Dynamical Astronomy, held in Cortina d'Ampezzo, Italy, August 9–12, 1972.* 1973, XIII + 468 pp.
40. A. G. W. Cameron (ed.), *Cosmochemistry. Proceedings of the Symposium on Cosmochemistry, held at the Smithsonian Astrophysical Observatory, Cambridge, Mass., August 14–16, 1972.* 1973, X + 173 pp.
41. M. Golay, *Introduction to Astronomical Photometry.* 1974, IX + 364 pp.
42. D. E. Page (ed.), *Correlated Interplanetary and Magnetospheric Observations. Proceedings of the 7th ESLAB Symposium, held at Saulgau, W. Germany, 22–25 May, 1973.* 1974, XIV + 662 pp.
43. Riccardo Giacconi and Herbert Gursky (eds.), *X-Ray Astronomy.* 1974, X + 450 pp.
44. B. M. McCormac (ed.), *Magnetospheric Physics. Proceedings of the Advanced Summer Institute, held in Sheffield, U.K., August 1973.* 1974, VII + 399 pp.
45. C. B. Cosmovici (ed.), *Supernovae and Supernova Remnants. Proceedings of the International Conference on Supernovae, held in Lecce, Italy, May 7–11, 1973.* 1974, XVII + 387 pp.
46. A. P. Mitra, *Ionospheric Effects of Solar Flares.* 1974, XI + 294 pp.
47. S.-I. Akasofu, *Physics of Magnetospheric Substorms.* 1977, XVIII + 599 pp.
48. H. Gursky and R. Ruffini (eds.), *Neutron Stars, Black Holes and Binary X-Ray Sources.* 1975, XII + 441 pp.
49. Z. Švestka and P. Simon (eds.), *Catalog of Solar Particle Events 1955–1969. Prepared under the Auspices of Working Group 2 of the Inter-Union Commission on Solar-Terrestrial Physics.* 1975, IX + 428 pp.
50. Zdeněk Kopal and Robert W. Carder, *Mapping of the Moon.* 1974, VIII + 237 pp.
51. B. M. McCormac (ed.), *Atmospheres of Earth and the Planets. Proceedings of the Summer Advanced Study Institute, held at the University of Liège, Belgium, July 29–August 8, 1974.* 1975, VII + 454 pp.
52. V. Formisano (ed.), *The Magnetospheres of the Earth and Jupiter. Proceedings of the Neil Brice Memorial Symposium, held in Frascati, May 28–June 1, 1974.* 1975, XI + 485 pp.

53. R. Grant Athay, *The Solar Chromosphere and Corona: Quiet Sun*. 1976, XI + 504 pp.

54. C. de Jager and H. Nieuwenhuijzen (eds.), *Image Processing Techniques in Astronomy. Proceedings of a Conference, held in Utrecht on March 25–27, 1975*, XI + 418 pp.

55. N. C. Wickramasinghe and D. J. Morgan (eds.), *Solid State Astrophysics. Proceedings of a Symposium, held at the University College, Cardiff, Wales, 9–12 July 1974*. 1976, XII + 314 pp.

56. John Meaburn, *Detection and Spectrometry of Faint Light*. 1976, IX + 270 pp.

57. K. Knott and B. Battrick (eds.), *The Scientific Satellite Programme during the International Magnetospheric Study. Proceedings of the 10th ESLAB Symposium, held at Vienna, Austria, 10–13 June 1975*. 1976, XV + 464 pp.

58. B. M. McCormac (ed.), *Magnetospheric Particles and Fields. Proceedings of the Summer Advanced Study School, held in Graz, Austria, August 4–15, 1975*. 1976, VII + 331 pp.

59. B. S. P. Shen and M. Merker (eds.), *Spallation Nuclear Reactions and Their Applications*. 1976, VIII + 235 pp.

60. Walter S. Fitch (ed.), *Multiple Periodic Variable Stars. Proceedings of the International Astronomical Union Colloquium No. 29, Held at Budapest, Hungary, 1–5 September 1975*. 1976, XIV + 348 pp.

61. J. J. Burger, A. Pedersen, and B. Battrick (eds.), *Atmospheric Physics from Spacelab. Proceedings of the 11th ESLAB Symposium, Organized by the Space Science Department of the European Space Agency, held at Frascati, Italy, 11–14 May 1976*. 1976, XX + 409 pp.

62. J. Derral Mulholland (ed.), *Scientific Applications of Lunar Laser Ranging. Proceedings of a Symposium held in Austin, Tex., U.S.A., 8–10 June, 1976*. 1977, XVII + 302 pp.

63. Giovanni G. Fazio (ed.), *Infrared and Submillimeter Astronomy. Proceedings of a Symposium held in Philadelphia, Penn., U.S.A., 8–10 June, 1976*. 1977, X+226 pp.

64. C. Jaschek and G. A. Wilkins (eds.), *Compilation, Critical Evaluation and Distribution of Stellar Data. Proceedings of the International Astronomical Union Colloquium No. 35, held at Strasbourg, France, 19–21 August, 1976*. 1977, XIV+316 pp.

65. M. Friedjung (ed.), *Novae and Related Stars. Proceedings of an International Conference held by the Institut d'Astrophysique, Paris, France, 7–9 September, 1976*. 1977, XIV+228 pp.

66. David N. Schramm (ed.), *Supernovae. Proceedings of a Special IAU Session on Supernovae held in Grenoble, France, 1 September, 1976*. 1977, X+192 pp.

67. Jean Audouze (ed.), *CNO Isotopes in Astrophysics. Proceedings of a Special IAU Session held in Grenoble, France, 30 August, 1976*. 1977, XIII+195 pp.

68. Z. Kopal, *Dynamics of Close Binary Systems*, forthcoming.

69. A. Bruzek and C. J. Durrant (eds.), *Illustrated Glossary for Solar and Solar-Terrestrial Physics*. 1977, approx. 216 pp.

70. H. van Woerden (ed.), *Topics in Interstellar Matter*. 1977, VIII + 295 pp.

71. M. A. Shea, D. F. Smart, and T. S. Wu (eds.), *Study of Travelling Interplanetary Phenomena*. 1977, XII+439 pp.